STUDY GUIDE
to accompany

CHEMISTRY
The Molecular Science

MOORE ▲ STANITSKI ▲ JURS

Kathleen E. Murphy, Ph.D.
Daemen College
Amherst, New York

BROOKS/COLE
✦
™
THOMSON LEARNING

Australia • Canada • Mexico • Singapore • Spain
United Kingdom • United States

Printed in the United States of America

1 2 3 4 5 6 7 05 04 03 02 01

ISBN 0-03-032401-7

For more information about our products, contact us at:
Thomson Learning Academic Resource Center
1-800-423-0563

For permission to use material from this text, contact us by:
Phone: 1-800-730-2214
Fax: 1-800-731-2215
Web: www.thomsonrights.com

Asia
Thomson Learning
60 Albert Complex, #15-01
Alpert Complex
Singapore 189969

Australia
Nelson Thomson Learning
102 Dodds Street
South Street
South Melbourne, Victoria 3205
Australia

Canada
Nelson Thomson Learning
1120 Birchmount Road
Toronto, Ontario M1K 5G4
Canada

Europe/Middle East/South Africa
Thomson Learning
Berkshire House
168-173 High Holborn
London WC1 V7AA
United Kingdom

Latin America
Thomson Learning
Seneca, 53
Colonia Polanco
11560 Mexico D.F.
Mexico

Spain
Paraninfo Thomson Learning
Calle/Magallanes, 25
28015 Madrid, Spain

Table of Contents

- Answers to Chapter Exercises in Appendix -

To the Student:

Each chapter in this study guide is organized into three parts, a summary of the important points for each objective, a key word exercise, and a section of practice test questions and exercises for each chapter in the textbook. The key word exercise is meant to help you understand the critical definitions or concepts and tie them together for an overview of the material. Once completed, it should also be useful as a review of the material and definitions before a test.

A variety of exercises have been provided that will also help you assess your ability to apply the concepts. The last two to three exercises in each chapter are multi-objective and you should expect that they will be more challenging and that it will take you more time to develop the correct approach. One of the major goals of this guide is to help you develop a systematic way to solve problems in chemistry, so you won't need to be dependent on the problem wording to define the solution to the problem. Most students in general chemistry find knowing where to start the problem is the hardest part for solving homework or test problems. It isn't as bad as it seems, once you develop a systematic approach, but you will have to think about the situation described and should not expect to take all your clues from the numbers given in the problem. My students have consistently told me that these types of exercises build their confidence and get them ready for the test. I hope they will do the same for you.

This book is dedicated to my husband, Bob, and daughter, Jennifer, for their support and patience through this project and to my parents who taught me the value of learning.

Chapter 1

The Nature of Chemistry

This chapter serves as an introduction to the unique view of chemistry, both to the science and to its importance to our lives. The chapter begins to introduce you both to the methods chemists use to solve problems and also to start to build the framework for the study of chemistry. The basic ideas developed in this chapter involve properties of the materials we constantly work with. How we can identify, separate, measure and classify these substances are key factors you will encounter. Practice in imagining and describing chemical transformations at the nanoscale level will also be employed to help visualize the principles introduced in this chapter.

At the end of each chapter in the textbook, a number of learning objectives for the chapter are listed. Achieving these objectives will be crucial to developing a full understanding of the concepts in the chapter. This study guide is structured to help you achieve the objectives of the chapter. Key points about each objective and some problem-solving tips will be given for each. Exercises and practice tests that will help you assess your ability to achieve the objectives are given after the discussion of the objectives.

Objectives:

1.1 Describe the approach used by scientists involving problems.

1.2 Understand differences among a hypothesis, a theory, and a law.

1.3 Define quantitative and qualitative observations.

Use of the scientific method has allowed chemists, as well as other scientists, to develop a keener understanding of the complex world around us and then to use that understanding to manipulate natural processes. The three basic parts of the process, developing a hypothesis, law or theory can be understood by keeping in mind that the process moves from the very specific **(hypothesis)**, using observations and data collection in carefully designed experiments, to a broader statement which has no exceptions **(law)**, to a comprehensive sweeping statement **(theory)**, which can be then be used to explain and predict the result of a large number of experiments. The number of hypotheses and experiments needed to develop a law will be very large so that the process narrows from many facts to a few laws. It will narrow even further as a theory is developed, since the theory will explain several laws and many hypothesis, new and old.

hypothesis ⇨ *[experiments]* ⇨	**law** ⇨	**theory**
• *many very specific statements*	• *few, broad statements*	• *very few comprehensive statements*
• *may be proven true or false*	• *must have no exceptions*	• *may be modified and tested by ability to predict the result of new experiments*

In conducting experiments, chemists expend great effort in collecting data. The data collected must be complete and accurate. There are two basic types of data- **quantitative**, which uses numbers to describe some property and **qualitative** which uses descriptions in words and not numbers.

Another point this chapter addresses refers to the open-endedness of chemical data. Information is continuously being updated and this may have significant impact on how we evaluate a chemical material. We may have to change our minds (and theories) if the data justifies it. A crucial requirement of all scientific data is that it be reproducible and withstand the test of our peers, which is why communication between all areas of science and scientists is very important.

1.4 Identify physical properties of matter or physical changes occurring in a sample of matter.

1.5 Estimate Celsius temperature for commonly encountered situations.

1.6 Calculate the mass, volume or density given any two of the three.

Identification of materials is commonly done through the use of physical properties. Any property that you can observe or measure without causing the substance to change its identity is called a **physical property**. Altering physical properties, such as a changing from a liquid to a gas represents a **physical change** in the substance.

Physical changes, as well as physical properties, may be tied to the temperature of the substance. Temperature is a measure of the relative hotness or coldness of the substance. The temperatures at which a substance changes state, melting points and boiling points, are physical changes that are often used to identify the substance. The temperature scale used in science is the **Celsius temperature** scale, represented by °C symbol.

In United States, the Fahrenheit scale is used to measure temperatures. To compare the temperatures on the Celsius and Fahrenheit scales, two features need to be taken into account, the proportion and the zeros of the two scales. The proportion between the two scales is 1.8 degrees °F = 1.0 degrees °C, based on the melting and boiling points of water, and the 0°C for the Celsius scale is equal to 32°F. Consequently, the temperature will appear as a larger number when expressed in T(°F) instead of T(°C).

To estimate one from the other:

T(°C) → T(°F) *smaller → larger*	Multiply the value in T°C by the proportionality factor, 1.8 (or approximately 2) and add 32° to the result (to adjust for the zero). A larger number should be the result.
T(°F) → T(°C) *larger → smaller*	First adjust for the zero by subtracting 32°, and divide the value in T(°F) by 1.8 (or approximately 2). A smaller number should be the result.

Density is the measured ratio between two different properties, the mass and the volume. No matter how large or small a volume is measured, the proportion of mass to volume will be the same as long as the temperature is constant. The density is a physical property that can then also act as a proportionality or conversion factor. Given any two of the three values - mass, volume and density for the same substance- you can always calculate the third.

1.7 Identify the chemical properties of matter or chemical change occurring in a sample of matter.

1.8 Explain the difference between homogeneous and heterogeneous mixtures.

1.9 Describe the importance of separation, purification and analysis.

Chemical changes are different from physical changes in that the substance does not retain its identity and its characteristic physical properties upon undergoing chemical change. After the change, the physical properties of the new substances will be measurably different than what they were before the change. Chemical changes are summarized by **chemical reactions**, in which the starting materials called reactants appear to the lefthand of the reaction arrow and the new substances, the products, appear on the right hand side of the arrow. **Chemical properties** describe the chemical changes a substance can undergo, such as whether or not it will dissolve in water, react with oxygen in air to produce new substances, or break apart into new substances when heated above a certain temperature.

If a material is not pure, it is a mixture. Mixtures can be separated by physical processes, such as solubility differences, whereas pure substances cannot be separated any further. Other common methods of separation include distillation, chromatography and filtration. If phase boundaries exist between different components of the mixture that can be detected by the human eye (such as in a salt and sand mixture) or under a microscope, the mixture is **heterogeneous**. Making a mixture by dissolving one material in another, such as adding only salt to water, with stirring, makes for a very uniform system, which is then defined as a **homogeneous mixture or solution**. A homogeneous mixture always shows the properties of its components and the properties will be the same no matter where you sample the mixture or solution.

Using techniques of separation and purification are the only way a chemist can tell for sure if a substance is pure or is a mixture. Chemical analysis describes the process of determining the physical and chemical properties of a substance focusing on properties that are unique characteristic properties to identify substances.

1.10 Understand the difference between a chemical element and a compound.

1.11 Classify matter.

1.12 Describe characteristic properties of he three states of matter- gases, liquids and solids.

When chemists talk about a pure substance they mean either an element or a compound. Pure substances, at the microscopic level, can not be physically subdivided any further and still retain the original properties. Pure substances have definite fixed properties and a composition, in the case of compounds.

A **chemical element** is a pure substance that cannot be further broken down by chemical reactions. Each chemical element has its own symbol, name and unique set of chemical and physical properties. The fundamental building blocks of all substances are the chemical elements. Currently 112 elements are known and the symbols summarized in the periodic table.

Chemical compounds result when two or more elements undergo chemical reactions to form a specific grouping or combination. Unlike mixtures, compounds have a single possible composition, which cannot vary.

All three forms of matter- mixtures, compounds and elements- may exist in three different physical states of matter: solid, liquid or gas. To differentiate between the states, one needs to look at the whether both the volume and shape is fixed. **Solids** have a fixed shape and volume. **Liquids**, in contrast, have a variable shape, but the volume is fixed. **Gases** have variable shape and volume and are generally invisible to the naked eye. The state that an element, compound or mixture will be found in largely depends on the strength of attractions between its particles. These attractions weaken at higher temperatures. Consequently, as the temperature of a sample of matter is increased, its state changes from solid to liquid to gas.

1.13 Identify relative sizes at the macroscale, microscale and nanoscale levels.

1.14 Describe the kinetic molecular theory at the nanoscale level.

1.15 Use the postulates of modern atomic theory to explain macroscopic observations about elements, compounds, conservation of mass, constant composition and multiple proportions.

We are used to seeing matter interact and chemical reactions occur at the macroscopic level. When matter is visible to our eyes without instrumentation and we can easily transfer, measure and handle it, we are working on the **macroscale** with matter. Using instruments, we can extend the level of measurement to micrograms and microliters (masses and volumes too small to see without magnification) and we are measuring on the **microscale**. No instrumentation will easily measure matter at the **nanoscale** level, however, where masses would be in nanograms and volumes in nanoliters.

It is most important to note that nanoscale dimensions are equivalent to the dimensions of <u>single</u> particles of elements and compounds and at this level we must imagine the particles interacting to make mixtures and compounds, to connect to what we actually see happening at the macroscopic level.

The energy imparted to matter will determine the physical state it is in. The **Kinetic Molecular Theory** describes matter at the nanoscale level. A basic premise of this theory is that all particles of matter are in constant random motion. The energy imparted to matter will determine its physical state. As the temperature of matter increases, the motion of its particles also increases, attractions between particles lessen, particles move farther apart, and the physical state of the material changes.

The four postulates of the modern atomic theory given in this chapter form the corner-stone of the chemist's view of the connection between the nanoscale world and the macroscale world of matter. It is key feature of the theory that all the chemical and physical properties of macroscopic substances can be explained by looking at the structure at the nanoscale level. Tiny particles called **atoms** are the smallest form of matter that exhibit the physical and chemical properties we connect with macroscopic substances. Elements can then be defined as being composed of atoms that all have the same chemical and physical properties. Linking atoms together in definite groups results in **molecules** being formed. If the atoms are all the same, the molecule is still an elemental form. A **compound** is a collection of molecules, that have the same chemical composition, but are made up of different chemical elements.

1.16 Distinguish metals, nonmetals and metalloids according to their properties.

1.17 Identify elements that consist of molecules and define allotropes.

1.18 Distinguish among macroscale, nanoscale, and symbolic representations of substances and chemical processes.

1.19 Appreciate the balance of benefits and risks in our technological society.

Metals and nonmetals have fundamentally different chemical and physical properties. A heavy black line, a "zig zag" line or "staircase", that appears on the right-hand side of the table can always be used as a guide for differentiating metals, nonmetals and metalloids. Knowing the list of physical properties common to metals will also help to define those for nonmetals since they are often opposite to those of metals.

• Chemical elements that appear on the left side of the staircase are metals. The elements that appear below the main body of the table are also metals.

• Elements that touch the staircase are mainly metalloids, which show some properties of metals and some nonmetal properties.

• Elements to the right of the staircase are the nonmetals.

Allotropes, molecular forms of elements that differ in the number of atoms in the molecule, only occur in the nonmetals. The elements that form diatomic molecules are also nonmetals and are grouped together in the same part of the periodic table, with the exception of H_2.

Chemical reactions are represented by equations where arrows, $\rightarrow$, are used to separate reactants and products. There may also be notes over the arrows that give more information. For example, 100°C over the arrow means to heat to that temperature. When molecules appear, the material is represented with the proper chemical formula. Ions are represented with charge symbols. The state of matter may also be indicated by using the letters in parentheses (s), (l) or (g) to indicate solid, liquid or gas, respectively.

Building an appreciation of the risks versus benefits of chemical substances will be also be a part of your study. Assessing risk of a chemical substance is often very difficult to do but the knowledge of the fundamental chemical properties is some of the best information to use in an assessment, since these will not change.

Chapter Review - Key Terms

Chemistry uses the language of science and many other terms that are part of its own unique language. It is important that these terms become part of your vocabulary so that you can interpret the implications of the major terms as they come up. For this you need to develop a real understanding, not just the ability to restate a definition. The key terms introduced in each chapter are summarized at the end of each chapter. Within this study guide an exercise titled " Chapter Review - Key Terms" appears at the end of the objectives summary to help you practice putting the definitions in context which should aid with your understanding.

This chapter has a very large number of key terms in comparison to other chapters in the text so the exercise will appear much longer in this chapter than it will in others. However, you will find that deciding what word to use from the context will help you build both your proficiency and your confidence interpreting the language of chemistry.

In order to make all the sentences below TRUE, insert the appropriate word or phrase from the list of key terms which best fits the context of the sentence.
NOTE: Any phrase or word from the list may used more than once.

LIST:		
allotrope	gas	model
atom	heterogeneous mixture	molecule
boiling point	homogeneous mixture	nanoscale
Celsius temperature	hypothesis	nonmetal
chemical change	kinetic molecular theory	physical changes
chemical element	law	physical property(ies)
chemical formula	Law of Conservation of Mass	product
chemical property(ies)	Law of Constant Composition	proportionality factor
chemical reaction	Law of Multiple Proportions	qualitative
chemistry	liquid	quantitative
compound	macroscale	reactant
conversion factor	matter	solid
density	melting point	solution
diatomic molecule	metal	substance
dimensional analysis	metalloid	temperature
energy	microscale	theory

The Way Science is Done:

To accomplish and define science, a specific method is used to determine the facts or currently accepted "truths" about our world, called the scientific method. The process starts with a tentative explanation of an observation, or series of observations, called an (1) _____ . Experiments are then developed to test its validity, in which information is carefully collected as experimental data. If the information collected involves numeric data, then (2) _____ data has been obtained. In contrast, if no numeric experimental data is collected, then the data is termed (3) _____ data From a number of experiments, a statement with no known contradictions called a scientific (4) _____ can be developed. The (5) _____ must also be shown to reliably predict the results for any new (6) _____ related to it.

As the next step, a unifying principle that explains a large body of facts or scientific (7) _____s called a (8) _____ can be developed. To make the scientific (9) _____ more concrete, easily visualized or to put it in mathematical or physical form, a (10) _____ is often constructed. Open communication between all members of the scientific community ensures that the scientific method continues to produce new knowledge, allowing a new (11) _____ , (12) _____ or (13) _____ to be tested by many different laboratories and the data, whether (14) _____ and (15) _____, to be analyzed by many scientists before becoming accepted as fact.

Chemistry and Matter- Properties and Transformations

(16) _____ is the branch of science that focuses on the understanding the transformations of (17) _____, which is anything that occupies space and has. Change in either one or both of the two primary types of properties, (18) _____ or (19) _____, indicates a transformation of (20) _____ has occurred. Properties that can be observed or measured without changing the composition of are called (21) _____ and are often used to identify substances. For example, the (22) _____ that relates the volume of a sample to its mass is called the (23) _____ and is characteristic of that substance. A process in which these properties change, but the same substance is present before and after the change, describes a (24) _____ in the substance. This type of change can often be induced by transferring heat to or from the substance, resulting in a (25) _____ change for the substance. (26) _____s are different from (27) _____s in that the substance does not stay the same, but instead is converted to a new substance with a different set of properties. Two examples of (28) _____s classified by the temperature are represented by the (29) _____ and (30) _____ of a substance. In scientific measurements, the (31) _____ scale, represented by the symbol, °C, is the one typically used.

Given that the state and temperature is constant, an increase in volume of a sample will increase its mass, but the proportion (or ratio) of mass to volume will not change, represented by the (32) _____ of sample. A (33) _____ or a (34) _____ defines a relationship between two different type of units for the same substance. Using the (35) _____ as a (36) _____ allows one to determine the mass from the volume of a substance or vice versa. Including the units is an essential part of the calculation, since the units both show how the calculation should be set up and act as valuable check on the set up of the calculation. The approach of using both the units and the values for mathematical problem-solving is called (37) _____.

Working with Chemical Transformations:

When a substance is altered and forms a new substance that has a different composition during the transformation, a (38) _____ or (39) _____ has occurred. The (40) _____ of a substance are used to describe the changes a substance can undergo, to indicate how it will behave when combined with air or water, exposed to sunlight or mixed with other substances. To describe the chemical transformation in writing, an equation is written in which the starting substances that appear to the left of the arrow are called the (41) _____s, and the new substance(s) that appear after the transformation is complete are called the (42) _____s. As one of its characteristic features, the total mass of the (43) _____s in a (44) _____ will always equal the total mass of the (45) _____s, within detectable levels. The law that describes this feature is called the (46) _____. (47) _____s are also always accompanied by a transformation of (48) _____, which is defined as capacity to do work. The relationship between (49) _____ and the transformations represented by (50) _____s is a very important one in the study of (51) _____.

Classification of Types of Matter

When substances as combined together but no chemical reaction takes place, then a

(52) _____ has been formed, which can be one of two types. (53) _____s have an uneven texture, usually indicating boundaries between phases within the sample, and varying composition throughout the sample, which can either be seen by eye or through the optical magnification of a microscope. In contrast, a (54) _____, also called a (55) _____, consists of two or more substances in the same phase, so that no boundaries are visible, even under a microscope. No matter where it is sampled, the composition and (56) _____ of a (57) _____ are always the same throughout the sample.

Much of what is known about chemistry has come from separating and purifying the components of a mixture to study the properties of the components. Once a variety of methods of purification produces a substance with the same, unique set of (58) _____ and (59) _____, a pure substance has been obtained. Using a (60) _____, like a decomposition reaction, can be very useful in distinguishing the two types of pure substances, an (61) _____ from a (62) _____. An (63) _____ is a single substance that cannot be separated into two or more new substances by a chemical reactions (such as decomposition). If a single pure substance can be decomposed into two or more new substances, then it must be a (64) _____. Although both are composed of two or more substances, a (65) _____ is different from a (66) _____ in that it has a composition and set of properties that cannot be varied.

The three most common states of matter are (67) _____, (68) _____ and (69) _____. Distinguishing between the three states requires looking at three attributes- mass, volume and shape. In the (70) _____ state a substance has a definite mass, volume and shape, whereas in the (71) _____ state it still has a definite mass and volume, but not shape. In contrast, when the substance is characterized by having no fixed volume or shape, it is a (72) _____. Because of the large separations between the particles, (73) _____es always mix to form (74) _____s.

Dealing with Matter on the Nanoscale Level:

Matter whose properties can be detected with the unaided human senses represent a view at the (75) _____ level, which is typically how we are used to seeing matter and its interactions. Matter that is so small that its properties can only be seen with a microscope represents the (76) _____ level. But when the scale gets so small that even the best microscopes cannot let us " see" the matter, we have to imagine the particles of matter. We then are "seeing" interactions on the (77) _____ level. This term was chosen since the actual dimensions of particles at this level would be measured in nanometers, one-trillionth of a meter. To connect the (78) _____ world to what we measure at the (79) _____ or (80) _____ level, we must use theories developed using the scientific method.

One such theory is the (81) _____ which explains the properties matter by describing the action of particles at the (82) _____ level. Its major premise is that all matter is composed of extremely tiny particles called (83) _____s that are in constant motion at any given temperature. This theory can then explain the properties and differences observed between the three states of matter at the typical, (84) _____ level.

Another theory, the Modern Atomic Theory, a modification of Dalton's Atomic Theory, says that the (85) _____ is the smallest particle that will show all the (86) _____ that define a (87) _____. (88) _____s are then explained as being distinct combinations of different (89) _____s generally in small whole number combinations. Dalton's atomic theory explained why the (90) _____ applied to

(91) _____s. The identity of an (92) _____ and its mass would not be altered during the (93) _____, but the arrangement of the atoms would change which resulted in a new set of (94) _____. Thus a (95) _____ can be visualized on the (96) _____ level by the joining, separating or rearranging of (97) _____s, which are then never lost nor created as a result. That a postulate of the Dalton's atomic theory describes a (98) _____ as a small whole number combination of two or more different elements also explains two other laws: the (99) _____ and the (100) _____.

Each (101) _____ has a unique name and symbol, which is composed of one or two letters in which only the first letter is capitalized. The vast majority of the (102) _____s are (103) _____s which are typically in the (104) _____ state at room temperature and are ductile, malleable and good conductors of electricity and heat. (105) _____s, much fewer in number, do not conduct electricity and show a much more diverse set of (106) _____, such as their physical state at room temperature. In the periodic table, elements that appear between (107) _____s and (108) _____s, but show some properties of both, are called (109) _____s.

Describing Chemical Substances:

In the elemental form, (110) _____s also frequently group together to form (111) _____s in which two or more (112) _____s are chemically bonded together. If the (113) _____ consists of two identical (114) _____s bonded together it is called a (115) _____. A (116) _____ uses symbols combined with subscripts to indicate how many of each type of (117) _____ bonded together in the (118) _____. In a (119) _____ two or more different (120) _____s will be represented in the (121) _____. (122) _____s are different molecular forms of the same (123) _____ in which the subscripts for the atom are different, such as in the molecules O_2, oxygen, and O_3, ozone.

- PRACTICE TESTS -

After completing your study of the chapter and the homework problems, the following questions can be used to test yourself on how well you have achieved the chapter objectives.

1. Tell whether the following statements are either **True or False**-

   A) With further testing hypothesis may become theories but not laws.

   B) The physical state of a substance depends on temperature.

   C) Solutions are examples of homogeneous mixtures that only occur when substances are mixed in the gaseous state.

   D) A chemical formula that describes a molecule must be made up of at least two different chemical symbols and have subscripts to represent the number of atoms.

   E) Energy changes accompany chemical changes.

   F) Atoms are rearranged in chemical and physical changes which is why both are considered transformations of matter.

   G) A product of a reaction may have the same physical properties as a reactant but not the same chemical properties.

2. Classify each of the following statements as a **theory, law or hypothesis**:

   (a) All matters is composed of atoms   (b) 10 X 10 = 100

(c) Black cats are bad luck   (d) It does not snow in the month of July
(e) the gravitational pull of the moon causes the ebb and flow of ocean tides

3. What does the phrase "established scientifically" mean?

4. Tell whether the following properties given is a **chemical or physical property**:
  (a) Gold has a relatively low melting point for a metal of 1064°C.
  (b) Pieces of gold can be hammered into thin sheets called gold leaf
  (c) Gold does not dissolve in water
  (d) Gold is often found naturally in a pure form, such as gold nuggets

5.   A) If a outside temperature is measured as 70 °F, what is the equivalent temperature in °C?
  (a) 70 °C   (b) 39 °C   (c) 21 °C   (d) 35 °C   (e) none of these

  B) If the temperature at the top of a mountain is -15°C, what would the temperature be in degrees Fahrenheit?
  (a) -4 °F   (b) + 5 °F   (c) -27 °F   (d) +27 °F   (e) none of these

6. Tell whether each of the following represent a **chemical or physical change**:
  A) gaseous bromine burn in an atmosphere of hydrogen gas to form a compound
  B) water evaporates from the ocean and returns to earth as rain
  C) a cube of cane sugar is dissolved in a cup of tea.

7. A chemist collected the following pieces of information as data as part of an experiment. Tell which would be **quantitative and which qualitative**:
  A) The material isolated had a yellow-green color
  B) The density of the material was 3.568 g/mL
  C) The material is a solid at 20°C and has a distinct odor
  D) At 20°C, 29.3 g of the material dissolved in 1.0L of water
  E) The material is slightly soluble in alcohol
  F) Crystals of the material had a distinct geometric shape.

8.   The following are the results of an analysis of two samples which contain only P and O. Tell whether the samples are from the same or different compounds. Explain the reason(s) for your choice.   Sample# 1   2.581 parts Phosphorus   3.322 parts Oxygen
  Sample# 2   3.718 parts Phosphorus   2.881 parts Oxygen

9. Given the following changes for substance A:
  (a) Substances A and B form a heterogeneous mixture
  (b) Substances A and C form a solution
  (c) Substances A and D are chemically combined to form a compound
  (d) Substance A(l) is boiled to produce gaseous substance A
  (e) Substance A is decomposed to give substances M and Z

  A) For which of the changes will the chemical properties of substance A stay the same?
  B) Is A and element or compound? Explain the reason(s) for your choice.

10.   For the following: (1) identify the substance(s) that is(are) undergoing change and (2) tell whether the substance is undergoing physical change or chemical change:
  A) Charcoal is ignited and burned and marshmallows are toasted over the burning coals
  B) Margarine is melted in a frying pan, the pan is heated, beaten eggs are added and then scrambled.
  C) Silkworms feed on mulberry leaves and produce silk
  D) Sheep are sheared and the wool is spun into fibers and knitted into a sweater

11.   A) A chemist received a material which could not be purified further. Can the chemist tell at this point whether the material is a mixture, compound or element? What further steps could be taken to be sure which of the three he had been given?

B) What theory would best explain the changes observed when water freezes? Explain your choice.

12.   A) The text tells you that ancient people knew of nine elements- silver, gold, copper, tin, lead, mercury, iron, sulfur and carbon. Give the correct symbol for each and indicate whether the element is a metal, nonmetal or metalloid.

B) Arsenic and thallium are both used as poisons in detective novels. Are the pure substances metals, nonmetals or metalloids?

13. Suppose an experiment takes place in four steps. First, the flask it is evacuated so that it contains no gas particles. Then dry ice, solid CO_2 is introduced into the evacuated flask. Thirdly, the flask is warmed and the dry ice begins to convert to a gas. At the end of the experiment, the flask is completely filled with $CO_2(s)$ gas. Put the four flasks pictured below in the sequence that would describe the four steps of the experiment on the nanoscale.   [● = carbon dioxide molecule]

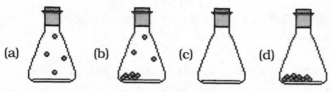

14. The density of bismuth metal is 9.8 g/mL and that of hafnium metal, 13.4 g/mL.
    A) Which would contain the greater mass, 5.0 mL of bismuth or 3.5 mL of hafnium?
    B) If you mixed together 5.0 mL of each metal, what would the total mass be?
    C) If the two metals reacted to form a compound would the total mass be the same
    or different than the mixture is formed in (B)? Which theory best explains your
    answer to this question, the kinetic molecular theory or the modern atomic theory?

15. Suppose 1.00 g of butter (density = 0.860 g/mL) is thoroughly mixed with 1.00 g of sand (density = 2.28 g/mL).
    A) Would the mixture be homogeneous or heterogeneous?
    B) Would the mixture have a density greater then, less than or equal to that for
    sand? Show calculations to support your answer.
    C) Describe what you assumed to be true about the volume of the mixture, in your
    answer to (C) and how this is explained on the nanoscale by the kinetic molecular
    theory of matter.

Chapter 2

Atoms and Elements

The purpose of this chapter is to introduce the fundamental particles within the atom, how the particles relate to the chemical properties of the elements, and the organization of the periodic table of elements. The primary counting unit for the chemistry, the mole, which is used in nearly all chemical calculations, is also defined. This counting unit lets chemists relate what is happening in chemical changes on the nanoscale to the macroscopic scale of grams and is essential in the study of chemistry. Common standard international (SI) units, and conversions between them and other systems, are also introduced and practiced, using dimensional analysis.

2.1 Describe radioactivity, electrons, protons, and neutrons and the general structure of the atom.

The study of radioactive elements, where an element spontaneously emitted invisible radiation detected by exposure of photographic plates, provided irrefutable evidence that atoms could be divided into smaller parts, called subatomic particles. Rutherford's experiments showed that subatomic particles exist in two fundamental regions of the atom, a positively charged **nucleus** and a surrounding region occupied by **electrons**, particles of negative charge. Further experiments proved the nucleus contains both **protons** and **neutrons**. The proton has a positive charge and is about the same mass as the neutron, which has no charge. The nucleus make up a very small part of the volume and is densely packed. The electrons occupy most of the volume of the atom, but contribute very little mass to the atom.

Evidence has shown that there are optimum values for the ratio of protons to neutrons in a nucleus. Nuclei that fall within these optimum values are stable. **Radioactivity** is the result of an unstable nucleus adjusting its ratio of protons to neutrons. There are three common types of nuclear decay: alpha (α), beta (β) and gamma (γ).

The number of protons determines the atom's **atomic number**, symbolized by **Z**. An atom's chemical identity is determined by the number of protons in the nucleus. If the number of protons in an atomic nucleus changes, that atom's identity changes.

In a neutral atom, the number of protons always equals the number of electrons. Since the electrons are in the region of contact where atoms interact, they play a fundamental role in chemical reactions and may be lost or gained by atoms. If the number of electrons becomes greater than or less than the atomic number, an **ion** results.

The **mass number** gives the total number of particles in the nucleus that have mass. The mass number, also called **atomic mass**, is then equal to the number of protons plus the number of neutrons and is symbolized by **A**. The terms **atomic weight** and atomic mass are used interchangeably. The notation for the full symbol for an atom is $_Z^A X$, where X is the one or two letter symbol for the element. The A term, the mass number, is written in the upper lefthand corner, and the Z term, the atomic number, in the lower lefthand corner, but is optional.

2.2 Use conversion factors for the units for mass, volume, and length common in chemistry.

Chemistry is a quantitative science. The use of measurement and units is a cornerstone of a chemist's work. Chemists employ the SI modified version of the metric system. You need to learn some of these units and conversions. Table 2.2 summarizes many of the conversions between units and will be an important resource for you.

The most frequently used units for general chemistry are:

Length: meter (m) centimeter (cm) millimeter (mm)
Volume: liter (L) milliliter (mL)
Mass: kilogram (kg) gram (g) milligram (mg)

Generally, the units involving powers of three - kilo, milli, micro, nano, and pico - are the most common and you will need to know what quantities they represent. Some guidelines for conversions are:

• When **converting between units within the metric system**, you can use your knowledge of the Greek prefixes to write the conversion factors to use between units.

• When **converting between the English system and metric units**, know the major conversions for each type:
length 1.0 in = 2.54 cm
volume 1.0 L = 1.057 quarts
mass 1.0 lb = 454 g

• Keep in mind that **some units can be interchanged**, for example 1.0 mL always equals 1.0 cm³.

Dimensional analysis, introduced in Chapter 1, is the primary method for unit conversion problem-solving. A major advantage is that the conversion factors can be used to guide you in the mathematical sequencing of the solution. To construct the plan for solving unit conversion problems by dimensional analysis, there are three basic question to ask:

First, what units are needed for the final answer?

Second, what units are stated in the problem?

Third, what is the relationship between the units stated in the problem and those needed in the answer?

Further guidance in problem-solving will result from asking the following questions:

• *Are the units I want and the one I know the same type, i.e . am I being asked to convert length to length? Volume to volume?*

• *Are the units I want and the one I know in the same system or am I asked to switch between systems, i.e. am I asked to convert meters to inches? Pints to milliliters?*

• *If the units are not the same type, for example being asked to convert mass to volume, then the density of the substance is needed as one of the conversion factors.*

2.3 Define isotope and give the mass number and number of neutrons for a specific isotope.

2.4 Calculate the atomic weight of an element from isotopic abundances.

2.5 Explain the difference between the atomic number and the atomic weight of an element and find this information for any element.

For a particular element, the number of neutrons in a nucleus can vary by several units even though the number of protons stay the same. Atoms of the same element that have different numbers of neutrons, and thus different atomic masses, are called **isotopes**. For an example, a carbon atom can have 6 protons and 6 neutrons, Carbon-12 or ^{12}C, or have with 8 neutrons with 6 protons, Carbon-14 or ^{14}C. The second isotope is radioactive and is used in the technique known as carbon dating. (There is also a Carbon-13 isotope.)

The atomic number, Z, of an element gives the number of protons in that element. In neutral matter, the number of protons equals the number of electrons. The mass number, A, or atomic mass is equal to the number of protons and neutrons in an element. Therefore, the mass number minus the atomic number, A-Z, will give you the number of neutrons in the nucleus of that atom.

Number of neutrons = A - Z = [mass number] - [atomic number]

The atomic mass of an element listed on the periodic table represents a weighted average of all isotopes found in nature for that element. This average mass depends on how many isotopes are present, the mass of the individual isotopes and the percent abundance of each isotope. The average atomic mass is determined by multiplying the mass number for each isotope times its relative abundance (% abundance/100) and adding the resulting values.

2.6 Relate masses of elements to the mole, Avogadro's number, and the molar mass.

2.7 Do gram-mole and mole-gram conversions for elements

The **mole** is the chemist's primary unit for measuring how many atoms, molecules or ions are present or reacting. It is an essential unit for chemical calculations involving gram quantities. A mole of any material has a fixed number of particles in the same way as a dozen describes a fixed number (12) of things. A mole contains 6.02×10^{23} particles of a substance. The numerical value of the mole (6.02×10^{23}) is known as **Avogadro's Number**.

The mass of each element given in grams will contain Avogadro's number of particles (or one mole). This means that 6.94 grams of lithium and 200.59 grams of mercury will contain the same number of atoms. If you need the number of atoms or molecules of substance Y present in a sample, or the reverse, Avogadro's number is used as the conversion factor.

$$\text{Number of particles Y} \times \frac{1 \text{ mole Y}}{6.02 \times 10^{23}} = \text{Number of moles Y}$$

It is imperative for you to be able to convert from gram amounts to mole amounts and the reverse. This mass-to-mole conversion is one of the basic steps in chemistry. When solving such conversion problems, look at the problem data and then use atomic mass values that have one more significant digit than the best data value.

2.8 Identify the periodic table location of groups, periods, alkali metals, alkaline earth metals, halogens, noble gases, transition elements, lanthanides and actinides.

One of the most useful organizational tools in chemistry is the Periodic Table. The original table was based on atomic weights of the elements. The present version is based on the atomic number (number of protons) in the nucleus which follows the law of chemical periodicity.

Groups, often called families, are vertical columns in the periodic table. Elements within the same group have similar chemistry. Some groups have been given names that are used to identify them by relative properties. Common group names are the **alkali metals** and **alkaline earth metals**, which appear on the left side of the table; the **halogens**, which appear on the right side of the table; and the **noble gases**, which is the last group in the table.

Periods are the horizontal rows. The elements in periods undergo periodic changes from metallic to nonmetallic to nonreactive. Notice that some periods are much longer than others. The significance of this will be discussed in later chapters.

The periodic table is commonly divided into major divisions known as the **main group elements**, usually identified with a letter A after the group number, and the **transition elements**, identified with a letter B after the group number. The 3A group contains main group elements and the 8B group is made up of the transition elements.

The two rows shown below the table are actually part of the sixth and seventh periods. The first of these, the **lanthanides** are found in the natural world in trace amounts and are also called the "rare earths". The second row, the **actinides**, are radioactive in all isotopic forms.

Chapter Review - Key Terms

In order to make _all_ the sentences below TRUE, insert the appropriate word or phrase from the list of key terms which best fits the context of the sentence.

NOTE: Any phrase or word from the list may used more than once.

LIST:		
actinides	ion	neutron
alkali metals	isotope	noble gases
alkaline earths	lanthanides	nucleus
atomic mass unit	Law of Chemical Periodicity	percent abundance
atomic number	main group elements	period
atomic structure	mass	periodic table
atomic weight	mass number	proton
Avogadro's number	mass spectrum	radioactivity
conversion factor	mass spectrophotometer	scanning tunneling microscope
electron	metric system	transition elements
group	molar mass	
halogens	mole	

Internal Structure of the Atoms:

The discovery that led to a major modification of Dalton's Atomic Theory that atoms could change their identity by emitting invisible radiation (only detected by the exposure of photographic film by the radiation) is a phenomena called (1) _____ . Experiments by scientists such as Madame Curie, Rutherford, Millikan and Thompson verified the existence of subatomic particles which revealed the inner (2) _____ . The atom is always composed of two regions. One exists at the center of the atom, is an extremely small dense form of matter called the (3) _____ while the other region is very diffuse, the second occupies nearly the all the space within the atom, and contains the (4) _____ . The (5) _____ is an instrument that allows us to see an image of single atoms or molecules on a surface.

Atoms are composed of subatomic particles, two of which have mass, the (6) _____s and (7) _____s, while the third, the electron, has virtually no mass in comparison. Both the (8) _____s and (9) _____s have charges, the first being positive and the second negative, but the (10) _____s have zero charge. The (11) _____s are essential for overcoming the repulsive force of the like charges of the (12) _____s in the nucleus. In the normal state, an atom has equal numbers of (13) _____s and (14) _____s producing atoms of zero charge. Atoms that lose or gain (15) _____s become (16) _____s, but they will retain their chemical identity. An instrument called a (17) _____ records the mass and abundance of (18) _____s on a graph called a (19) _____.

In order to classify atoms, the (20) _____, represented by the symbol Z, is used to give the number of (21) _____s in the nucleus which sets the unique chemical properties that identify each element. The (22) _____ for the atom represents the sum of the (23) _____ and (24) _____s in the (25) _____ of the atom. The mass of a single atom or molecule can be given in whole numbers if units of (26) _____ are used.

Symbols and Organization of the Elements:

Within the atomic symbol $^A X$, A represents the (27) _____, which is also called the (28) _____of the element. Because of the existence of (29) _____s, an average atomic mass is given on the periodic table for the elements. (30) _____s are atoms that have different numbers of (31) _____s but the an equal number of (32) _____s which have the same chemical properties. A normal sample of an element will contain all (33) _____s of the element and the fractional amount of each within the sample is called the (34) _____ and can be tabulated.

In the (35) _____, atoms are organized into rows called (36) _____s and columns called (37) _____s. The (38) _____ tells us that we can expect elements within a (39) _____ to have similar chemical properties. Special names have been given to different parts of the table. The two rows that appear outside the main body of the table are the (40) _____ and (41) _____, that are relatively rare elements in nature. The elements which appear in the middle section of the table and have a "B" after the group number are the (42) _____. The elements with an A included in the group number are called the (43) _____ and are present in larger quantities in the natural world. The 1A group of metals is called the (44) _____, whereas 2A metals are called (45) _____. A group of elements that appears on the right side of the table that are very unreactive are called the (46) _____. Next to this group are elements that exist as diatomic molecules and are very reactive called the (47) _____.

Units for Chemistry

Chemists and other scientists typically use a combination of standard international (SI) units, and the (48) _____ for the recording and reporting of measurments. The defined relationship between these units (and other systems of units) function as a (49) _____ within a calculation, and used to change the unit to a smaller or larger unit in the same or different system. To relate the (50) _____ of a single atom or molecule, which is on the nanoscale, to a more convenient, macroscale counting unit, a quantity called the (51) _____ is defined by chemists to track amounts of atoms or molecules in calculations. The (52) _____ of one (53) _____ is called the (54) _____ of the atom or molecule. (55) _____, the exact number of atoms in 12.00 grams of carbon, is the number of particles that must be contained in one (56) _____ of anything_. The mass of a single atom or molecule has the same value when it is expressed in the nanoscale unit of (57) _____ as does when expressed in the macroscale unit of grams as the (58) _____.

- PRACTICE TESTS -

After completing your study of the chapter and the homework problems, the following questions can be used to test yourself on how well you have achieved the chapter objectives.

1. Tell whether the following statements are either **True or False**-
   A)   All radioactivity involves a change in the nucleus of the atom.
   B)   The main group elements in the periodic table include all elements that are metals, metalloids and nonmetals, but not the lanthanides and actinides.
   C)   If you know the atomic number of an element, but not its mass, you can still identify it.
   D)   A mass spectrometer can separate isotopes of an element.
   E)   Most of the mass and size of the atom are accounted for by the nucleus.
   F)   Carbon-14 and Carbon-12 would have the same chemical properties, but different physical properties.

1.   G)   Scanning tunneling microscopy is a method used to show the position of single atoms by interacting with the nucleus.

           H)   The Law of Chemical Periodicity says that the observed physical and chemical properties of an element varies periodically with its atomic mass.

2. Match the element in the box below with one of the properties (A-L) in the list below.
   NOTE: Each element (a) - (l) can be used only once.

____	(A)	Not a main group element		(a) N (Z = 7)
____	(B)	Is an alkaline earth element		(b) O (Z = 8)
____	(C)	A generally unreactive, noble gas		(c) Si (Z =14)
____	(D)	Forms a gaseous diatomic molecule at room temperature		(d) Ar (Z = 18)
____	(E)	Is a halogen in period 4		(e) Br (Z = 35)
____	(F)	Is a metal in group 6A		(f) Rb (Z = 37)
____	(G)	A group 5A nonmetal with the smallest value for Z		(g) Sr (Z = 38)
____	(H)	Is a transition metal		(h) Rh (Z = 45)
____	(I)	Is an alkali metal		(i) In (Z = 49)
____	(J)	Is a metal in group 3A that has 66 neutrons.		(j) Bi (Z = 83)
____	(K)	Is a metalloid whose A value is double its Z value		(k) Po (Z = 84)
____	(L)	Is a metal in group 5A		(l) Pu (Z = 94)

3. Consult Table 2.2 in the text (p. 51) as needed to answer the following:

     A)   If you had a beaker that contains 5.0 liters, would it be bigger, smaller or the same size as a beaker that held 5.0 quarts?

     B)   Which is the longer distance, 10 kilometers or 5 miles?

     C)   In Ireland a person's weight is measured in "stones" where 1 stone is equal to 14.0 pounds. Convert your own weight, in pounds, to (a) stones and (b) kilograms.

4. Complete the following conversions (Consult Table 2.2 in the text (p. 51) as needed)

     A)   Convert 3010 cm^3 to m^3.

     B)   Convert 50 yards to km.

     C)   Convert 10 cubic feet (ft^3) to m^3

5. A water bed has the dimensions 8.0 ft X 7.0 ft X 8.0 inches.

     A) How many gallons of water are needed to fill the waterbed?

     B) If the density of water is taken as 1.0 g/mL, what would be the weight of water used in pounds, to fill the waterbed?

6. Answer the following questions:

     A) For the following list of atoms: ^{31}Si, ^{35}Cl, ^{33}P, ^{28}Si, ^{29}Al, ^{33}S

         (a) which are isotopes of each other?

         (b) which would have the same number of neutrons?

     B) Identify the element and give the symbol for an atom, using the $^A X$ notation:

         (a) that has a total of 60 for the sum of its protons, neutrons and electrons, with equal numbers of each.

         (b) with a mass of 234 where there are 8 neutrons for every 5 protons.

7. Briefly but completely answer the following:

A) What accounts for the fact the atomic number (Z) are always whole numbers, but atomic weights for elements on the periodic table are not whole numbers?

B) Which statement would be true concerning the masses of individual copper atoms, that *all, some or none* have a mass of 63.546 amu and why?

C) Which of the 3 symbols given below convey the same information and why?
$$^{57}_{26}\text{Fe}, \quad _{26}\text{Fe}, \quad ^{57}\text{Fe}$$

8. An element is found to have two isotopes that have masses of 78.92 amu and 80.92 amu, with 49.31% of the atoms being the heavier isotope. What is the average atomic weight and the identity of this element?

9. Sulfur has an average atomic mass of 32.064. There are four natural isotopes with masses of 31.972, 32.971, 33.968, and 35.967. Given only this information, which of the isotopes do you expect to have the greatest natural abundance and which the least?

10. A) Arrange the following in order of increasing mass:

(a) 2 K atoms   (b) a Br_2 molecule   (c) a Cesium atom

B) If two atoms of Neptunium have masses of 3.836542×10^{-22} g and 3.869792×10^{-22} g, what are the molar masses of the two atoms?

11. Given the following isotopes and sample weights:

Isotope:	^{24}Mg	^{52}Cr	^{124}Sn	^{59}Co	^{35}Cl
Sample weight:	12.0 g	20.8 g	42.5 g	25.0 g	12.0 g

A) Considering the isotopes:

(a) which isotope(s) have an equal number of neutrons and electrons?

(b) which isotope has the greatest ratio of neutrons to protons in the nucleus?

B) Considering the sample weight of each isotope:

(a) which sample weights contain the same number of moles of atoms?

(b) which sample weight contains the largest number of moles of atoms?

12. Chalcopyrite a bronze-colored mineral is 34.67% Cu, 30.43% Fe, and 34.94% S by mass.

A) How many moles of Cu would be in 125 g chalcopyrite?

B) How many grams of chalcopyrite would contain 3 moles of S?

13. A box of aluminum foil contains a single piece of Al that is 1.0 m long, 304 mm wide and 0.60 mm thick and the density of Al is 2.70 g/mL.

A) How many grams of Al are in the piece of foil?

B) How many moles of Al are in the piece of foil?

C) If you have a piece of gold, Au, of equal volume, how much would the foil weigh? (density Au = 19.31 g/mL)

D) If the two pieces of metal could be melted and form a homogeneous mixture, how many atoms of Au would there be for each Al atom in the mixture? (State in simplest whole number ratio)

Chemical Compounds

There are millions of different chemical compounds, yet each can be classified as belonging to one of two major classification systems. Either a compound is **organic**, which means it contains carbon, or it is **inorganic**, which means it does not contain carbon. Alternatively, all compounds can be classified as being either **ionic** or **molecular**. The latter system is the most useful for the study of general chemistry, since many of the properties we study and the naming system used for compounds is based on this classification.

Through the study of this chapter you will learn how to recognize the two types of compounds, ionic and molecular, their basic structures, learn to name compounds and know what to expect for some of the chemical and physical properties of the compounds. In addition, the mole concept introduced in Chapter 2 will be extended to compounds which will increase your ability to interpret chemical changes. You will also become familiar with describing the composition of compounds, either through a molecular formula or percent composition by mass, which are both extremely useful tools in the mathematics of chemical conversions. Lastly, the importance of ions (from ionic compounds) and carbohydrates (molecular compounds) in biological systems is discussed.

3.1 Interpret the meaning of empirical formulas, condensed formulas, and structural formulas.

3.2 Name binary compounds, including straight chain alkanes.

3.3 Write structural formulas for and identify straight- and branched-chain alkane constitutional isomers.

Formulas for molecular compounds can be given in several forms. The **empirical formula** represents the simplest ratio of atoms in the compound. It may not give a true picture of the molecule in a molecular compound, but is always the correct combination in ionic compounds. The **molecular formula** shows the actual number and kinds of atoms in one molecule of the compound and is a whole number multiple of the empirical formula.

In **structural formulas**, chemists use lines between atoms to represent the bonds that result from sharing of electrons. Structural formulas are not used to represent ionic compounds since the binding force results not from sharing, but from the attraction of opposite charges. In **condensed formulas**, no lines are drawn, but the atoms bonded together are indicated in groups.

Binary molecular compounds are formed from the reaction of two nonmetals. The compound is named using the element name of the first nonmetal and naming the second by adding an "ide" suffix to the root of its name. The number of each nonmetal is placed in the name as a prefix to the element name. The mono prefix is not used for the first element. For example, CS_2 is named carbon disulfide, not monocarbon disulfide, and CO is named carbon monoxide. It is important to know the common Greek prefixes given below:

1	mono-	4	tetra-	7	hepta-	10	deca-
2	di-	5	penta-	8	octa-	12	dodeca-
3	tri-	6	hexa-	9	nona-		

The largest group of binary molecular compounds are the **hydrocarbons**, which contain only carbon and hydrogen. The **alkanes**, represented by general molecular formula, C_nH_{2n+2}, are an important class of hydrocarbons. The naming system for alkanes provides the basis for naming all hydrocarbons. In order to name an alkane, the parent chain (the longest continuous chain of carbons in the molecule) must first be identified. The alkane name consists of a prefix indicating the number of carbons in the parent chain followed by an "ane" suffix. For example, C_8H_{18} is called octane since the prefix for eight is "octa". The prefixes used for hydrocarbon names are given below. Note that chains with 1 to 4 carbons have different prefix names than the Greek prefix for that number.

Carbons in chain	1	2	3	4	5	6	7	8
Prefix	meth-	eth-	prop-	buta-	penta-	hexa-	hepta-	octa-

Isomers are very common in hydrocarbons and are the result of being able to put the same number of atoms together in different arrangements which produces isomers with different chemical and physical properties. **Constitutional isomers** appear in hydrocarbons with 4 carbons or more. The carbon atoms may be arranged in a straight chain or as a shorter chain with branches attached to it. The branches are **alkyl groups** which are fragments of an alkane molecule minus a hydrogen atom. Alkyl groups are named using the prefix indicating the number of carbons in the group followed by an "yl" suffix. For example, a branch having the structure C_2H_5- or CH_3CH_2 is called ethyl.

Functional groups are the chemically active parts of organic molecules. In this chapter you will encounter several functional groups. The first of these is the **alcohol** functional group, an OH group. Simple alcohols are named by replacing the "e" in the appropriate name for the parent chain with an "ol" suffix. For example, CH_3CH_2OH is called ethanol.

The naming of alkanes with side chains and functional groups follows very specific rules. For our purposes only a few rules are needed:

(1)   Using the structural or condensed formula, identify the parent chain. The name of this chain will appear <u>last</u> in the molecule name.

(2)   Numbers are used to identify where an alkyl or functional group is located on this chain. The numbering is always set so the smallest numbers are used.

(3)   The branches are named as alkyl groups.

3.4 Predict the changes on monatomic ions of metals and nonmetals.

3.5 Know the names and formulas of polyatomic ions.

Some elements gain or lose electrons when forming compounds. Ionic compounds, formed when a metal reacts with a nonmetal, are an example of this behavior. Ionic compounds contain **cations** (positive ions resulting from electron loss by metals) and **anions** (negative ions that result when nonmetals gain electrons). The position on the periodic table can be used to predict the charges on monatomic ions.

*   **Nonmetals in group 4A-7A** form negative ions and the charge is normally 8 minus the group number, e.g. O (group 6A) forms O^{2-}.
*   **Group 1A-3A metals** have a positive charge equal to group number, e.g., K (group 1A) forms K^+, Mg (group 2A) forms Mg^{2+} ions
*   **Metals from 4A-6A and the transition metals (B groups)** have several possible positive charges and the group number represents the maximum positive charge. Predicting the charge of these elements is not possible, so it is best to be very familiar with the charges the monatomic ions can have (Figure 3.2 in text). Often, the charge on the anion in the compound will help you derive the charge on these metals.

Polyatomic ions, groups of atoms bonded together with a charge, are typically formed from nonmetals. The group reacts as a single unit, stays together in the crystal lattice or when the compound is dissolved in water. There are about 30 common polyatomic ions (Table 3.7 in the text) and it is important to know their names and charges. Polyatomic ions do have some features to help you remember the names. Most are **oxoanions**, which are nonmetals combined with one or more oxygen atoms. These features are:

*   The name for the ion uses the root of the non-oxygen atom's name with an "ate" or "ite" ending, SO_4^{-2} is sulfate.
*   If two oxoanions exist, the one with the more oxygens is given an "ate" ending and the one with fewer oxygens, an "ite" ending, e.g., NO_3^- is nitrate and NO_2^- is nitrite.

- Those ions with an H at the beginning have hydrogen in the name; HCO_3^- is called hydrogen carbonate. If two H atoms are present, dihydrogen is used.

- The oxohalogen ions fall into a series of four ions with the same charge. The ion name consists of three parts: a prefix (except where none is indicated), the root name for the halogen and a suffix as given below:

Ion:	Prefix:	Suffix:	Example:
XO_4^-	per	ate	IO_4^- is periodate
XO_3^-		ate	IO_3^- is iodate
XO_2^-		ite	IO_2^- is iodite
XO^-	hypo	ite	IO^- is hypoiodite

- There is only one common positive polyatomic ion, the ammonium ion, NH_4^+.

3.6 Describe the properties of ionic compounds and compare them with the properties of molecular compounds.

3.7 Write the formulas of ionic compounds.

3.8 Name ionic compounds.

3.9 Describe electrolytes in aqueous solution and summarize the differences between electrolytes and nonelectrolytes.

Ionic compounds are not distinct molecules. They are aggregates of cations and anions held together by their opposite charges. In an ionic solid, the ions are in a **crystal lattice** where each positive ion is surrounded by negative ions and the negative ions are surrounded by positive ions. A **formula unit** is used to represent the formula for ionic crystals. A formula unit is the simplest ion ratio that has the properties of the material in the lattice and has zero charge.

Ionic crystals have several characteristic properties. They are hard, brittle, have a regular geometric shape, and also have distinct cleavage planes. Ionic solids will have a high melting points and when molten conduct an electric current. In contrast, molecular compounds have much lower melting points, are not electrically conducting when molten, and form brittle but soft crystals.

All formulas for ionic compounds must be electrically neutral. The total sum of the charges in the compound must equal zero. This does not mean that the number of positive and negative ions must be equal. When writing formulas for ionic compounds a simple cross-over technique is useful.

- Write the cation symbol first and note its charge and do the same for the anion. The cation charge then becomes the anion subscript (with no sign) and the anion charge, the cation subscript. For example the ionic compound formed by barium with nitrogen would be: $(Ba^{2+}) + (N^{3-}) \rightarrow Ba_3N_2$
- If each subscript can then be divided by 2 or another number, you must reduce to the simplest whole number ratio, since only empirical formulas are written for ionic compounds. For example: $(Ca^{2+}) + (O^{2-}) \rightarrow Ca_2O_2$ reduces to CaO.

Ionic compounds are named using a different system than that for molecular compounds. The guidelines are:

(1)   When naming ionic compounds, the cation is named first, followed by the anion. If a monatomic ion is combined with the cation, an "ide" ending is added to the nonmetal's name. For example, K_2O, is potassium oxide, and CaF_2, calcium fluoride. Greek prefixes are <u>never</u> used in names of ionic compounds to indicate the number of ions.

(2)   If more than one charge on the metal is possible, the charge of the metal is indicated in parentheses using Roman numerals.

- The transition metals have several possible charges. The anion can be used to determine the charge on the metal. For example in $FeCl_2$, the charge of chloride is (-1) so the iron ion has a +2 charge and the name given would be Iron (II) chloride.

- The number in parenthesis is the metal ion charge, not the number of anions present. For example, the formula for cobalt (II) sulfide describes Co^{+2} combining with S^{-2} and is CoS, not CoS_2.

(3)   When a metal forms a compound with a polyatomic anion the same rules apply as in (1) and (2). The anion name is added to the name of the metal so that KCN is potassium cyanide and $NiCO_3$ is nickel (II) carbonate.

Remember that ionic compounds and molecular compounds are named differently. It is necessary to pay very close attention to the type of compound you are trying to name.

Ionic crystals often are water soluble. When they dissolve, they break up into the charged ions that are surrounded by water molecules, and the resulting solution conducts an electrical current. An ionic compound that completely dissociates into ions is considered a **strong electrolyte**. Many molecular compounds dissolve in water but do not dissociate, so they are **nonelectrolytes**.

3.10   Thoroughly explain the use of the mole concept for chemical compounds.

3.11   Calculate the molar mass of a compound.

3.12   Calculate the number of moles of a compound given the mass, and vice versa.

3.13   Explain the formula of a hydrated ionic compound and calculate its molar mass.

When compounds participate in chemical changes, a chemist still needs to be able to scale up from the single ion or molecule that is reacting or dissolving on the nanoscale to the macroscale of grams. Consequently, being able to use the counting unit of moles of the compound or ion is still very important. Avogadro's number can also be applied to molecules or formula units in the same way it could in Chapter 2 for elements.

The **molar mass of a compound** can be easily calculated from the molecular formula. The atomic masses for each atom must be added together to get the total mass of the compound. For each element in the compound, the subscript in the molecular formula is multiplied by its atomic mass (from the periodic table) and these are added together. For example, the molar mass of $Mg(IO_4)_2$ would be the sum of the mass of 1 mole of magnesium atoms, 2 moles of iodine atoms, and 8 moles of oxygen atoms and is calculated as [1× (24.3 g) + 2× (126.9 g) + 8× (16.0 g)] = 406.1g. The molecular weight, the weight of a single molecule, would be 406.1 amu. The molar mass of the compound is 406.1 grams. The term molar mass will be used to indicate that the macroscale scale unit of grams per mole is to be used. The molar mass once calculated, can be used as a conversion factor for changing mass of a compound into moles and vice versa, using the same techniques introduced in Chapter 2.

When ionic compounds are recrystallized from water, **ionic hydrates** are often formed. These compounds have water molecules present as part of their crystal structure. The molecular formula indicates how many water molecules there are per formula unit of the compound. When named, the normal name is given to the ionic compound and the number of water molecules indicated with a Greek prefix followed by the word hydrate, so that $CaCl_2 \cdot 2H_2O$ is named calcium chloride dihydrate. Hydrates vary in their stability toward heat but water can be driven off to leave the just the ionic compound. The molar mass for calcium chloride dihydrate, $CaCl_2 \cdot 2H_2O$, is equal to the sum of the masses of 1 mole of calcium, 2 moles of chlorine, and two water molecules which is [1×(40.0 g) + 2×(35.5 g) + 4×(1.0 g) + 2×(16.0 g)] = 147.0 grams per mole.

3.14   Express molecular composition in terms of percent composition.

3.15 Use percent composition and molar mass to determine the empirical and molecular formulas of a compound.

The **percent composition by mass** compares the mass of the element to the molar mass of the compound. As shown in the schematic below, it can be calculated one of two ways.

$$100 \times \frac{(wt.\ X,g)}{total\ wt.\ compound} = \boxed{\begin{array}{c}\textbf{\% X in}\\\textbf{compound}\end{array}} = \frac{(no.\ atoms\ X) \times (atomic\ mass\ X)}{molar\ mass\ \ compound} \times 100$$

know masses: know formula:

The **empirical formula** for an unknown compound can be calculated from percent composition data, using the left side of the equation above. If the formula is unknown, but the percent by mass of each element in a total weight of the compound is given, then the mass of the elements in the compound can be calculated. Those masses can then be converted to moles, using the atomic masses as conversion factors and the simplest ratio between the elements determined.

①
Convert %'s to mass of element, assuming 100 g of compound →

②
Use atomic mass to convert grams to moles of each element →

③
Divide each by smallest number of moles. If fractions result, multiply by factor to produce whole numbers.

3.16 Identify biologically important elements.

3.17 Identify the important functional groups in carbohydrates and fats.

The biological periodic table (Figure 3.7) summarizes elements of biological importance.

Carbohydrates all have the same general formula, $C_x(H_2O)_y$. They are not truly hydrates, since there are no H-OH bonds in the molecules. The simplest carbohydrates are sugars containing five or six carbons called **monosaccharides**. Monosaccharides can exist in two constitutional isomers, as either the ring or straight chain forms, and have names ending in "ose".

Disaccharides are molecules composed of two sugar units bonded together. They have common names that do not indicate which sugars have been bonded together. For example, sucrose is a disaccharide made from bonding glucose with fructose. **Polysaccharides**, such as cellulose or amylose, are molecules made up of large numbers of monosaccharides bonded together to form a single chain. These polymers act as chemical storehouses of energy or provide structural material in biological systems.

Lipids or **fats** differ from carbohydrate materials in several ways. Although they contain a large number of carbon atoms, fats do not form polymers like carbohydrates. Three fatty acids, long chain alkanes with a **carboxylic acid** functional group, COOH on one end, combine with glycerol (which has alcohol groups) to form an triester. It is the type of fatty acid bonded that determines the physical properties of the fat. As an energy source, fats have much higher caloric content than carbohydrates and are good fuel molecules, much like other long chain alkane molecules.

Chapter Review - Key Terms

This chapter has a large number of key terms in comparison Chapter 2, so this exercise is longer than it will be in most chapters. However, you will find that being able to distinguish between the terms will be important for the study of later chapters. Also, once completed, this exercise, will serve as a good review tool for the chapter concepts before a test on the material.

In order to make all the sentences below TRUE, insert the appropriate word or phrase from the list of key terms which best fits the context of the sentence.

NOTE: Any phrase or word from the list may used more than once.

LIST:

alcohol	electrolyte	molecular weight
alkane	empirical formula	monatomic ion
alkyl group	formula unit	monosaccharide
anion	formula weight	nonelectrolyte
binary molecular compound	functional group	organic compound
carbohydrate	halide ion	oxoanion
carboxylic acid	hydrocarbon	percent composition by mass
cation	inorganic compound	polyatomic ion
chemical bond	ionic compound	polysaccharide
condensed formula	ionic hydrate	strong electrolyte
constitutional isomer	isomer	structural formula
crystal lattice	major mineral	trace element
dietary mineral	molecular compound	water of hydration
disaccharide	molecular formula	

Formulas and Types of Compounds

Atoms in a compound are linked into a single unit through (1) _____ s. To describe the molecule, a chemist can use several types of formulas, each of which can communicate different kinds of information. If the formula only gives the number and kinds of atoms in the compound, it is called a (2) _____, which is the shortest notation that can be used. If the actual (3) _____ s are drawn out as lines or dashes between atoms, then a (4) _____ has been written. A (5) _____ also gives information about the way the atoms are linked, but shows the atoms as groups within the molecule. Any of the three formulas can be used to determine the (6) _____ of the molecule, which is the sum of all the weights of the atoms in a molecule, but the (7) _____ is the most convenient for the computation.

As a very general classification, compounds of two or more different atoms can be classified as either (8) _____ s, which do not contain C, or (9) _____ s, which do contain C and usually H. Differences in water solubility and physical properties are used to further classify compounds as either (10) _____ s, which form ions when dissolved in water, or (11) _____ s, in which most of the molecules stay intact when the substance is dissolved in water. The (12) _____ s for (13) _____ s indicate the distinct unit, whereas those for (14) _____ s are the simplest whole number combinations of ions that will produce zero charge.

The simplest (15) _____ s composed of two different nonmetal elements are the (16) _____ s, which incorporate Greek prefixes before the element name to indicate the numberof atoms of an element in the molecule. (17) _____ s formed from only C and H are called (18) _____ s. The (19) _____ s are a family of hydrocarbons having the (20) _____, C_nH_{2n+2}, where the number of carbons is indicated in the name of the compound.

Molecules that share the same (21) _____, but have different bonding arrangements of the atoms are (22) _____ s of each other. (23) _____ s with more than 4 carbons can have several different forms, such as straight chain, branched, or ring structures which are examples of (24) _____ s. Removal of a H atom from a C atom of an

(25) _____s creates a fragment of a molecule called an (26) _____. The name of an (27) _____ is the same as the (28) _____ having the same number of carbons, but the "ane" ending is replaced with an "yl" ending.

   If a hydrogen is replaced by a OH group, the molecule becomes an (29) _____. The OH group is an example of a (30) _____. (31) _____s are responsible for the characteristic chemical behavior of the molecule and are easily recognized when either the (32) _____ or (33) _____ is written.

Comparing Ionic to Molecular Compounds

   Metals reacting with nonmetals typically form (34) _____s, which are composed of (35) _____s (positively charged ions) and (36) _____s (negatively charged ions). When in the solid state, the ions form an arrangement called a (37) _____, where each ion is surrounded by ions of opposite charge. (38) _____s are also characterized by high melting points, whereas (39) _____s have much lower values. When single atoms become ions, they are called (40) _____s. When negative ions are formed in this way, an "ide" ending is added to the element name in the compound. As a general name, a (41) _____ formed by an element in Group 7A is called a (42) _____. When a group of 2 or more atoms becomes an ion that has a net electrical charge, it is a (43) _____. When the (44) _____ contains oxygen, it is an (45) _____ and typically has names that end or "ate" or "ite", depending on the number of oxygen atoms.

   A major difference in behavior occurs when the two major types of compounds are dissolved in water. (46) _____s produce (47) _____s when dissolved in water because the solution then conducts electricity. If the compound completely dissociates into ions, it is a (48) _____. (49) _____s also can dissolve in water, but often, show no dissociation into ions and are then (50) _____s.

   The formula for an (51) _____ is written as the smallest whole number ratio of (52) _____ to (53) _____ and is the (54) _____ of the compound. The (55) _____ is calculated as the sum of the atomic weights of ions in the compound, based on the formula, and is used, as is the (56) _____ for a (57) _____, to convert mass to moles.

   When water molecules are trapped and then incorporated in the (58) _____ of an (59) _____, often after recrystallization from water, an (60) _____ has been formed. The water molecules, called the (61) _____, are given in the (62) _____ as the number of water molecules for every one (63) _____ of a compound. The (64) _____ can be lost from the compound upon heating.

   The (65) _____ is one way of conveying information about the composition of the compound in mass units, unlike formulas that give the information in moles or atoms. The (66) _____ is obtained by taking the weight of a particular element in the compound, dividing it by the total weight of the compound and multiplying the result by 100. The (67) _____ gives the smallest possible whole number ratio of atoms in a compound and is the same as the (68) _____ in an (69) _____. For a (70) _____, the actual (71) _____ is a whole number multiple (1,2,3....) of the (72) _____. For example the molecular unit for benzene, C_6H_6, is six times the simplest ratio of one C to one H (or CH). The (73) _____ of a either type of compound can be used to calculate the (74) _____ of the compound.

Biologically Important Compounds and Elements

Four elements, C, H, N, and O make up the major portion of the human body on the basis of weight, and also in numbers of atoms. The essential elements other than these four required for healthy functioning of the human body are the (75) _____ which are efficiently recycled within the body, so only small amounts are required. The (76) _____ are divided into two classes, based on the percentage of body weight, the first of these, (77) _____ are present in quantities less than 0.01% of body weight and second, (78) _____, an even smaller percentage.

In one of the major types of molecules, the C,H and O atoms are combined into (79) _____, which are a major source of energy for the body and have the general formula $C_x(H_2O)_y$, where x and y are whole numbers. Simple sugars, hexoses and pentoses, are (80) _____s that contain 6 or 5 carbon atoms, respectively. The sugar molecules also always have (81) _____ or OH groups bonded to the carbons in the molecule A (82) _____ is formed when two (83) _____s are bonded together to form a single molecule, such as in the molecule sucrose. When hundreds or thousands of the sugar monosaccharides are bonded together, a (84) _____ is formed, such as in starch and cellulose. Fats differ from (85) _____ in that they are formed from three long chain (86) _____molecules that have a (87) _____ as a (88) _____ on one end that bond with the (89) _____ groups on glycerol, to form a glyceride and water molecules. Consequently, fats retain some of the properties associated with high molecular weight (90) _____s, such as being solid at room temperature. Fats also have about two times the caloric or energy values of (91) _____.

- PRACTICE TESTS -

After completing your study of the chapter and the homework problems, the following questions can be used to test yourself on how well you have achieved the chapter objectives.

1. Tell whether the following statements would be always **True or False**:

    A)   Grey tin and white tin are compounds of tin and oxygen that have two different crystal lattices so they are isomers of each other.

    B)   Although a small percentage of body weight, the halide ions belong to the class of major minerals and must be obtained from the daily diet to remain healthy.

    C)   A functional group is often the reactive group of an organic molecule.

    D)   The molecular formulas of organic compounds can be represented several ways, but there is only one representation of an ionic compound.

    E)   If the percent composition of two organic compounds is the same, they must have the same molecular and structural formulas.

    F)   The location on the periodic table is very useful in predicting the charge an atom will have as a monatomic ion, but is not useful for predicting the charge for polyatomic oxoanions.

    G)   Larger ions tend to form stronger bonds, as seen by the physical properties such as melting point, than smaller ions of the same charge.

    H)   When metalloids react with nonmetals an ionic compound forms.

    I)   Ionic compounds must be electrolytes and molecular compounds must be nonelectrolytes when dissolved in water.

    J)   Alkanes are a class of hydrocarbons, as are alcohols, so they have similar names.

K) The molar mass for an ionic compound is calculated the same way as for a molecular compound, from the empirical formula.

L) Hydrates are molecules that contain H_2O as part of the molecular formula, therefore carbohydrates, $C_x(H_2O)_y$ and $CuSO_4 \cdot 5H_2O$ would have similar characteristics in their crystal lattices for the solid state.

2. Tell whether the following information would be sufficient information to determine:
A) the molecular formula of a molecular compound, if you know the:
  (a) number of moles of each type of atom in a given sample of the compound
  (b) mass % of each element and the actual number of one of the atoms in a molecule of the compound
  (c) mass % and the molecular weight of the compound
  (d) structural formula of the compound
B) the formula unit for an ionic compound, if you know the:
  (a) name of each ion in the compound.
  (b) the mass % and identity of the metal in the compound and the name of the other elements.

3. Briefly answer the following:
A) Explain why the element sodium can exist by itself as pure sodium in the solid state, but sodium ion cannot.

B) All alkanes have the same empirical formula but some are solid at room temperature while others are a gas or liquid. What appears to be the main factor for determining the physical state of the alkanes?

C) If 14.50 grams of $CuCl_2$ was dissolved in a certain amount of water, would the solution contain more, the same number or fewer moles of copper ions than if 14.50 grams of $CuCl_2 \cdot 2\,H_2O$ was dissolved in an equal amount of water?

D) Fluoride toothpaste converts the mineral apatite in tooth enamel to fluorapatite, $Ca_5(PO_4)_3F$. Given that calcium and phosphate ions have their usual ion charges, is the charge on the F ion different from its most common value in $Ca_5(PO_4)_3F$? Explain your reasoning.

4. Draw the correct structural and condensed formula for:
  A) 3-methylpentane   B) 2,4 dimethyldecane

5. Give the proper name from the choices for the compound structure shown below.

CH_3
$CH_3CHCH_2CH_2CHCH_3$
$CH_2CH_2CH_3$

Choices:
(a) 2-methyl-5-propylpentane   (b) 4,7-dimethyloctane
(c) 2-methyl-5-propylhexane   (d) 5-methyl-2-propylhexane
(e) 2,5-dimethyloctane   (f) none of these

6. For the four alkane molecules (a) - (d):
A) give the proper name for each molecule
B) tell which are constitutional isomers, if any

(a) CH_3
$CH_3CHCH_2CH_2CH_3$

(b) CH_3
$CH_3CH_2CCH_3$
CH_3

(c) $CH_3CH_2CH_2CH_2CHCH_3$
CH_3

(d) CH_3
$CH_3CH_2CHCHCH_3$
CH_3

7. For each of the following pairs of molecules, tell which of the three descriptions apply to the pair:
  (a) The molecules are constitutional isomers
  (b) The molecules are not constitutional isomers
  (c) The molecules represent the same compound

A) $CH_3CHCH_2CHCH_3$ and CH_3
   CH_3 CH_3 CH_3CHCH_2
                       $CHCH_3$
                       CH_3

B) CH_3 and OH
$CH_3CH_2CH_2CH$ $CH_3CH_2CHCH_2CH_3$
OH

8. Given the following ions: Mg^{2+}, NH_4^+, HS^-, NO_3^-, ClO_3^-, $H_2PO_4^-$, S^{2-}

   (Note there may be more than one right answer for the following)

   A) which would: (a) be an oxoanion?   (b) be a cation?

   (c) have hydrogen included in its name.

   B) which are two ions that could form a ionic compound:

   (a) made with only two elements?

   (b) that has 4 hydrogens in the formula unit?

9. For the compound K_3PO_4:

   A) what would be the cation and anion in the compound, the name and charge on each?

   B) In the beaker on the right draw a nanoscale picture of two molecules of K_3PO_4 dissolved in water, showing the cation and anion with the proper charge and in the correct proportion.

10. For the compound $Ge[S(CH_2)_4CH_3]_4$ determine the:

    A) molar mass of the compound.

    B) percent by mass of S in the compound

    C) number of moles of compound in 2.0 g of the compound.

11. For the following compounds:

    A) Write either the proper formula or name for the compound, whichever is not provided.

    B) Tell whether the compound would be considered an ionic or molecular compound.

    C) <u>If the compound is ionic,</u> tell what ions are in the compound and their charges.

Compound:	Formula	Type:	Ions:
(a) Iron (III) sulfate	_____	_____	_____
(b) Nitrogen trichloride	_____	_____	_____
(c) Lithium phosphide	_____	_____	_____
(d) _____	C_7H_{16}	_____	_____
(e) _____	$Cr(NO_3)_2$	_____	_____
(e) _____	S_2O_3	_____	_____

12. For the following amounts of compounds:

    (a) 3 moles of C_2H_5OH (b) 2 moles of CO_2,   (c) 2 moles of KIO_3 (d) 3 moles of Na_2O

    A) Which contains the greatest mass of Oxygen?

    B) Which compound has the largest %O?

    C) In which of the compound(s) would O be a monatomic ion with -2 charge?

13. For the following substances, calculate the value for the:

    A) number of moles in 0.585 g lithium bromide

    B) number of grams in 5.85 moles of ethylene glycol, CH_2OHCH_2OH

    C) volume of 5.85 moles of chloroform, $CHCl_3$, d = 1.48 g/mL

14. Calculate the atomic weight and identity of a metal that can form an oxide with the formula unit MO_2 that is 66.5% M by mass.

15. A white crystalline compound is found by analysis to contain 35.0% N, 5.0 %H and 60.0% O and a solution of the compound in water conducts electricity very well..

    A) What is the empirical formula of the compound?

    B) Based on the conductivity, is the compound molecular or ionic?

    C) Given that the molar mass of the compound is 80.0 grams/mole and considering answer for (B), what would be the molecular formula for the compound.

16. Compound X was isolated as a minor component of liquefied fuel gas. By analysis, X was determined to be 85.69 % C and 14.31% H by mass and its molecular weight was 55.9.

    A) What is the molecular formula for X?

    B) Is X a hydrocarbon and is it an alkane?

Chapter 4

Quantities of Reactants and Products

This chapter discusses how to write and interpret chemical equations in order to develop quantitative relationships between reactants and products. The reactions are classified by type and the criteria are identified. This chapter develops your ability to write chemical conversion factors (also called stoichiometric factors), which mathematically relate two different chemical substances. Chemical stoichiometry involves using calculations to predict the amount of a substance isolated or used based on the stoichiometric factors. Common types of stoichiometric calculations, such as identifying the limiting reactant, calculation of the theoretical yield and percent yield for a reaction are introduced. The empirical formula calculation, introduced in Chapter 3 is extended with the use of chemical reaction data.

4.1 Interpret information conveyed by a chemical reaction.

4.2 Recognizing the general types of reactions.

4.3 Balancing simple chemical equations.

Any **balanced chemical reaction** contains much useful information. The reactants and products are all identified with formulas. The numbers in front of the substances, called the **stoichiometric coefficients**, contain useful information regarding the amounts of substances reacting. Relationships between the reactants and/or product can be read on either the macroscale (molar ratios) or nanoscale (atom to atom, molecule to atom) directly from the balanced chemical reaction. You CANNOT read mass relationships directly from the balanced chemical reaction. The respective molar masses would be needed to convert the stated molar relationship into a mass relationship.

For the general reaction: $aA + bB \rightarrow cC + dD$
where a, b, c, d are the stoichiometric coefficients determine the ratios

Mole ratio: Mass Ratio:
$$\frac{a \text{ moles A}}{b \text{ moles B}} = \frac{a \times (\text{molar mass A})}{b \times (\text{molar mass B})}$$

Being able to recognize the type of reaction occurring can be useful when predicting products or performing stoichiometric calculations. The important features that distinguish the four types of reactions discussed in this chapter are:

Combination reaction:
- Only one product formed, which must be a compound.
- Reactants are either elements or compounds.
- Formation reactions are one example of this reaction type.

Decomposition reaction:
- Always only a single reactant, which must be compound.
- Metal carbonates decompose to CO_2 and the metal oxide.

Displacement reaction:
- Two reactants form two products in one-sided exchange.
- Reactants are typically a compound and an element

Exchange reaction:
- Two reactants (compounds) make two products, but in a two-sided exchange.
- Forming a gas, solid or molecular substance, such as H_2O, is the driving force for this reaction.

Many reactions that are combination, displacement, decomposition or exchange are typically balanced through a trial and error method that can be very frustrating if not approached systematically. Using the steps outlined in the text will get you quickly and easily to the balanced reaction. Some pointers on using the first three steps are:

- When writing the correct formula apply the rules for naming compounds introduced in Chapter 2
- To start the balancing, pick an atom that appears only once in the products and only once in the reactants.
- Change only the coefficients in front of the substance, never change the subscript to try to balance the equation.

4.4 Use molar ratios to calculate the number of moles or mass of one reactant or product from the number of moles or mass of another reactant or product by using the balanced chemical reaction.

4.5 Use stoichiometric principles in the chemical analysis of a mixture.

The **stoichiometric factor or ratio** is a <u>chemical</u> conversion factor that is used to convert from one chemical substance to another, not from one unit to another.   The key point to remember is that the chemical conversion is defined by deciding what chemical substance you want and which chemical substance you have, or will to start from.

A chemical reaction often serves as the starting point for stoichiometric calculations. Therefore it is essential that the reaction used for the stoichiometric factors be a <u>balanced</u> chemical reaction. Employing the stoichiometric ratio between the substance you have and the one you want will be the <u>critical</u> step in setting up the mathematical route to solving a stoichiometry problem. Other unit conversions can then be added before or after the chemical conversion to complete the mathematical solution.

The chemical conversion needed for problem-solving is found in one of three places:
(1)   **In the balanced chemical reaction** A chemical reaction will always take precedence as the best indicator of the relationship between the substances.

(2)   **In the subscripts for an atom in a molecular formula**, which defines the relationship between the molecule and the atom (the second chemical substance).

(3)   **In other stated relationships that are given in the problem**, such as a percentage or concentration units. Analysis for a certain substance in a chemical mixture is often completed using this type of relationship.

4.6 Determine which of the two reactants is the limiting reactant.

A limiting reactant situation is recognized using two features:
- the reaction described has **two or more reactants** <u>and</u>
- you **have starting amounts of at least two of the reactants**.

This normally means that only one of the reactants is completely consumed. Only if you have the stoichiometric amount of each reactant, i.e. have amounts that match the needed mole ratio in the balanced reaction, would both be completely reacted. You cannot tell by just looking at the starting amounts- either in mass or moles- which reactant will be limiting. **To determine which reactant is limiting**:
(1) **Convert** the starting amounts of each reactant to moles.

(2) **Compare the reactants** by applying the stoichiometry of the reaction to the reactants using one of two methods.

<u>Method I: Comparing Amount of Product from Each Reactant</u>
- Convert the starting amounts of each reactant to number of moles of each reactant.

- Then calculate the amount of product that can be made from each reactant, using the mole ratios.

- Whichever reactant produces the least amount of product is limiting

- Although smaller, this amount of product is the maximum amount that can be made.

REMEMBER: Never add the calculated amounts of product together. It's an either/or situation- you get one OR the other, but NOT BOTH!

<u>Method II: Comparing Reactant Ratios</u>
- Convert the starting amounts of each reactant to number of moles of each reactant.

- Calculate the molar ratio of reactants in the balanced chemical reaction and compare it to the reactant ratio you have.

a A + b B → products   **Need:**   **Have:**
                                    a moles A   ? moles A
                                    b moles B   ? moles B

- If the "**have**" **ratio is smaller than needed, A is limiting** since there is not enough A to react with all of B. *Use moles A to calculate product.*
- If the "**have**" **ratio larger than needed, B limiting** since there is more than enough A to react with B. *Use moles B to calculate product.*
- If **ratios equal, neither limiting** and have the exact stoichiometric amounts of each. *Use either moles A or B to calculate product.*

Method I is best when you are only asked to determine the amount of product from the combination. Method II is more useful when then problem you are trying to solve asks for something other than just the amount of product.

4.7 Explain the difference among actual yield, theoretical yield and percent yield and calculate theoretical and percent yields

The **theoretical yield** is the calculated amount of product, based on stoichiometry, and is always the maximum amount of product you can make. Various experimental factors will prevent the theoretical yield from actually being achieved. The **actual yield** needs to be measured in the laboratory. The **percent yield**, a measure of reaction efficiency, is calculated using the formula:

$$\% \ yield \ = \frac{Actual \ yield}{Theoretical \ yield \ (calculated \ value)} \times 100$$

If the % yield and actual yield of a product are known for a reaction, the equation for percent yield can also be used to calculate how many moles of limitng reactant are needed for reaction from the theoretical yield. For example, the calculation for moles of A needed to produce a certain actual yield of product C from the reaction aA + bB → cC + dD would be written as:

$$\left[\text{actual moles product C} \times \frac{100}{\% \ yield} \right] \times \frac{\text{a mole A}}{\text{c mole C}} = \text{Moles reactant A needed}$$

4.8 Use stoichiometric principles to find the empirical formula of an unknown compound using combustion analysis.

The last section of stoichiometric calculations introduced in this chapter extends the discussion of empirical formula calculations begun in Chapter 3 to include reaction data, in particular, combustion analysis. The key points to remember for combustion analysis are:

(1)   In combustion reaction, a compound is burned in (reacts with) O_2 to form the oxides of the elements in the compound.

(2)   The amount of the element in the compound can be determined from the oxide of the element produced using stoichiometric ratios.

(3)   For a hydrocarbon or carbohydrate, the products of combustion will be CO_2 and H_2O, so that:
     • *The amount of C in the compound is calculated from the amount of* CO_2 *produced.*
     • *The amount of H in the compound is calculated from the amount of* H_2O *produced.*

(4)   For a compound containing oxygen, O, the amount of O in the compound should be calculated last and can be determined from the calculation:

wt. O = (wt. of compound burned) - (sum of wt. of all <u>other</u> elements in compound)

Learning Extension: Developing a Technique for Problem-solving:

It is important to develop a systematic way to solve problems in chemistry. The full guidelines for the approach are given in the "Appendix for Problem Solving" in the textbook, but the following steps will guide you in this process. Several examples are also produced to illustrate how to apply these steps to analyze, and then solve, different problems using the concepts developed in Chapter 4.

1. Define the problem.

First carefully read through the question and then determine the following:

A) What is the problem asking you to find?

Are you being asked for moles or grams of product or reactant, an empirical formula, or something defined by a specific formula (such as molarity, % yield, dilution, titration, etc.) ? If formula-based, write out the formula needed with all the terms specified.

B) What information is given in the problem? Making a list may be very helpful.

Useful information may include the balanced chemical reaction, molecular formulas, masses or volumes of reactants or products. Definitely take note of terms like molarity and density, but keep in mind that these are often intended for conversions, not as starting points.

2. Develop the plan.

A) Compare the known information to what you want to find. Determine what other information is needed to solve the problem.

What chemical conversions are needed? (Need mole ratios from formula or balanced reaction, molarity?) What unit conversions are needed? (Need density, molar mass, other?)

B) Write out a plan in the form of a map or series of steps. Gather the information needed to solve the problem.

3. Execute the plan.

Set up the mathematical equation and solve it, checking that the units cancel appropriately.

4. Check your answer.

The following questions will aid you in determining if the problem has been correctly solved and avoid some common errors.

- *If a chemical reaction was used, is it balanced?*
- *Estimate the size of the answer from the values given. Does the answer make sense?*
- *Does the answer have the correct units and number of significant figures?*

Example 1: Given the combination reaction $P_4O_{10} + 6\,H_2O \rightarrow 4\,H_3PO_4$ what would be the number of moles of P_4O_{10} that must be reacted in order to make 300. g of phosphoric acid, H_3PO_4?

1. Define the problem:

>**Looking for:** Number of moles P_4O_{10} reacted
>
>**Know:** 300. g H_3PO_4 produced

2. Develop the plan:

Chemical conversion needed:

>Mole ratio P_4O_{10}/ H_3PO_4
>(from balanced reaction)

Unit conversion needed:

>Molar mass H_3PO_4

$$P_4O_{10} + 6\,H_2O \rightarrow 4\,\mathbf{H_3PO_4}$$
$$300.\ g$$

$$? \text{ mol } P_4O_{10} \xleftarrow[P_4O_{10}/H_3PO_4]{\text{Mole ratio}} ? \text{ mol } H_3PO_4 \quad\Big\downarrow\ \begin{array}{l}\textit{Need molar}\\\textit{mass } H_3PO_4\end{array}$$

3. Executing plan

$$300\ g\ H_3PO_4 \left[\frac{1\ mole\ H_3PO_4}{98.0\ g\ H_3PO_4}\right]\left[\frac{1\ mole\ P_4O_{10}}{4\ mole\ H_3PO_4}\right] = \underline{\underline{0.765\ mole\ P_4O_{10}}}$$

4. Checking answer:

> 300. g H_3PO_4 is about 3.0 moles, so based on mole ratio, you should need about 0.75 moles P_4O_{10} to react, and the answer is reasonable. The units and significant figures (limited to 3) in answer also correct.

Example 2: A mixture containing Cu_2O and other inert substances was analyzed by heating a sample of mixture with H_2 gas to produce gaseous H_2O and $Cu(s)$. If 4.351 g of the mixture reacted with excess H_2 produce 1.152 g Cu,

>A) what is the balanced reaction?   B) what is the % Cu_2O in the mixture?

A) $Cu_2O(s) + H_2(g) \rightarrow 2\,Cu(s) + H_2O(g)$

B) 1. Define the problem:

>**Looking for:** % Cu_2O in mixture, $\%\ Cu_2O = \dfrac{wt.\ Cu_2O\ in\ mixture}{total\ wt.\ mixture} \times 100$
>
>**Know:** wt. mixture = 4.351 g, wt. Cu produced = 1.152 g
>**Needed to solve for %:** wt. Cu_2O in mixture = wt. Cu_2O reacted = ?

2. Develop the plan:

Chemical conversion:

>Mole ratio Cu/Cu_2O
>(from balanced reaction)

Unit conversions:

>Molar masses Cu_2O, Cu

$$\mathbf{Cu_2O}(s) + H_2(g) \rightarrow 2\,\mathbf{Cu}(s) + H_2O$$
$$?\ g \qquad\qquad 1.152\ g$$

$$\begin{array}{l}\textit{Need molar}\\\textit{mass } Cu_2O\end{array}\Big\uparrow \qquad\qquad \begin{array}{l}\textit{Need molar}\\\textit{mass Cu}\end{array}\Big\downarrow$$

$$? \text{ mol } Cu_2O \xleftarrow[Cu_2O/Cu]{\text{Mole ratio}} ? \text{ mol Cu}$$

3. Executing plan:

$$1.152\ g\left[\frac{1\ mole\ Cu}{63.55\ g\ Cu}\right]\left[\frac{1\ mole\ Cu_2O}{2\ mole\ Cu}\right]\left[\frac{143.1\ g\ Cu_2O}{1\ mole\ Cu_2O}\right] = 1.297\ g\ Cu_2O$$

$$\%\ Cu_2O = \frac{1.297\ g\ Cu_2O}{4.351\ g\ mixture} \times 100 = \underline{\underline{29.81\%}}$$

4. Checking answer:

> Based on the mole ratio, the mass ratio of Cu_2O/Cu should be about 143/127, slightly more than 1.0, so the weight of Cu_2O should be slightly larger than the Cu isolated. So the answer for wt. Cu_2O seems reasonable. The units and significant figures (limited to 4) in answer also correct.

Example 3: Hot Cl_2 gas will combine with gold to form gold (III) chloride, by the reaction:
$2\ Au(s) + 3\ Cl_2(g) \rightarrow 2\ AuCl_3(s)$ Suppose 20.0 g Au and 10.0 g of Cl_2 are sealed together in a container and heated until complete reaction takes place.

     A) Which reactant is limiting?
     B) What weight of $AuCl_3$ is formed?
     C) What weight of excess reactant should remain after complete reaction?

A) 1. Define the problem:

    Looking for: Limiting reactant
     • *Since Part B asks for amount of product, Method II will be most convenient to use.*

  2. Develop the plan: $2\ Au(s) + 3\ Cl_2(g) \quad\rightarrow\quad 2\ AuCl_3(s)$

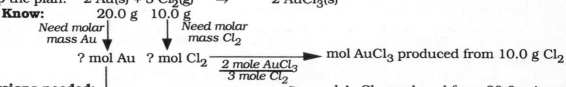

Unit conversions needed:
    Molar masses Cl_2, Au

The smaller of the two is used to calculate the mass of product formed

  3. Executing plan:

$$20.0\ g\ Au \left| \frac{1\ mole\ Au}{197.0\ g\ Au} \right|\left| \frac{2\ mole\ AuCl_3}{2\ mole\ Au} \right| = 0.102\ mol\ AuCl_3$$

$$10.0\ g\ Cl_2 \left| \frac{1\ mole\ Cl_2}{71.0\ g\ Cl_2} \right|\left| \frac{2\ mole\ AuCl_3}{3\ mole\ Cl_2} \right| = 0.0939\ mol\ AuCl_3 \quad \boxed{Cl_2\ \textbf{is limiting}}$$

B) 1. Define the problem: **Looking for:** wt. $AuCl_3$ produced
  2. Develop the plan   **Know:** 0.0939 moles $AuCl_3$ produced,
                       mol $AuCl_3 \rightarrow$? g $AuCl_3$
               Need: Molar mass $AuCl_3 = 1(197.0) + 3(35.45) = 303.4$

  3. Executing plan: $0.0939\ mol\ AuCl_3 \left| \dfrac{303.4\ g\ AuCl_3}{1\ mole\ AuCl_3} \right| = \underline{\underline{28.5\ g\ AuCl_3}}$

C) 1. Define the problem: **Looking for:** wt. Au left after reaction
  2. Develop the plan: **Know:** 0.0939 moles $AuCl_3$ produced, wt. Au at start = 20.0 g
                          • wt. Au left = (wt. Au at start) - (wt. Au reacted)
                          • 0.0939 mol $AuCl_3 \rightarrow$? mol Au reacted $\rightarrow$? wt. Au reacted
               Need: Molar ratio Au/$AuCl_3$, Molar mass Au

  3. Executing plan: $0.0939\ mol\ AuCl_3 \left| \dfrac{1\ mole\ Au}{1\ mole\ AuCl_3} \right|\left| \dfrac{197.0\ g\ Au}{1\ mole\ Au} \right| = 18.5\ g\ Au\ reacted$

        (20.0 g Au at start) – (18.5 g Au reacted) = **1.5 g Au left**

  4. Checking Answers:
    A) 10.0 g Cl_2 is about 0.14 moles and 20 g of Au is about 0.10 moles. So the ratio we have of mole Cl_2/mole Au is .14/.1 = 1.4/1 which is less than the needed ratio, 3/2= 1.5/1, so Cl_2 is limiting and answer is correct..
    B) Since weight of $AuCl_3$ is about 300 g/mole, producing slightly less than 0.10 mol would give a value slightly smaller than 30.0 g. So the answer seems reasonable. The units and significant figures (limited to 3) in answer also correct
    C) Have about 0.006 moles of Au left (0.102-0.094), so should be about 0.008(200) = 1.6 g left, and answer seems reasonable. The units and significant figures (limited to 3) in answer also correct

Chapter Review - Key Terms

In order to make all the sentences below TRUE, insert the appropriate word or phrase from the list of key terms which best fits the context of the sentence.

LIST:

actual yield	combustion reaction	oxide
aqueous solution	decomposition reaction	percent yield
balanced chemical reaction	displacement reaction	stoichiometric coefficent
coefficient	exchange reaction	stoichiometric factor
combination reaction	limiting reactant	stoichiometry
combustion analysis	mole ratio	theoretical yield

NOTE: Any phrase or word from the list may used more than once.

Types of Chemical Reactions:

Chemical reactions have distinguishing features that allow them to be classified. For example, a reaction is called a (1) _____ when a single compound is converted to two or more products in the reaction. The substances produced can either be elements or compounds. In contrast, a (2) _____ produces a single product from two or more reactants. The product of a (3) _____ must always be a compound. When a metal carbonate undergoes a (4) _____, it typically produces an (5) _____ of the metal and carbon dioxide. A formation reaction is an example of a (6) _____ where all the reactants are in elemental form. If a metal reacts with oxygen the (7) _____ produced will be an ionic compound. If a nonmetal or metalloid reacts with oxygen, by a (8) _____, a molecular compound is formed. Hydrocarbons or carbohydrates react with oxygen in (9) _____s to produce carbon dioxide and water. The experimental technique that uses this type of reaction to determine the percent composition of a hydrocarbon or carbohydrate is called (10)_____.

A reaction that involves an element reacting with a compound to produce two products, a compound and element different from the reactants, is called a (11)_____. In contrast, a reaction that results from an interchange, or two-sided switching, of ions or atoms between two reactant compounds is called an (12)_____.

Writing Chemical Reactions:

Because the number of atoms on each side of the reaction arrow must be equal, (to obey the Law of Conservation of Mass) a chemical reaction must always be written as a (13)_____. The relationship between the mass of chemical reactants and products is called chemical (14)_____.

The numbers that appear in front of the chemical substances on either side of the reaction arrow are called (15)_____s. These are used to construct the (16)_____s or (17)_____s needed to calculate the amount of product made, or reactant consumed, in a reaction. The amount of product calculated in this way is called the (18)_____ for the reaction. The (19)_____ from a reaction or synthesis would have to be measured in the laboratory and would be expected to be less than the (20)_____. To evaluate how efficient the reaction is at converting reactants to products the (21)_____ is divided by the (22)_____ and the resulting fraction is multiplied by 100 to give the (23)_____ for the reaction.

In a reaction that has two or more reactants, the (24)_____ is the one that determines the maximum amount of product that can be formed and is expected to be completely consumed. To determine which reactant is the (25)_____, the (26)_____ of the reactants to each other or the reactants to one of the products must be used in a calculation with the starting amounts of reactants. Without calculation, it is very difficult to correctly tell which reactant is the (27)_____ from the starting amount of reactants in either grams or moles.

- PRACTICE TESTS -

After completing your study of the chapter and the homework problems, the following questions can be used to test yourself on how well you have achieved the chapter objectives.

1. Tell which of the following statements is always **TRUE or FALSE**.
   - A) In every reaction that involves two reactants, one reactant is limiting and the other in excess.
   - B) The compounds $C_3H_6O_3$ and $C_6H_{12}O_6$ have the same empirical formula so that the mass of CO_2 produced per gram of the compound in a combustion reaction will be the same.
   - C) The number of moles of individual atoms must always be balanced on both sides of the equation in a chemical reaction.
   - D) The percent yield measures the efficiency of the reaction by comparing the actual yield of product to the amount of limiting reactant.
   - E) There must always be at least two products in a decomposition reaction.
   - F) There must always be two reactants and two products in a combustion reaction.

2. For the reaction: $2\,H_2S + SO_2 \rightarrow 3\,S + 2\,H_2O$ which of the following are TRUE?
   - A) 3 grams S are produced for each gram of SO_2 consumed
   - B) 3 mol S are produced for every 1 mol of H_2S consumed
   - C) The total number of moles of product formed will equal the total number of moles of reactant consumed
   - D) 1 mol of H_2O is produced for every 1 mol of H_2S consumed
   - E) Two-thirds of the S produced comes from the H_2S molecules.

3. Balance the following reactions and indicate whether the reaction is a combination, decomposition, displacement or exchange reaction.

   A) $H_2O_2 \rightarrow O_2 + H_2O$ B) $HgCl_2 + Al \rightarrow AlCl_3 + Hg$
   C) $UO_2 + HF \rightarrow UF_4 + H_2O$ D) $PI_3 + H_2O \rightarrow H_3PO_3 + HI$
   E) $Cl_2 + F_2 \rightarrow ClF_3$ F) $Br_2 + H_2O \rightarrow HBr + HBrO$
   G) $Cl_2O_7 + H_2O \rightarrow HClO_4$ H) $HAuCl_4 \rightarrow Cl_2 + Au + HCl$
   I) $Zn(NO_3)_2 + Na_2S \rightarrow ZnS + NaNO_3$

4. Fill in the table for the reaction : $(CH_3)_2N_2H_2 + 2\,N_2O_4 \rightarrow 2CO_2 + 3\,N_2 + 4\,H_2O$ and then answer the following questions concerning the relationships from the table.

   A) For which two products would there be 2 moles of product made for 1 mole of reactant consumed? State the ratio to prove your choice.

   B) For which reactant and product would there be 1.2 g of product made for every 1.0 g of reactant consumed?

$(CH_3)_2N_2H_2$	N_2O_4	CO_2	N_2	H_2O
	2 molecules			
			3 moles	
		88.0 g		

5. A reaction used to convert exhaled CO_2 to O_2 in closed systems, such as spacecrafts is:

$$4\ KO_2(s) + 2\ CO_2(g) \rightarrow 3\ O_2(g) + 2\ K_2CO_3(s)$$

A) Write all the molar stoichiometric ratios for the reaction.

B) In order to function well as an oxygen recharging system, the reaction must be able to consume 1.38g of CO_2 for every 1.00 g of O_2 produced. Does this reaction meet this criteria? Prove your answer.

6. Predict the likely product(s) and then balance the reactions for the following:

A) The decomposition of barium carbonate

B) The displacement reaction that occurs between ZnS and O_2

C) The formation of magnesium nitride from magnesium metal and N_2 gas

D) The displacement reaction of H_2 gas with tungsten (VI) oxide

E) The exchange reaction that occurs between ZnO and sulfuric acid, H_2SO_4.

7. Silver carbonate, Ag_2CO_3 decomposes to produce Ag(s), $CO_2(g)$ and $O_2(g)$.

A) Balance the decomposition reaction $Ag_2CO_3 \text{ Æ } Ag(s) + CO_2(g) + O_2(g)$.

B) What mass of Ag_2CO_3, in grams, must have decomposed if 56.2 g of Ag(s) is obtained from the decomposition reaction?

8. Which of the following reactions produces the largest amount of O_2 per gram of reactant decomposed:

A) $2\ NH_4NO_3(s) \rightarrow 2\ N_2(g) + 4\ H_2O(l) + O_2(g)$

B) $2\ N_2O(s) \rightarrow 2\ N_2(g) + O_2(g)$

C) $2\ Ag_2O(s) \rightarrow 4\ Ag(s) + O_2(g)$

9. H_2S (g) reacts with oxygen to form sulfur dioxide and water: $H_2S + O_2 \rightarrow SO_2 + H_2O$

A) What are the coefficients in the balanced the reaction?

B) How many grams of O_2 will be consumed when reacted with 0.985 g of H_2S?

10. Calcium nitride is formed when calcium metal and nitrogen gas react.

A) Predict the formula for the product, balance the chemical reaction and put in the appropriate symbols for the physical states: $Ca\ (\ ) + N_2\ (\ ) \rightarrow \underline{\ \ ?\ \ }\ (\ )$

B) Calculate the mass of calcium nitride solid that can be prepared from 54.9 grams of Ca metal and 43.2 grams of N_2 gas.

11. A glimpse of the "before" and "after" reaction between molecule A_2 and molecule B_2 to form A_3B is pictured in the box on the right (with the symbols indicated). Use the diagrams to answer the following questions:

A) What are the coefficients for A_2, B_2 and C in the balanced reaction?

B) Which reactant appears to be limiting? Use the method II type of calculation for a limiting reactant to show prove your choice.

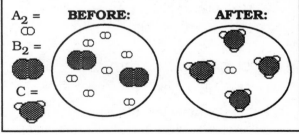

12. A can of butane lighter fluid contains 1.20 moles of butane, C_4H_{10}. If the lighter fluid burns in air,

A) write the balanced chemical reaction occurring and identify the type of reaction.

B) how many grams of CO_2 will be produced when all the butane is burned?

13. When strontium bromide, $SrBr_2$, is heated in a stream of Cl_2 gas, it is completely converted to $SrCl_2$ and another product.

    A) Identify the second product and write the balanced reaction.

    B) What type of reaction is this?

    C) Will the mass of $SrCl_2$ produced always be greater than, less than, or equal to the mass of $SrBr_2$ consumed?

    D) How many grams of $SrCl_2$ will be produced when 14.5 g of $SrBr_2$ is completely converted?

14. Anhydrous Na_2SO_4 absorbs water from the atmosphere until it is completely converted to the decahydrate, $Na_2SO_4 \bullet 10\ H_2O$. In an experiment, a student placed 3.50 g of anhydrous (without water of hydration) Na_2SO_4 out on a lab bench. At the end of one hour, the student reweighed the now hydrated Na_2SO_4 and found the weight had increased to 6.93 g.

    A) What general type of reaction is taking place in this situation?

    B) Write the balanced chemical reaction occurring with the Na_2SO_4.

    C) Had the Na_2SO_4 been completely converted to the decahydrate in that hour? Prove your answer.

15. Para-cresol is used as a disinfectant and in the manufacture of herbicides. A 0.4039 g sample of para-cresol containing only C, H and O, yielded 1.152 g CO_2 and 0.2691 g H_2O in a combustion analysis.

    A) What is the empirical formula of the compound?

    B) Write the balanced combustion reaction based on the empirical formula.

16. For the following reaction $CaCN_2 + H_2O \rightarrow CaCO_3 + NH_3$

    A) Balance the reaction.

    B) If 1.0 kg of $CaCN_2$ is combined with 1.0 kg of water and 360 g of NH_3 was obtained, what is the % yield for the reaction?

17. A sample of white solid was suspected of being either pure cocaine, $C_{17}H_{21}NO_3$, pure sucrose, $C_{12}H_{22}O_{11}$, or some combination of both.

    A) It is possible to tell which of the three possibilities is true of the white solid if the sample is analyzed by combustion. Explain why combustion analysis can be used.

    B) If a 18.32 mg sample of the powder produced 51.31 mg of CO_2 and 12.99 mg of H_2O, was the powder pure cocaine, pure sucrose or some combination of both?

18. The rusting of iron metal in air results in the formation of solid Fe_2O_3. If a 13.263 g iron sample is reacted with O_2 in the air and 2.381g of O_2,

    A) what is the balanced reaction taking place?

    B) what mass of Fe_2O_3 has been formed?

    C) what is the percent by mass of the iron sample that has been converted to rust?

Chapter 5

Chemical Reactions

This chapter has two major areas of focus: developing an understanding of typical chemical reactions that occur in aqueous solution and the usefulness of the concentration unit, molarity, for stoichiometric calculations. The prediction of whether or not a reaction will occur and the likely products that result depends on knowledge of the criteria for types of reaction discussed in this chapter. Reducing a reaction to a "net" ionic reaction, provides a clearer picture of the atoms or ions actually involved, and the similarities and differences between the types of reactions become more apparent. Molarity, one of the most commonly encountered units in chemistry allows the stoichiometric calculations begun in Chapter 2 to be extended to include solutions.

5.1 Predict products of common types of chemical reactions: precipitation, acid-base, and gas-forming.

5.2 Write a net ionic equation for a given reaction in aqueous solution.

5.3 Recognize common acids and bases and predict when neutralization reactions will occur.

Exchange reactions occur between electrolytes and are chemical changes where a precipitate (ppt), a gas or a molecular substance, such as water, is formed as a product of the reaction. Producing any of these three types of substances will remove that product from the reaction system so the conversion from reactants to products will be nearly complete. Only one of the products has to meet this criteria to have the chemical change occur.

Precipitation is a reaction that produces an insoluble ionic compound as a product, and typically occurs when at least one of the reactants is an ionic compound. If the reactants are both ionic, the anions in the two compounds switch places, i.e. compounds AD and XZ become AZ and XD. The switch of anions still occurs when one reactant is ionic and the other molecular. The anion part of the molecular compound will be based on the common poly-atomic or monatomic ions of nonmetals introduced in Chapter 4. For example, H_2S is seen as $(H^+)_2 \, S^{2-}$ for the purposes of exchange, while H_2O is $(H^+)(OH^-)$. If either the cation or anion can form a soluble compound, then precipitation cannot occur. Prediction of ionic compound solubility requires knowledge the general rules of solubility given in Table 5.1 of the text.

Spectator ions from the soluble compounds are not part of the precipitate, molecular compound or a gas formed and do not participate in the chemical change. To identify the spectator ions, the balanced overall equation is rewritten to show all the ions that exist freely in solution on both the reactant and product sides. Ions that appear in equal numbers on both sides of the reaction are then eliminated. This process gives the **net ionic reaction** that shows only the ions, or molecular substances, that are participating in the chemical change. Several key points to keep in mind that when rewriting the reaction are:
  • The stoichiometry of the balanced reaction must be retained
  • Separate into ions only if compound is a *strong electrolyte*:
    • *if ionic*, must be soluble compound
    • *if molecular*, must be a strong acid
  • Pure solids, liquids and gases are always kept intact
  • Weak electrolytes are always kept intact
  • Only ions that appear in equal numbers on both sides of the reaction can be deleted.

Acids and **bases** act as chemical opposites because when put together they neutralize the properties of each other, often by forming water. The acid produces **hydronium ion**, written as H^+ (aq), or H_3O^+ and the base produces **hydroxide ion**, OH^-. **Strong acids** or **strong bases** are strong electrolytes and dissociate nearly completely in solution. The molecular formula for a strong acid has the hydrogens that can be donated written first, as in **HCl**, **HNO$_3$**, or **H$_2$SO$_4$**. Strong bases are soluble metal hydroxides. The number of strong acids is very limited and the most common are given in Table 5.2 in the text.

Weak acids do not dissociate completely and form a smaller number of free $H^+(aq)$ in solution. Organic acids, which are weak acids, contain the carboxylic acid functional group -CO_2H or -COOH, such as in acetic acid, $CH_3CO_2\mathbf{H}$. The best way to tell the strong from weak acids is to know formulas of the strong acids.

Weak bases are typically amines, which have NH_3 or -NH_2 groups in the molecular structure Like weak acids, only small percentage of the weak base molecules produce ions in solution. Weak electrolyte dissociation is a dynamic situation and is represented using a double reaction arrow to indicate that although the reactants form ions, the product ions can recombine to reform reactants.

In a neutralization reaction, an acid, either strong or weak, reacts with a base to form water. The second product, made from the anion of the acid and cation of the base, is called a **salt** and is always an ionic compound. The neutralization is a one-way conversion since the product water molecules and salt do not reform the reactants. Therefore only single reaction arrows are used for neutralization reactions.

5.4 Recognize oxidation-reduction reactions and common oxidizing and reducing agents.

5.5 Assign oxidation numbers to reactants and products in a redox reaction, identify what has been oxidized or reduced, and identify oxidizing agents and reducing agents.

5.6 Use the activity series to predict products of displacement redox reactions.

Oxidation-reduction (redox) reactions involve the exchange of electrons and can be a source of electricity or energy. Redox reactions include battery reactions, burning, metabolism, and corrosion. **Oxidation** describes the loss of electrons by one substance and **reduction** the gain of electrons by another. Oxidation-reduction terminology can be confusing because the context of the change is important. The main difference is that:

• the terms **oxidizing and reducing agents** are used when you want to talk about the participants in terms of what substance causes another to change.

• the terms **oxidation and reduction** are used when the discussion is about the change itself.

Electron gain and loss is fundamental to redox, but may also be seen as the gain or loss of oxygen, halogens or hydrogen (Table 5.4). These are convenient ways to recognize that electron transfer has occurred without determining oxidation numbers. Another clear indication of an oxidation-reduction reaction is when an element is converted to a compound in the chemical reaction or vice versa.

Using the system of **oxidation numbers** is the most reliable method for recognizing redox reactions. The system is arbitrary, but as long as it is used in a consistent manner, the correct answers will result. The general rules for assigning oxidation numbers are outlined on Section 5.4 of the text. Three key points are:

(1) The sum of the assigned oxidation numbers for any ion or molecule must always equal the overall charge on the ion or molecule.

(2) When H and O appear in compounds or ions, they are usually assigned oxidation numbers of +1 and -2 respectively. These values can then be used to assign the oxidation numbers for other elements in the ion or molecule.

(3) The group numbers of an element, is a good indicator of the maximum positive oxidation state for the element. The minimum negative oxidation number for a nonmetal is generally equal to (group number) - 8.

When the oxidation number for an element increases, the element has undergone oxidation, lost electrons and acted as a reducing agent. When the oxidation number decreases for an element, reduction has occurred which indicates the element has gained electrons and acted as a oxidizing agent.

Some generalizations can be made to help you predict whether a substance will act as an oxidizing or reducing agent (Table 5.3). Oxidizing agents gain electrons (they are reduced), so nonmetals, such as oxygen or the halogens, are good oxidizing agents. Reducing agents lose electrons (to form cations), so metals are and $H_2(g)$ are good reducing agents. However, although a metal can be a reducing agent, it could act as either a reducing or oxidizing agent. For example, Fe^{2+} is a reducing agent when it is converted to Fe^{3+}, but acts as an oxidizing agent when it is reduced to Fe^0.

Not all metals are oxidized (lose electrons) with equal ease. The **metal activity series**, given in Table 5.5, lists the metals in decreasing order of ability to lose electrons. The major dividing line in chemical behavior is marked with H_2. Metals above H_2 are oxidized to their ions by acids, and possibly water, while those below H_2 are not. Any metal in the series can be oxidized by the ion of a different metal below it in the series through a displacement reaction. It is important to remember that the ion of the second metal is required as the reactant. Combining two elemental metals cannot result in a redox reaction since neither one can be reduced, only oxidized. Also note that the most reactive metals tend to be from groups 1A and 2A, while the least reactive are from groups 1B and 2B in the transition metals.

5.7 Define molarity and calculate molar concentrations.

5.8 Determine how to prepare a solution of a given molarity from the solute and water or by dilution of a more concentrated solution.

A very large percentage of chemical reactions occur in solution, in particular, aqueous solutions. The major component of the solution is the solvent, which is water in an aqueous solution. The substance dissolved in the solvent is called the solute. The concentration unit, **molarity**, M, is the ratio of moles of solute per liters of solution and gives the composition of the solution in a chemically useful form.

Molarity = (moles solute)/Volume solution, L = (moles solute) / (1.0 L solution)

Once the molarity of the solution is known, it can act as a conversion factor between the volume of a solution and moles of solute:

$$\text{Moles of solute X} \overset{\times\,1/M}{\underset{\times\,M}{\rightleftarrows}} \text{volume of solution, L}$$

To prepare solutions with a specific molarity, one of two methods can be used to determine the amount of solute and solvent needed.

Method I: Add the correct amount of solute to a certain volume of solvent and mix thoroughly. To determine the amount of solute needed, you must know two things:
 (1)   the molarity of the solution you want to prepare and
 (2)   the final volume of solution needed.

Then the mass of solute needed is calculated from:

(Final volume Soln , L) $\times$ (Molarity X) $\times$ (Molar mass X) = grams of X needed

Method II: A more concentrated solution, having a higher molarity of solute, is diluted to the needed value by adding solvent. In this situation, the dilution formula can be used. In a dilution method, the number of moles of solute is not changed, but is just distributed over a larger volume of solvent, so the molarity decreases. Since the number of moles of solute in the two solutions are the same, the following relationship is always true for dilution:

$V_{conc} \times M_{conc} = V_{dil} \times M_{dil}$ where V_{conc}= smaller, INITIAL volume taken
 M_{conc} = initial solution of HIGHER concentration
 V_{dil}= larger, FINAL volume of solution
 M_{dil} = final solution of LOWER concentration

Also in a dilution it is useful to remember that:

• the volume units used are not limited to liters and can be whatever is convenient, as long as both volumes have the same units.

• the volume of solvent needed to dilute is always: $V_{solvent} = V_{dil} - V_{conc}$

5.9 Solve stoichiometry problems using solution molarities.

5.10 Understand how aqueous solution titrations can be used to determine the concentration of an unknown solution.

In earlier chapters, stoichiometry was introduced as a way to describe mass relationships between reactants and products in chemical reactions. If a reaction takes place in solution, these calculations can be extended to include molarity. Figure 5.17 in the text is a useful representation of the extensions that molarity provides and diagrams the common pathways that apply to solving problems involving solutions. Note that the critical chemical conversion is still the same, but now an alternative way exists to calculate moles of A or B as needed.

Titrations use this pathway specifically for exchange reactions. In a **titration**, solutions of two reactants, A and B, that undergo a one-way, exchange reaction in a procedure whre one solution is carefully added to the other. At the **equivalence point** of the titration, the stoichiometric ratio for A and B needed is <u>exactly met</u>, so that it will be true for the reaction:

$$aA + bB \rightarrow \textbf{products} \quad \text{and} \quad \frac{\textbf{No. moles A used}}{\textbf{No. moles B added}} = \frac{a}{b}$$

Experimentally, finding the equivalence point of a titration is the most important and also the most difficult part of a titration since the solutions themselves rarely signal the equivalence point. For that reason, either indicators, which are highly colored compounds, or instrumentation techniques that can locate the equivalence point are used to identify it in the laboratory procedure. A **standard solution** which is a solution with a known molarity, prepared by using either of the techniques described previously, are frequently used in a titration to provide reactant A or B.

Chapter Review - Key Terms

In order to make <u>all</u> the sentences below TRUE, insert the appropriate word or phrase from the list of key terms which <u>best fits</u> the context of the sentence.

NOTE: Any phrase or word from the list may used more than once.

LIST:		
acid	oxidation number	solute
base	oxidation-reduction reaction	solvent
concentration	oxidized	spectator ions
equivalence point	oxidizing agent	standard solution
hydronium ion	precipitate	strong acid
hydroxide ion	redox reactions	strong base
metal activity series	reduced	titration
molarity	reducing agent	weak acid
net ionic equation	reduction	weak base
oxidation	salt	weak electrolyte

Concerning Exchange Reactions:

When a chemical reaction occurs in an aqueous solution, the reactants can exchange ions to form products. Chemical reaction only occurs if a (1) _____, molecular compound that is not a strong electrolyte or a gaseous molecular compound is formed as a product. These chemical reactions can be written as (2) _____ where only the symbols and formulas of the ions that participate in the chemical change are included. Missing from these reactions are the (3) _____, those ions that remain free in solution and not part of the chemical change. An exchange reaction, specifically a neutralization reaction occurs when an (4) _____ reacts with a (5) _____ and water and a (6) _____ are produced. The extent of ionization of a compound determines whether the substance will be classified as a (7) _____, strong electrolyte or nonelectrolyte, when dissolved in water.

(8) _____s are substances that produce (9) _____s, represented by the symbol $H^+(aq)$ or $H_3O^+(aq)$, in solution. (10) _____ ionize completely in aqueous solution and are molecular compounds like HCl or HNO_3. Many other molecular compounds that contain an acidic hydrogen are (11) _____s, since they produce (12) _____ in aqueous solution, but do not undergo complete ionization. Ionic compounds that contain (13) _____ and a metal cation act as a (14) _____ when dissolved in water. Molecular compounds such as NH_3 react with water to produce (15) _____ and are also (16) _____s, but are (17) _____s since the molecules ionize to only a very limited extent in water.

A neutralization reaction in which a (18) _____ reacts with a (19) _____ always has the same (20) _____, namely $H^+(aq) + OH^- \rightarrow H_2O$. The same type of reaction occurring between a (21) _____ and a (22) _____ always contains the molecular form of the acid and the anion of the (23) _____ as well as water.

Concerning Oxidation and Reduction Reactions:

A reaction in which electrons are directly transferred between two substances is an (24) _____ or in shorter form, a (25) _____. The two processes (26) _____ and (27) _____ must always occur together in the same reaction since they are two halves of the same process. A substance that gains electrons undergoes (28) _____ and acts as the (29) _____ in the reaction. Simultaneously, the second substance participating loses electrons and undergoes (30) _____. The (31) _____ substance is then the (32) _____, causing (33) _____ to occur. To keep track of what has happened to a substance, each element must be assigned an (34) _____. If an increase in the (35) _____ for an element is observed when comparing the reactant form to the product form, (36) _____ of that element has occurred. If the reverse is true, the (37) _____ has decreased, then the element has been (38) _____. Chemcial species that gain oxygen or halogen atoms when converted from reactants to products are (39) _____, which will also signal that the (40) _____ has increased. (41) _____ can be signaled by the gain of H atoms, or loss of O or halogen atoms, during the chemical reaction.

In exchange reactions, the (42) _____s of the elements involved do not change, so these are never (43) _____s. However, reactions that are displacement or combustion reactions are always (44) _____s, while decomposition and combination reactions, may or may not be. Recognition of substances that typically act as a (45) _____ or (46) _____ helps to predict the type of reaction that will occur and possibly the products. A pure metal reacting with an (47) _____ to produce H_2 gas is an example of a (48) _____ that is also a displacement reaction in which the (49) _____ is the (50) _____. The ranking of the relative reactivity of metals towards acids, led to the development of the (51) _____. Metals that are listed in the topmost part of the (52) _____ are very reactive and can act as (53) _____s towards water to produce highly flammable H_2 gas, while the metal itself undergoes (54) _____. The metals at the bottom of the (55) _____ are not (56) _____ by (57) _____s.

Concerning Concentration Units:

A solution is always composed of at least two components, a (58) _____ and a (59) _____. The (60) _____ is always the one that is thought of as being dissolved in the (61) _____, so the (62) _____ particles can be visualized as being surrounded by (63) _____ molecules in the solution on the nanoscale level. To describe the composition of the solution, the (64) _____ of the solute in the solvent must be given. The unit preferred by chemists is (65) _____, which is defined as the moles of (66) _____ per in one liter of solution. The (67) _____ of a solution is a chemical conversion factor since it can be used to convert from moles of (68) _____ to the volume of solution, or vice versa.

In the laboratory, the (69) _____ of an unknown solution can be determined using a technique called a (70) _____, that involves using a measuring device called a buret. The reaction used for a (71) _____ must always involve two reactants, substances A and B, that undergo complete conversion to products. Two separate solutions are prepared, once containing A and the other B. One of the two solutions must have a known (72) _____ and is called the (73) _____. In a (74) _____ a solution containing B is carefully added to the solution of A using a buret. The point where just enough moles of B has been added to completely convert the moles of A to products is the (75) _____ of the (76) _____. Since the (77) _____ is the only point in the (78) _____ where the exact stoichiometric ratio of moles of B to A in the chemical reaction was accomplished, the volume needed to achieve it can be used to calculate the unknown (79) _____.

- PRACTICE TESTS -

After completing your study of the chapter and the homework problems, the following questions can be used to test yourself on how well you have achieved the chapter objectives.

1. Tell whether the following statements would be always **True or False**:

   A) A strong electrolyte is one that completely decomposes in aqueous solution so that virtually none of the molecular form remains.

   B) Acids and bases can be thought of as chemical opposites and when put together neutralize the properties of each other in a neutralization reaction.

   C) It is possible for two acid solutions with different concentrations, one a weak acid and the other a strong acid, to have the same concentration of hydronium ions in solution.

   D) Diluting a solution will not change the number of moles of solute.

   E) Salts containing sodium and potassium ions are always soluble in water.

F) A salt is an ionic compound formed from the cation of an acid and the anion of a base in a neutralization reaction.

G) Calculations for an unknown concentration in with a titration can be done if you know only, the balanced chemical equation and the molarity of the standard solution.

H) Combustion reactions are always oxidation reduction reactions in which oxygen is reduced.

I) A net ionic reaction is required only if there are spectator ions in the reaction.

J) A strong base solution is a solution of high concentration in which the base completely ionized.

K) Noble metals are pure metals that are poor oxidizing agents with respect to acids.

2. Based on the molecular formula and information given in the chapter, tell which of the following would be: (a) weak acid (b) strong acid (c) strong base (d) weak base (e) none of these

A) $Al(OH)_3$ B) CCl_4 C) HBr   D) HCO_3^- E) HCN
F) $CH_3CH_2CO_2H$ G) CH_3NH_2 H) NaBr   I) HNO_3

3. A few compounds that contain N are given below:

(a) KNO_3 (b) NH_4Cl (c) KNO_2 (d) NO (e) N_2H_4 (f) N_2O_4

A) Indicate the oxidation number on N in each compound.
B) Which compound(s) could be reduced to N_2?
C) Which compound(s) could be oxidized to N_2?
D) In which compound(s), if any, is N completely oxidized?
E) In which compound(s), if any, is N completely reduced?

4. Fill in the blanks below using one of the following:

(a) Oxidation   (b) Reduction (c) An oxidizing agent (d) A reducing agent

A) _____ is a chemical change in which substance gain electrons.
B) _____ is always reduced.
C) _____ is always accompanied by reduction.
D) _____ contains at least one element whose oxidation number is increased.
E) _____ accompanies the loss of O atoms.
F) _____ accepts one or more electrons.

5. Tetraphosphorus decaoxide reacts with water in a combination reaction to produce an acid where P has an oxidation number of +5. Listed below are the possibilities for the product(s)

(a) H_3PO_4 (b) H_3PO_3 (c) $H_3PO_4 + O_2$ (d) $H_3PO_3 + O_2$

A) Which of the above could be the products?
B) What is the name of the acid?
C) Is the acid a strong or weak acid?
D) Write the appropriate balanced chemical reaction.
E) Is the reaction an example of an acid-base reaction, oxidation-reduction reaction or neither?

6. Using the list of reagents below, write the balanced reaction and net ionic reaction for the production of $ZnCl_2(aq)$ by:

Zn(s), Cl_2 (g), $ZnSO_4$, HNO_3, $Zn(OH)_2$, NaCl, HCl, $Zn(NO_3)_2$, $CuCl_2$, $CaCl_2$

A) a neutralization reaction   C) a redox reaction
B) an exchange reaction (not acid- base)

7. Which of the following reactions are redox reactions? If redox, identify the species reduced and oxidized and the oxidation numbers for each form.

    A) $3 Zn(s) + 2 CoCl_3 (aq) \rightarrow 3 ZnCl_2 (aq) + 2 Co(s)$

    B) $PCl_3(l) + 3 H_2O(l) \rightarrow 3 HCl (aq) + H_3PO_3 (aq)$

    C) $C_3H_8(g) + 5 O_2(g) \rightarrow 3 CO_2 (g) + 4 H_2O(l)$

8   Using the metal activity series (Table 5.5), tell whether the following reactions would occur:

    A) $Fe(s) + MgCl_2 (aq) \rightarrow FeCl_2(aq) + Mg(s)$

    B) $Mg(s) + H_2O(l) \rightarrow MgO(s) + H_2(g)$

    C) $Ca(s) + 2 HCl(aq) \rightarrow CaCl_2(aq) + H_2(g)$

9. For the following reactant combinations, determine the likely products and give the:

        (a) balanced overall (molecular) reaction,

        (b) overall reaction *with all ionic species clearly indicated*

        (c) net ionic equation, with spectator ions removed.

    A) $CuBr_2 (aq) + Na_2S(aq) \rightarrow$?

    B) $Ba(OH)_2(aq) + HNO_3(aq) \rightarrow$?

    C) $AgNO_3(aq) + HCl(aq) \rightarrow$?

    D)   Sodium hydrogen carbonate acting as an acid when mixed with KOH(aq)

10. A solution is prepared by dissolving 245.0 g H_2SO_4 in enough water to make 800. mL of solution.

    A)   What is the molarity of H_2SO_4 in the solution?

    B)   What volume of the solution would contain 1.50 mol of H_2SO_4?

    C)   If 20.0 mL of the solution was titrated with KOH(aq) and 30.15 mL of the KOH(aq) was needed to reach the equivalence point, what was the molarity of the KOH?

    D)   If 50.0 mL of the solution is mixed with 100 mL of H_2O, what would be the molarity of H_2SO_4 in the diluted solution?

11. For the reaction: $HBr (aq) + Pb(OH)_2 (s) \rightarrow$?

    A)   Determine the products and balance the chemical reaction

    B)   What volume of 3.50M HBr(aq) is needed to react completely 80.0 g of $Pb(OH)_2$?

12. An antacid containing $CaCO_3$ is reacted with HCl and the following reaction occurs.

        $2 HCl(aq) + CaCO_3(s) \rightarrow CO_2(g) + H_2O + CaCl_2(?)$

    A)   Name $CaCl_2$ and identify its physical state in the reaction.

    B)   If 29.47 mL of 0.430 M HCl is required to react completely all of the $CaCO_3$ in the tablet, how many milligrams of $CaCO_3$ does the tablet contain?

13. For the following UNBALANCED reaction: $Te(s) + HNO_3(aq) \rightarrow H_2O(l) + NO(g) + TeO_3(s)$

    A)   Write the balanced overall (molecular) reaction.

    B)   Write the balanced net ionic reaction, if appropriate.

    C)   If 100 mL of 6.12 M $HNO_3(aq)$ and 50.0g of Te(s) are allowed to react and 50.0 g of $TeO_3(s)$ is produced, what is the % yield for the reaction?

14. A solution of $KMnO_4$ reacts with oxalic acid, $H_2C_2O_4$, and sulfuric acid as shown in the following: $2 KMnO_4 + 5 H_2C_2O_4 + 3 H_2SO_4 \rightarrow 2 MnSO_4 + CO_2(g) + H_2O + K_2SO_4$

    A)   What are the proper coefficients for $CO_2(g)$ and H_2O in the balanced equation?

    B)   Since this an oxidation-reduction reaction, identify the **reducing agent** and **oxidizing agent**.

    C)   What is the molarity of a $H_2C_2O_4$ solution when a 50.0 mL sample of the solution requires 26.50 mL of 0.203 M $KMnO_4$, in excess sulfuric acid, H_2SO_4, to reach the equivalence point of the titration?

15. If a drop of 12.0M HCl(aq), having a volume of 0.05 mL, is allowed to spread over a sheet of Aluminum foil.

    A)   Will a reaction occur and if so what is the complete, balanced reaction that occurs?

    B)   If all of the HCl reacts, what volume of Al metal has reacted? (The density of the aluminum foil is 2.70 g/cm^3)

    C)   If the foil is 0.10mm thick, how many drops would produce a hole, created by the reacted aluminum the size of a penny, about 1.8 cm^2?

16. A student is given a bottle that has not been opened since it was received from a commercial supplier, but most of the label has been torn off. All that can be read is the molar mass of the compound (171.3) and that the compound is a hydroxide salt. Dissolving 6.84 g of the solid in 500.0 mL of water produced a solution with an hydroxide concentration of 0.160M, found by titration. Reaction of the solution with a solution of H_2SO_4 produced a white precipitate. What is the likely identity of the compound in the bottle?

Chapter 6

Energy and Chemical Reactions

In this chapter the two major classes of energy, kinetic and potential energy are defined and related to changes in chemical systems. Changes in temperature ΔT indicate changes in kinetic (motion) and are defined by the heat capacity of a substance. Chemical potential changes, defined as enthalpy, ΔH, accompany phase change (physical change) or chemical changes of substances and act as the major energy source for all living systems. The scientific laws, the Law of Conservation of Energy stated in the First Law of Thermodynamics and Hess's Law will be used to quantitatively determine the thermal energy changes occurring in the chemical systems. The connection of enthalpy to changes in bond energies of reactants and products, an important relationship, is used to illustrate why only certain compounds can act as chemical fuels.

6.1 Understand the difference between kinetic energy and potential energy

6.2 Be familiar with the typical energy units and be able to convert from one unit to another.

6.3 Understand the conservation of energy and energy transfer by heating and working.

In chemistry, **kinetic energy** is connected with the motion of nanoscale particles. **Potential energy** is stored energy and represents energy waiting to be released as work or thermal energy. The primary unit used to measure energy is joules (J), the standard international (SI) unit. The calorie, an older unit, is a larger unit of thermal energy that equals 4.184 joules. Most energy changes associated with chemical or physical changes are measured in kilojoules, kJ (1 kJ = 1000 J). The nutritional calorie (Cal), used to define the energy from diets or food, is actually a kilocalorie equal to 4.184 kilojoules (or 4184 joules) of energy.

The Law of Conservation of Energy states that energy can be transformed, stored or transferred between substances, but it cannot be created nor destroyed in the process. The first law of thermodynamics applies this law to the internal energy of closed systems and divides energy into two types, thermal energy or **heat** (kinetic), symbolized as q, or **work**, as w, (potential), so that E = q + w.

6.4 Recognize and use thermodynamic terms: system, surroundings, heat, work, temperature, thermal equilibrium, exothermic, endothermic and state function.

6.5 Use specific heat capacity and the sign conventions for transfer of energy.

To describe the movement of energy, two regions called the system and surroundings must be developed which may be separated by real or imagined boundaries. Generally it is only necessary to define the system precisely within the situation considered, then everything else becomes the surroundings. Changes in **internal energy**, ΔE, associated with working, (w) or heating (q), can be accurately measured, even if it is not possible to measure the total internal energy. Thermodynamic values, like heat and work, have a sign, as well as a magnitude, indicating the direction of energy change. Thermal energy flow into a system, from the surroundings, is always positive, or **endothermic**, while flow out is always negative, **exothermic**.

Temperature is a measure of the kinetic energy of nanoscale particles Spontaneous transfer of heat always occurs in the direction of the higher to lower temperature. Two objects at the same temperature are at **thermal equilibrium** and no net transfer of heat occurs. Solids, liquids and gases change temperature when heat is transferred to or from the substance. **Heat capacity**, c, is a proportionality factor relating the quantity of thermal energy (q) transferred to the temperature change (ΔT). Heat capacity has units of joules/°C. **Specific heat capacity** (J/g-°C) gives the heat capacity per gram of the substance and allows the thermal energy, q, to be calculated for any mass changing temperature. The **molar heat capacity** (J/mol-°C) relates the thermal energy change to the number of particles undergoing change. The specific heat capacity is different for each substance and is given in reference tables, such as Table 6.1 in the text. The molar heat capacity can be nearly the same for different substances, as in the case of metals.

6.6 Distinguish between change in internal energy and the change in enthalpy for a system.

Changes in state or phase changes also cause heat to be transferred. No temperature change is observed while the phase change is occurring, however, since only potential energy is affected. **Enthalpy**, ΔH, describes the thermal energy change at a constant pressure and temperature and can be either exothermic or endothermic.

The two major changes of state for all substances, melting and boiling then have endothermic enthalpy values. The enthalpy for melting, the **heat of fusion**, ΔH_{fus}, and that for boiling, the **heat of vaporization**, ΔH_{vap}. The values are given in either kJ/mol or kJ/g of a substance. If a substance freezes, the sign of the ΔH_{fus} becomes negative, to indicate an exothermic phase change. The same is true for ΔH_{vap} during condensation. The mass of substance undergoing change will determine the actual quantity of thermal energy, q, for the change. Figure 6.10 in text shows a graph of temperature as a function of the quantity of heat added. If a substance changes state while being heated or cooled, the heat transferred cannot be calculated from the temperature change alone. Instead the heat transferred for each region must be calculated separately and then added together. It is important to remember that:

- the **thermal energy, q, for a temperature change within a state** is calculated from **the specific (or molar) heat capacity.**

- **q for change in state** is calculated from the ΔH **for the phase change** (Make sure the sign of ΔH is positive if heating and negative, if substance is cooling.)

- **Convert the units of q for each region to all the same units**, kilojoules or joules, **before adding** the values together.

Experiments show that ΔE and ΔH are examples of state functions. The values don't depend on how the change occurs but just on the initial and final states. ΔE and ΔH for a process differ by the work involved in the change. Measurements of a thermal energy change under constant volume conditions means that work due to expansion is zero. Consequently the change internal energy equals the thermal energy change, or $\Delta E = q + 0 = q_v$. However, the same energy change measured under a constant pressure conditions equals ΔH and includes work, since $\Delta H = q_p = \Delta E + w$ from the first law of thermodynamics.

6.7 Use thermochemical equations and derive thermostoichiometric factors from them.

6.8 Use the fact that the standard enthalpy change for a reaction, $\Delta H°$, is proportional to the quantity of reactants consumed or products produced when the reaction occurs.

6.9 Understand the origin of the enthalpy change for a reaction in terms of bond enthalpies

Chemical reactions also occur at a single temperature and also have an associated ΔH either exothermic or endothermic. The **standard enthalpy change for a reaction, $\Delta H°$,** is the thermal energy change measured under conditions of 1 bar of pressure and at 25°C. The **thermochemical equation** gives the $\Delta H°$ and the balanced chemical reaction, with the physical state (solid, liquid, aqueous solution) for all reactants and products.

The standard enthalpy change, $\Delta H°$, for reactions are used to write **thermostoichiometric factors**, the ratio between the moles of a reactant (or product) in the balanced reaction and the energy transferred during the reaction. The thermostoichiometric factors are stoichiometric conversion factors which are used to relate the actual thermal energy change, q, to moles of reactant consumed or product made.

Experiments show that the $\Delta H°$ for a chemical reaction is mainly due to the difference in energy used to break bonds in reactants and the energy released when new bonds are formed. The **bond enthalpy (bond energy)** is the thermal energy needed to break a bond between to atoms for a substance in the gas state and are always positive values. A strong bond has a high bond energy. Exothermic reactions have more strong bonds formed in the products (or more bonds formed) then broken in the reactants. The reverse situation produces endothermic reactions.

6.10 Describe how calorimeters can measure the quantity of thermal energy transferred during a reaction.

6.11 Apply Hess' Law to find the enthalpy change for a reaction

6.12 Use standard molar enthalpies of formation to calculate the thermal energy transfer when a reaction takes place.

A **calorimeter** is a device used to measure an exchange of thermal energy (q) between two parts, or substances, within the calorimeter. Within the calorimeter, one part (substance) loses while the other gains and $q_{gain} + q_{lost} = 0$. There are two basic types of calorimeters, closed (bomb) calorimeters which can measure internal energy changes, since $\Delta E = q_v$ and open (coffee cup) calorimeters which can measure ΔH $(= q_p)$ directly. Either a ΔT or ΔH process can define the loss or gain of the substances in the calorimeter. Both parts (substances) can be using same type of process or one of each type can occur.

Hess's Law is an alternative way to determine the ΔH for a chemical reaction It involves adding together reactions with known ΔH to produce a specific overall reaction. Since ΔH is a state function, the type of reactions used do not affect the net energy change involved as long as the reactions can be shown to add up to the same overall reaction. A method that can be used to develop the correct sequence for adding the listed reactions is to:

(1)   Identify the substances that appear only <u>once</u> in the listed reactions and are also in the overall reaction.

(2)   Add the listed reaction so that the substance is on the correct side of the reaction arrow compared to overall reaction. Then multiply the listed reaction by the fraction or whole number that will make the coefficient in front of the substance the same as in the overall.

(3)   If any listed reaction(s) remain they will be used to cancel out the unwanted substances.

The equation: $\mathbf{\Delta H^0 = \Sigma \Delta H_f^0 (products) - \Sigma \Delta H_f^0 (reactants)}$ summarizes another form of Hess's Law, using the formation reactions. It can be used when all the standard heats of formations are known for each substance in the balanced chemical equation. The key points to remember when using this equation are:

• Use the ΔH_f^0 on the table for the correct physical state of the substance

• Multiply the ΔH_f^0 by the coefficient in the balanced chemical equation.

• If the $\Delta H°$ for the reaction is known, then this equation can also be used to solve for the ΔH_f^0 for a substance in the reaction.

6.13 Be able to define and give examples of some chemical fuels and to evaluate their abilities to provide heating.

6.14 Describe the main components of food and evaluate their contributions to caloric intake.

Substances that are useful as **chemical fuels** have combustion reactions that are exothermic, are substances that are available in large amounts and at reasonable cost. The **fuel value** of substance is the energy released per gram of the fuel and the **energy density** is the energy released per liter. The higher the fuel value or the energy density, the better the better the fuel. Fossils fuels, which are similar to hydrocarbon molecules, have a much higher density than fuels that contain more carbohydrate molecules, such as wood or biomass.

Three classes of compounds in foods, fats, carbohydrates and proteins are important fuels that act as foods for biological systems. The **caloric value** of these compounds is equal to the enthalpy of the combustion reaction measured with bomb calorimetry since combustion and metabolism have the same end products. Fats have a higher energy density (9 Cal/g = 38 kJ/g) and are used to store chemical energy, whereas carbohydrates, with about half the caloric value of fats, are more quickly utilized fuel sources. Carbohydrates and proteins have about the same caloric values (4 Cal/g = 17 kJ/g). The minimum energy needed per day achieved by the conversion of these substances is estimated by the **basal metabolic rate** (BMR) which is about 1 Cal (4 kJ)/kg of body weight per hour for resting and normal activities.

Summary of Important Equations for Thermal Energy Changes (q):

ΔT processes:
- **Objects** (like a bomb calorimeter)

Heat capacity (J/°C):

$$q_{object} = C_{object} \times \Delta T$$

- **Pure substances**

Specific heat capacity, c (J/g-°C):

$$q = c \times mass \times \Delta T$$

Molar heat capacity, c_m(J/mol-°C):

$$q = c_m \times (no.\ moles) \times \Delta T$$

ΔH processes:
Change in state:

$$q = \frac{\Delta H_{chnage\ of\ state}}{1.0\ g\ X} \times (mass\ X\ changing\ state)$$

$$q = \frac{\Delta H_{chnage\ of\ state}}{1.0\ mol\ X} \times (no.\ moles\ X\ changing\ state)$$

Chemical reaction:

$$q = \frac{\Delta H^0\ reaction}{no.\ mol\ X\ (balanced\ reaction)} \times (no.\ moles\ X\ reacted)$$

Chapter Review - Key Terms

In order to make all the sentences below TRUE, insert the appropriate word or phrase from the list of key terms which best fits the context of the sentence.

NOTE: Any phrase or word from the list may used more than once.

LIST:		
basal metabolic rate	fuel value	standard enthalpy change
bond enthalpy(bond energy)	heat/heating	standard molar enthalpy of formation
caloric value	heat capacity	standard state
calorimeter	heat of fusion	state function
change of state	heat of vaporization	surroundings
chemical fuel	Hess's Law	system
(Law of) conservation of energy	internal energy	thermal equilibrium
endothermic	kinetic energy	thermochemical equation
energy density	molar heat capacity	thermodynamics
enthalpy change	phase change	thermostoichiometric factor
exothermic	potential energy	work/working
first law of thermodynamics	specific heat capacity	

Describing Energy Changes:

The study of the interconnection between chemical change and energy transformations is called (1) _____. Energy associated with the motion of nanoscale particles is often thermal energy, a form of (2) _____, and is connected to the temperature of the substance. Chemical (3) _____, which is different from (4) _____, results from the attraction and repulsions of the electrons and nuclei of atoms or molecules in a substance. One of major scientific laws, the (5) _____, says that for any chemical or physical change, energy may be transformed or moved, but it cannot be created or destroyed. To be able to follow the movement of energy, the universe has to first be divided into two parts, the (6) _____ and the (7) _____. The (8) _____ is the defined, primary region of concern enclosed by real or imaginary boundaries. The (9) _____ is everything else in the universe.

The (10) _____ of a system, ΔE_{sys}, is the sum of all the (11) _____ and (12) _____ of all the nanoscale particles contained within the system. The law represented by the equation: $\Delta E_{sys} = q + w$ is the (13) _____ which gives the changes in (14) _____, ΔE_{sys}, where w, is defined as (15) _____ and is a process that transfers (16) _____ to or from the system while q, (17) _____ , describes the thermal energy transfer that occurs when samples at different temperatures are brought into contact. Heat will naturally flow from the region of high temperature to low temperature, until (18) _____ has been established and the temperature of two samples is the same. The signs of q or w are important indicators of the direction of energy flow. Negative values for either indicate that the (19) _____ of the (20) _____ has been lowered and energy has been transferred to the (21) _____, whereas positive values indicate the opposite process has occurred.

A change in temperature for a substance also indicates that thermal energy has been lost or gained. To relate the observed temperature change, ΔT, to the thermal energy transferred, in joules (J), the proportionality factor called the (22) _____ for the substance is defined. Like density, (23) _____ relates two major units, joules and degrees, and has a numerical value that is different for each substance. Unlike density, its value depends on the mass since the (24) _____ needed for a 1°C change varies with the mass of the object or substance. To make it single-valued, the (25) _____, represented by the symbol, c, is defined as the amount of heat needed to change the temperature of 1.0 gram of a substance by 1°C and has the units, J/g-°C. The (26) _____, c_m, is the (27) _____ defined in terms of the number of particles changing, instead of the mass and has the units = J/mol-°C. The observation that the (28) _____ of metals are nearly the same, while the (29) _____ are different, can be explained by the fact that since the forces between metal atoms in the solid state are very similar, the energy needed to increase the motion of a single atom is also similar.

Concerning Enthalpy Changes:

Plotting the temperature changes during heating (or cooling) of pure substances, shows that (30) _____s always occur with either the gain (or loss) of thermal energy, but the temperature of both phases remains the same during the change. Heat must be added during changes such as melting, which is a (31) _____ , yet the temperature of the substance is not observed to change until the conversion from solid to liquid is complete. This type of thermal energy transfer, that occurs independent of a ΔT, is called an (32) _____, ΔH. Since ΔH is a measure energy change due to the interactions of the particles changing, not their motion, it is an example of a (33) _____ change, not a change in (34) _____. The quantity of heat needed to melt 1.0 mole of a substance, at its melting point, is called the molar (35) _____, and is represented by the symbol, ΔH_{fus} and is always a positive value. Similarly the heat needed to boil or vaporize 1.0 mole of a substance is called the (36) _____, ΔH_{vap}. When freezing of the 1.0 mole of substance occurs instead of melting, the quantity of heat would still equal the (37) _____, but the sign would be changed to a negative value. Negative values for (38) _____s are (39) _____ values and always indicate the system (substance) has lost thermal energy independent of its temperature. When the (40) _____ is positive, the change is called (41) _____ and the energy of the system increases. Since the magnitude does not depend on how the change is accomplished, but only on what the initial and final states are, an (42) _____ is an example of a (43) _____ .

Chemical reactions also have either (44) _____ or (45) _____ ΔH values which are usually much larger than those associated with a (46) _____. The combination of the balanced chemical reaction with the reaction (47) _____, ΔH_r, for the reaction produces the (48) _____. A (49) _____ is the ratio of the enthalpy of reaction, ΔH_r, to the moles of one of the reactants or products in the balanced chemical reaction. The (50) _____ can be used to calculate the exact quantity of heat lost or gained when a specific amount of reactant or product, not equal to the stoichiometric amount, is involved in the conversion. The source and magnitudes of the (51) _____ for a chemical reaction can be explained by changes in the (52) _____ of the products versus the reactants. The (53) _____ measures the energy needed to break bonds in gaseous molecules into gaseous atoms. Generally, if strong bonds in the reactants are converted into weaker bonds in the products, or into fewer strong bonds, the reaction will be (54) _____. If the reactants have fewer strong bonds, or have weaker bonds than those in the products, the reaction is (55) _____.

Measuring and Classifying Enthalpy Changes:

In the laboratory, a (56) _____ is the device used to measure heat transfers between two or more substances and can be used to determine the (57) _____ for a reaction or a (58) _____ for substance undergoing a change in temperature . If an (59) _____ for a reaction is measured under (60) _____ conditions, 25°C and 1.0 bar(atm) pressure, a superscript appears with the symbol, $\Delta H°$ and defines the (61) _____. An alternative way to calculate the (62) _____ for an overall reaction, is to use (63) _____ and sum reactions with known $\Delta H°$'s in a certain sequence to produce the desired overall reaction.

The (64) _____ $\Delta H°_f$, has been measured and tabulated for many compounds, assuming the enthalpy of formation of all elements is zero. It is defined as equal to the (65) _____, $\Delta H°$, for the formation reaction in which one mole of the compound is made from elements in their standard states. These defined values are useful since the (66) _____ for any reaction can then be calculated directly from the sum of the of all the (67) _____s of products minus that of the reactants, as long as the value for each reactant and products is known.

A (68) _____ is any substance that has an (69) _____ value for $\Delta H°$ when it reacts with $O_2(g)$ which is also available at a reasonable cost and in reasonable quantity. Currently fossil fuels and biomass are the major (70) _____s in use. Since much smaller volumes of fossil fuels are needed to provide the same amount of energy than foods or biomass, fossil fuels have a higher energy per unit volume or (71) _____. The number of kilojoules released per gram of the substance, or (72) _____ of fossil fuels, is also highest of all the (73) _____.

The (74) _____ is the minimum energy intake for humans required to maintain body activity whether awake or asleep. In humans, three classes of compounds, fats, carbohydrates and proteins, provide this energy acting as (75) _____s since metabolism is the same type of reaction as combustion. Therefore the $\Delta H°$ for the combustion reaction of either fats, carbohydrates or protein with $O_2(g)$, produces the (76) _____ of foods. Fats have the higher (77) _____ or (78) _____ than carbohydrates or proteins which is directly related to the higher number of weaker C-C or C-H bonds in the fat molecules than the stronger C-O or O-H bonds, which are more prevalent in carbohydrates and proteins.

- PRACTICE TESTS -

After completing your study of the chapter and the homework problems, the following questions can be used to test yourself on how well you have achieved the chapter objectives.

1. Tell whether the following statements would be always **True or False**:

    A. Chemists need a fixed set of conditions to use for comparison purposes. The set of conditions known as the standard state is a temperature of 0°C and pressure of 1 bar.

    B . Any allotropic form of an element can be considered as a standard state and have a standard enthalpy of formation of zero.

    C. Unless zero in value, thermodynamic values should always have a sign associated with them.

    D. Properties such as melting point and density are characteristic properties because they help identify one material from others. The molar heat capacity values also vary from material to material and thus can be used as chemical identifiers.

    E. A calorie is a larger unit of thermal energy than a joule.

    F. The change to the gaseous state demands more energy than melting because the bonds between the atoms must be completely broken to produce gaseous atoms.

    G. Changes in thermodynamic values, such as enthalpy, heat and work are written with a delta (Δ) in front of them to indicate they are not absolute values.

    H. Endothermic reactions occur when stronger bonds are formed in the product than the bonds present in the reactant substances, assuming the number of bonds is about the same.

    I. Calorimetry can be used to determine either enthalpy changes or heat capacities, while Hess's law can only be used to determine enthalpy changes.

    J. A bomb calorimeter is used to measure the thermal energy change for decomposition reactions, which are often exothermic reactions.

    K. $\Delta H°f$ values for compounds are determined from reactions having only elements in the standard states as the starting materials.

    L. Two important criteria for fuels, energy density and fuel values, can both be determined from the enthalpy of combustion and the molecular formula for the fuel.

    M. A thermochemical equation tells us how much energy is transferred as a chemical process occurs and is always given in kilojoules per mole.

    N. If the surroundings were to do work on the system under study, the sign of w in the equation, $\Delta E = q + w$, would be positive and the energy of the system would have increased.

    O. The typical daily energy requirement for basal metabolic rate of humans is in millions of joules per day.

2. Tell whether the following statements are **True or False**:

A. The $\Delta H°$ for a combustion reaction equals the $\Delta H°_f$ for the oxide.

B. Since energy is always conserved, it is impossible to have an "energy shortage".

C. There is really no such thing as "cold" only more or less heat. That means that refrigerators don't make cold, they move heat.

D. You have two containers of water both at 30°C. One container is filled to a volume of 1.0 liter and the other to 10.0 liters. The larger container must have a higher molar heat capacity, since it has more mass.

E. Using Hess's Law, one can explain why $CO_2(g)$ is formed instead of $CO(g)$ when burning a fossil fuel in excess oxygen conditions.

3. Suppose you had two test tubes, one containing 50.0 g of C_6H_6 and the other 50.0 g H_2O. Both test tubes were placed in a freezer and brought to -10°C. If you then remove the test tubes and placed them in the same hot water bath so they heated at the same rate, which substance would:

A) begin to melt first?

B) be the first to be completely converted to liquid?

	melting point T (°C)	ΔH_{fus} (kJ/mol)	c (s) (J/g-°C)
C_6H_6	5.0	10.56	1.74
H_2O	0.0	6.01	4.18

4. Given the properties for liquid and gaseous Freon, CCl_2F_2, in the box below, suppose 2.0 kg of CCl_2F_2 is heated from - 40°C to + 40°C

Properties for CCl_2F_2

Boiling point, °C = -29.8°C
c (liquid) = 0.598 J/g-°C
c (gas) = 0.969 J/g-°C
$\Delta Hvap$ = 20.11 kJ/mol

A) Draw a diagram similar to Figure 10, showing the three regions of the heating curve for heating liquid CCl_2F_2 from - 40°C to the boiling point, the phase change at the boiling point, and heating the gas to 40°C.

B) How many kilojoules of heat can be removed from the air when 2.0 kg of CCl_2F_2 is heated from - 40°C to + 40°C?

5. The specific heat of ethanol is 2.46 J/g-°C. A sample of 100. grams of ethanol in an insulated container was at 24.0°C and a second sample of 75.0 grams of ethanol at an unknown temperature was added. If the final temperature of the mixture was 28.0°C, what was the temperature of the ethanol added?

6. When 44.0 mL of water at 23.80°C is added to a metal cup that weighs 48.93 g and has a temperature of 98.50 °C, the final temperature of the metal cup and water is 28.37°C. What is the specific heat of the metal?

7. When a certain mass of KI(s) is added to 200 mL of water in a calorimeter at 23.5°C the temperature of the resulting solution decreases to 19.3°C. If the enthalpy change for KI(s) dissolving is +21.3 kJ/mol KI, density of solution is 1.21 g/mL and the specific heat capacity of solutionis 4.184 J/g-°C, what mass of KI was added?

8. Silver metal undergoes a displacement reaction with $Zn(NO_3)_2$ in aqueous solution. When excess silver metal at 20.00°C is added to 100 mL of 0.274 M $Zn(NO_3)_2$ also at 20.00°C, the temperature of the solution rises to 43.85°C.

A) Complete and balance the reaction: $Ag (s) + Zn(NO_3)_2(aq) \rightarrow$ __?__(aq) + Zn(s)

B) Is the reaction endothermic or exothermic?

C) Assuming the density and specific heat capacity of solution is the same as that of water, what is the enthalpy change for the reaction per mole of $Zn(NO_3)_2$?

D) Write the net ionic reaction occurring in solution. Should the ΔH for the net ionic reaction be different than that for the balanced overall reaction? Explain.

9.   A 0.50 gram of asparagus was burned in a bomb calorimeter whose heat capacity was 5.24 kJ/°C and the temperature of the calorimeter rose from 22.45°C to 23.17°C.

A) What is the quantity of thermal energy was released per gram of asparagus from the combustion?

B) What is the number of nutritional calories that can be obtained from eating 4 ounces of asparagus (1 pound = 16 ounces), based on the answer to (A)?

10.   When 1.0 mole of NO(g) forms from its elements, 90.29 kJ of heat is absorbed.

A) Write the thermochemical equation for the formation reaction

B) Is the reaction endothermic or exothermic?

C) How much thermal energy is transferred when 1.50 g of NO(g) is formed from the elements?

11.   A fermentation reaction that occurs in many natural juices produces ethanol, $C_2H_5OH(l)$, from glucose, $C_6H_{12}O_6(s)$.

$$C_6H_{12}O_6(s) \rightarrow 2\ C_2H_5OH(l) + 2\ CO_2\ (g) \quad \Delta H° = -68.4\ kJ$$

A) If 18.0 g of glucose were fermented how many kilojoules of heat would be transferred by this fermentation at 25°C, 1 bar pressure?

B) Compare the percentage of heat released in (A) to what would be released by the complete combustion of 18.0 g glucose, $\Delta H_{comb} = -2801.6\ kJ/mol\ C_6H_{12}O_6(s)$.

C) Considering the change in the number and type of bonds in the products of the two reactions, explain why the second reaction has a lower enthalpy.

12. Dissolving of BaSO4(s) in water is an endothermic process:

$$BaSO_4(s) \rightarrow Ba^{+2} + SO_4^{-2} \quad\quad\quad \Delta H° = +19.3\ kJ$$

A) What would be the $\Delta H°$ for the precipitation of one mole of $BaSO_4(s)$?

B) What would be the heat produced in solution when $BaSO_4$ precipitated from the reaction of 50.0 mL 0.105 M $Ba(NO_3)_2$ solution with excess H_2SO_4 solution?

C) If 50.00 mL of 0.105M $Ba(OH)_2$ were substituted for $Ba(NO_3)_2$ would the heat released be the same or different? Explain.

13. Using Hess' Law, determine the ΔH of reaction for: $2\ S(s) + 2\ OF_2(g) \rightarrow SO_2(g) + SF_4(g)$

Reactions: $OF_2(g) + H_2O(l) \rightarrow O_2(g) + 2\ HF(g)$ $\Delta H = -276.6\ kJ$

$SF_4(g) + 2\ H_2O(l) \rightarrow SO_2(g) + 4\ HF(g)$ $\Delta H = -827.5\ kJ$

$S(s) + O_2(g) \rightarrow SO_2(g)$ $\Delta H = -296.9\ kJ$

14. Rust, Fe_2O_3, is formed from the series of reactions given below.

STEP 1: $2\ Fe(s) + 6\ H_2O(l) \rightarrow 2\ Fe(OH)_3 + 3\ H_2(g)$ $\Delta H_1 = +321.0\ kJ$

STEP 2: $2\ Fe(OH)_3 \rightarrow Fe_2O_3(s) + 3\ H_2O(l)$ $\Delta H_2 = ?$

STEP 3: $2\ H_2(g) + O_2(g) \rightarrow 2\ H_2O(l)$ $\Delta H_3 = -571.6\ kJ$

Using the standard heats of formation, $\Delta H°_f$ as needed:

A) Calculate the $\Delta H°_f$ for $Fe(OH)_3$ from the first step

B) Calculate ΔH_2 for the second step.

C) Show that the 3 steps added together produce the formation equation for one mole of Fe_2O_3 and that the resulting ΔH also equals the heat of formation for $Fe_2O_3(s)$.

15. One way to remove NO(g), a primary air pollutant, is to react it with NH_3(g). The unbalanced reaction is;   ___ NH_3(g) + 6 NO(g) → ___ N_2(g) + 6 H_2O(l)

   A) What are the coefficients for NH_3 and N_2 in the balanced reaction?

   B) What is the $\Delta H°$ for the reaction, using standard enthalpy of formation values?

   C) Is the reaction endothermic or exothermic?

   D) How many kilojoules of heat would be lost or gained when 60.0 g of NO(g) reacts?

16. Propene and cyclopropane have the same molecular formula, C_3H_8, and are isomers. If the $\Delta H°_f$ of propene is +20 kJ/mol and that of cyclopropane is +53.3 kJ/mol,

   A) which would have the largest fuel value when combusted?

   B) Compare the bond strengths in propene to that in cyclopropane, assuming the number of bonds involved in the combustion to be the same.

17. Hydrogen sulfide, H_2S(g) is a potential chemical energy source used by bacteria, instead of hydrocarbons, since its oxidation is also exothermic.

   A) Write the thermochemical equation for the oxidation of H_2S(g) to SO_2(g), assuming gaseous water is the second product.

   B) If 100g H_2S(g) reacted with 100 g O_2, what is the maximum amount of heat that can be obtained?

   C) Given the density of H_2S(g) is 1.39 g/L, calculate its fuel value and energy and compare to the values for H_2(g) and CH_4(g) listed below. Would H_2S make a good chemical fuel? Explain your answer.

	H_2(g)	CH_4(g)	H_2S(g)
Fuel Value (kJ/g)	120.5	79.0	_____
Energy density (kJ/L)	9.86	51.7	_____

Chapter 7

Electron Configurations and the Periodic Table

In this chapter, the relationship between electromagnetic radiation (or light energy) and electron energies is explained. The quantum nature and organization of electrons into energy levels and orbitals within the atom is described. The quantum model is then used to build the electron configurations for atoms which strongly influences both the chemical and physical properties. The relationship of position on the periodic table and chemical properties of an element can be easily correlated and used to explain the similarity of properties for elements in a group and the periodic trends observed for atomic radii, ionic radii, ionization energy and electron affinity. This description will be fundamental to understanding the next chapters that deal with the types of chemical bonding an atom will undergo and the molecules it forms.

7.1   Use the relationships among frequency, wavelength, and the speed of light for electromagnetic radiation

7.2   Explain the relationship between Planck's quantum theory and the photo-electric effect

The major characteristic used to classify **electromagnetic radiation** is either **frequency**, ν, or **wavelength**, λ. Frequency has units of cycles per second (cps or Hertz) and describes the number of peaks or wave crests that pass by a fixed pint in one second. In chemistry, frequency is used with units of 1/sec or s^{-1} only. The wavelength is the distance the wave travels in one second which is the same as the distance between the crests in the wave. Throughout the electromagnetic spectrum, the wavelength varies from meters for radio waves, to nanometers for visible radiation, and to as small as picometers for cosmic (alpha, beta and gamma) radiation. If either the wavelength, l or frequency, ν, changes for a wave, the other must also change since $\nu \times \lambda = c$, where c equals the speed of light, 3.00×10^8 meters/sec.

The energy of the wave varies with frequency since the more waves passing a point, the more energy transmitted. Short wavelengths mean high frequency. The exact relationship between energy and frequency (or wavelength) was defined by Max Planck as $E = h\nu = hc/\lambda$, where h equals **Planck's constant**, 6.62×10^{-34} J-sec, and wavelength has units of meters. The "**quantum**" of energy, hν, described by Planck, is a fundamental "packet' or particle that makes up all energy, much like atoms as fundamental particles make up all matter. Einstein redefined the quantum as a "**photon**", a fundamental particle of light. An individual quantum of radiation has an extremely small amount of energy, but we generally deal with very large quantities of photons in the macroscopic world. For example, a burning 100 watt light bulb is producing 100 joules of visible radiation per second and emitting approximately 10^{21} photons of visible light per second, producing a mole or more of photons every 2 minutes.

The **photoelectric effect**, which is the production of electrical current by light absorbed by the surface on a metal, depends on frequency of light. Each metal requires a certain minimum frequency of light, called the threshold frequency, to free the electrons. This minimum frequency is unique to the metal. The effect could only be explained if electrons in metals were in "quantized" energy levels, where the energy states are like steps on a ladder. The energy transmitted from the frequency of light has to match the exact spacing between levels or the electron can not move to a higher step to be eventually be removed from the metal. That different threshold energies are required by different metals means electrons are starting out in different positions on the energy "ladder".

7.3   Use the Bohr model of the atom to interpret the emission spectra and the energy absorbed or emitted when electrons in atoms change energy levels.

7.4   Calculate the frequency, energy, or wavelength of an electron transition in a hydrogen atom and determine in what region in the electromagnetic spectrum the emission would occur.

Bohr defined the electron states for an atom as orbits that had an quantized energy related to n^2, where n was a whole number greater than zero and gives the **principal energy level**. The energy for a Bohr orbit is - (2.179 X 10^{-18} J)/n^2 where the negative value indicates how far the electron is below the free electron state, given a zero energy (see diagram below). The diagram of the n levels with the Bohr energy also shows that:

(1)   Large values of n indicate that the electron exists further out from the nucleus and will require less energy to be removed.

(2)   The spacing between the levels is not constant and gets smaller as the n values get larger.

(3)   Transitions between energy states are given by: $\Delta E_{transition} = E_{upper} - E_{lower}$ where:

$$\Delta E = h\nu = 2.179 \times 10^{-18} J \left[\frac{1}{n^2_{lower}} - \frac{1}{n^2_{upper}} \right]$$

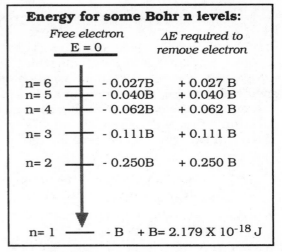

Energy for some Bohr n levels:

	Free electron E = 0	*ΔE required to* *remove electron*
n= 6	- 0.027B	+ 0.027 B
n= 5	- 0.040B	+ 0.040 B
n= 4	- 0.062B	+ 0.062 B
n= 3	- 0.111B	+ 0.111 B
n= 2	- 0.250B	+ 0.250 B
n= 1	- B	+ B= 2.179 X 10^{-18} J

The observed lines in the emission spectra of hydrogen are correctly predicted by this equation. This is proof the energy levels are quantized. Normally the electron is in the lowest available n level, its **ground state**. If an electron absorbs energy of the correct energy ΔE, it can move to a higher n level and produce an **excited state**. **Line emission spectra** result from electrons moving from excited states to the ground state, giving off the energy (ΔE) as light. The wavelength for the transition, λ = hc/$\Delta E_{transition}$, is compared against a published electromagnetic spectrum, such as in Figure 7.1 or Table 7.1, to determine the region it will appear in. Emission and absorption lines between identical n values have the same energy and wavelength. Consequently, the same equation is used to calculate ΔE for either and the direction of the change is not indicated.

7.5 Explain the use of quantum mechanical model of the atom to represent the energy and probable location of electrons.

7.6 Apply quantum numbers

7.7 Understand the spin properties of electrons and how they affect electron configurations and the magnetic properties of atoms.

7.8 Describe and explain the relationships among shells, subshells and orbitals.

Another physicist, de Broglie, expanded the Bohr concept by proving electrons have a dual nature and could have both particle and wave properties. In the De Broglie relation, the **momentum**, mass × speed (mv), determines wavelength of the particle, λ= h/$mv_{particle}$.

The Quantum Mechanical Model is based on the wave-particle duality in which the electrons occupy **orbitals** that have the characteristic of a wave. The electron orbitals are described with a **boundary surface** representing a probability (usually 90%) of finding the electron within a region. The use of the boundary surface, or probability, is necessary. According to the **Heisenberg Uncertainty Principle** it is impossible to know the exact position and exact momentum of the electron simultaneously. Thus, given its wave properties, we can only specify a region in space as the orbital where the electron is likely to be at any time.

Four quantum numbers give the individual characteristic of each electron on the atom. The **principal quantum number**, n, is used to specify energy and distance from the nucleus. The type or shape of the orbital is given by ℓ, the azimuthal quantum number, and the orientation in space by the m_ℓ quantum number. According to the **Pauli exclusion principle**, two electrons can have the same n, ℓ, and m_ℓ numbers and occupy a single orbital, but must differ in a fourth quantum number called the spin quantum number, m_s. The spin quantum number describes the opposite response the two electrons have to a magnetic field. Therefore, each electron has its own unique set for the four quantum numbers. Table 7.3 in the text shows the exact relationship between the four quantum numbers and the organization of electrons into shells and subshells. Overall the organization can be summarized as:

	shells →	subshells →	orbitals →	single electron spin
has quantum number	n	ℓ	m_ℓ	m_s
sets major property:	energy	shape	orientation	up (↑) or down (↓) spin
has possible values:	1,2,3...	0,1,2..n-1	$-\ell,..0,..+\ell$	+1/2 , - 1/2

There are as many **shells** as there are n levels on an atoms. The shells represent major changes in energy for an electron, so that generally, a shell must be completely filled, starting with n=1, before higher n levels can be occupied. **Subshells** divide the shell into different types of orbitals, where the number of different types equals n, so that there are 3 types in n = 3 and only one type in n= 1. The number of orbitals within a subshell varies, but is always an odd number. When ℓ = 0, there is only one orbital, an "s" orbital, when ℓ = 1, there are 3 orbitals of equal energy and similar shapes called "p" orbitals and when ℓ = 2, there are 5 "d" orbitals. Each orbital has its own orientation in space, so they coexist by being in different regions of space at any given time.

The **electron configuration** is a description of the occupied orbitals on an atom. The condensed form of electron configuration gives the n and ℓ values, represented by its letter designation (s, p, d, or f) for the occupied subshells starting with the 1s subshell. The number of electrons in the subshell is given by a superscript. The configuration $1s^2 2s^2 2p^3$ means there are 2 electrons in the 1s subshell (with n = 1, ℓ = 0), 2 electrons in the 2s subshell (with n = 2, ℓ = 0), and 3 electrons are in the 2p subshell (with n = 2, ℓ = 1).

To specify all four quantum numbers, the orbital box diagram form of the electron configuration is written, where each orbital is shown as a box (or line), and the electron spin as either an up or down arrow, so that $1s^2 2s^2 2p^3$ becomes 1s_↓↑_ 2s _↓↑_ 2p _↑_ _↑_ _↑_. As a compromise, the expanded electron configuration shows the separate subshell orientations, but does not show the electrons as arrows and $1s^2 2s^2 2p^3$ becomes $1s^2 2s^2 2p_x^1 2p_y^1 2p_z^1$. **Hund's Rule** stipulates that electrons must have the maximum number of <u>unpaired</u> spins in any unfilled subshell.

7.9 Use the periodic table to write the electron configuration of atoms and ions of main group and transition elements.

7.10 Explain variations in valence electrons, electron configurations, ion formation, and paramagnetism of transition metals.

The position on the periodic table can be used to write the electron configuration for an element. Each period (row) in the periodic table corresponds to a different n level or shell being filled by electrons as shown in Figure 7.13 in the text. There are basically two types of electrons in any configuration:

(1)   Electrons in filled shells become **core electrons**, which are very difficult to remove and do not participate in bonding or exchange with other atoms.

(2)   The electrons in unfilled shells are **valence electrons** which are more easily removed and the ones used for bonding or exchange.

(3)   Given that n, the principal quantum number equals the period (row) number, the number and placement of valence electrons can be directly correlated to the group number for the element:

• Groups 1A (ns^1) and 2A (ns^2) form the **s-block elements**, where the s subshell is filled.

• Groups 3A ($ns^2 np^1$) → 8A ($ns^2 np^6$) form the **p-block elements** where the p subshell is filled, after the s subshell.

- Groups 3B ($ns^2(n-1)d^1$) → 2B ($ns^2(n-1)d^{10}$) form the **d-block elements**, or transition metals, where the d subshell with an n value of (row number-1) is filled after the s subshell.
  - The lanthanides and actinides are the **f-block elements** where the 4f and 5f subshells are being filled, respectively.

Experimental evidence indicates that the ns^2np^6 configuration, observed for the noble gases (group 8A), is a particularly stable configuration which corresponds to the core electrons for an atom. In the **noble gas notation** the symbol of the noble gas that appears at the end of row <u>before</u> the element replaces the core electrons in the electron configuration. A filled d subshell also act as core electrons and is listed after the noble gas symbol for p-block elements in the fourth or higher rows. Half-filled as well as filled d or f orbitals have enhanced stability. Electron "borrowing" of an electron from the ns orbital to the d (or f) subshell occurs in the groups 6B and 1B groups of the transition metals and at similar positions for the f-block elements. For example, the position of copper, Cu, gives it the configuration $[Ar]\, 3d^9 4s^2$ but its actual configuration is $[Ar]3d^{10}4s^1$.

To form ions, the s and p-block metals lose electrons from the subshell last filled. When p-block nonmetals gain electrons, they add electrons to the unfilled p subshell. When the s and p block elements gain or lose electrons they will also generally do so to achieve the same electron configuration as the nearest noble gas element and become isoelectronic with the noble gas. The common ions of the nonmetals that appear before argon, Ar, in the periodic table and the metals after, which all have 18 electrons, show this general trend:

$$Cl^{-1}, S^{-2}, P^{-3} \rightarrow [Ar] = 1s^2 2s^2 2p^6 3s^2 3p^6 \leftarrow K^{+1}, Ca^{+2}$$

Transition metals form ions by losing valence electrons from the s subshell first and then the d subshell. Most elements have only paired electrons and are not attracted to a magnetic field and are called **diamagnetic**. Elements have unpaired electron spins are attracted to a magnetic field and are **paramagnetic**. The groups headed by Fe, Co and Ni in the d-block, also have unpaired spins, and can become permanent magnets, indicating large scale alignment of the spins, to form **ferromagnetic** materials.

Valence electrons for neutral atoms and ions can be represented using dots around the four sides of the element's symbol in a **Lewis Dot symbol**. All elements in the same family will have exactly the same number of dots. For example the 1A elements; such as H, Li, Na, and K all have the Lewis dot symbol: Na•, K•, Li• and H•.

7.11 Explain how nuclear magnetic resonance works and how it is used in chemical analysis and medical diagnosis.

Nuclear spin is a phenomena similar to electron spin where the separation between two energy states becomes apparent in a magnetic field. Only certain nuclei exhibit nuclear spin, but one of the most common is ^1H, with a single proton as the nucleus. When subjected to a magnetic field, the spin of a proton can be changed when it is influenced by tuned radio wave frequencies in an **nuclear magnetic resonance** (NMR) instrument. Protons are influenced by the number and types of atoms near them, so that the NMR signal can give information about the environment around the H atom. The fact that the human body has so much material containing hydrogens makes the body a good subject for analysis by this method. To avoid the possibly misleading "nuclear" designation, the use of a magnetic field and radio frequencies to image human tissue is called magnetic resonance imaging or MRI.

7.12 Describe trends in atomic radii, based on electron configuration.

7.13 Describe trends in ionic radii and explain why ions differ in size from their atoms.

7.14 Use electron configurations to explain trends in the ionization energies of the elements.

7.15 Describe electron affinity.

Atomic radii, the size of neutral atoms, has two important trends within the periodic table. Progressing down a group (column) in the periodic table causes valence electrons to occupy energy levels with higher n values. This results in an increase in atomic radii. Progressing from left to right in a row (period) results in atomic radii gradually decreasing. The electrons are being added to the same energy level or shell while more protons are added to the nucleus. The resulting increase in attractive force of the nucleus results in the decrease of the radii.

For cations (positive ions) formed by the loss of electrons, the **ionic radius** decreases dramatically from the neutral atom, showing an increased nuclear pull. Anions (negative ions) have much larger radii than the neutral atom because adding more electrons produces less attraction per electron. The anions will also have much larger radii than cations which are isoelectronic with it.

Ionization energy is the energy needed to remove an electron and produce a cation from a neutral gaseous atom. Ionization energy has trends opposite to that for atomic radii, it increases going across a row (period) and decreases going down a group (family) in the periodic table. As electrons are removed, the number of protons remains constant and the effective pull of the nucleus increases, so the second ionization energy is always greater than the first ionization energy. On the same atom, the ionization energy required to remove a core electron is very high compared to removing valence electrons indicating that the core electrons are more tightly bound to the nucleus.

Electron affinity is the energy associated with the reaction: $X(g) + e^- \rightarrow X^-(g)$. It is a measure of the attraction of a nucleus for additional electrons. The trend for electron affinity within the periodic table is not as regular as for atomic radii or ionization energy. The highest electron affinities, which are exothermic values, occur with p-block nonmetals, where the radii and ionization energy values also indicated the nuclear attraction was strongest.

Chapter Review - Key Terms
In order to make all the sentences below TRUE, insert the appropriate word or phrase from the list of key terms which best fits the context of the sentence.
NOTE: Any phrase or word from the list may used more than once.

LIST:		
atomic radius	ionic radii	photons
boundary surface	ionization energy	Planck's constant
continuous spectrum	isoelectronic	principal energy level
core electrons	Lewis dot symbol	principal quantum number
diamagnetic	line emission spectra	quantum
electromagnetic radiation	momentum	quantum theory
electron affinity	noble gas notation	s-block elements
electron configuration	nuclear magnetic resonance (NMR)	shell
excited state	orbital	spectrum
ferromagnetic	paramagnetic	subshell
frequency	Pauli exclusion principle	transition elements
ground state	p-block elements	uncertainty principle
Hund's rule	photoelectric effect	valence electrons
		wavelength

Defining Light Energy and Connection to Atoms:
(1) _____ is composed of waves of oscillating electric and magnetic fields that move through space at the same speed, the speed of light. The waves differ in
(2) _____, the distance between adjacent crests in the wave and (3) _____,

the number of waves that pass a certain point in one second, given in units of hertz. A
(4) _____ is the term used to describe a distribution of all (5) _____s or
(6) _____s, of light or radiation. The (7) _____, λ, is always equal to the
speed of light, c, divided by the (8) _____, v, so either can be used to characterize the
wave. High energy is transmitted by waves that have short (9) _____s.

   Planck was the first to correlate the observed (10) _____ of light waves emitted
by a hot object to energy observed by defining a (11) _____ of energy as the smallest
packet of energy that could be emitted. A constant, h, called (12) _____, multiplied by
the (13) _____ defines the value of the (14) _____ of energy produced.
Planck's (15) _____, which said that light was composed of packets of energy, was not
well accepted until it could be shown to account for other unexplained phenomena. One of
these was the (16) _____, where a metal emits electrons and produces an electrical
current when light of a certain (17) _____ (or (18) _____) is absorbed by the
surface of the metal. Einstein renamed the (19) _____, a (20) _____ of energy. The connection of (21) _____s to electron transition energies within the atom led
to the application of (22) _____ to the model of the atom.

   When gaseous atoms are heated, only certain energies of visible light are emitted and
produce a discontinuous (23) _____, characteristic of the element, not a
(24) _____ like a rainbow, when all the (25) _____s, from 400 to 700 nm, of
visible light are included. The same lines appear to be missing from the absorption
(26) _____ where electrons are undergo transitions between the (27) _____
energy, the lowest energy state for the electron, to a higher,(28) _____ . Neils Bohr is
credited with the first model that explained the connection between allowed energy transitions
and electron states. Bohr defined the (29) _____s as electron orbits characterized by
an integer value, n, the (30) _____, which determined the energy of the electron and
its distance from the nucleus.

The Quantum Mechanical Model for Electrons:

   In the quantum mechanical model of the atom, the electrons are viewed as waves. De
Broglie related the (31) _____ of the electron to its wavelike properties. The
(32) _____ says that as waves, it is impossible to determine both the exact
(33) _____ and position of the electron at the same time, unlike particles. Consequently, the electron wave can only be described by a (34) _____, defining the region in
which the electron resides 90% of the time. Three quantum numbers are needed to completely
describe the (35) _____ of the electron wave, called an (36) _____. The first n,
the (37) _____ is needed to define both the (38) _____ and size; the second,
the azimuthal quantum number, ℓ, defines the shape and third, m_{ℓ}, which defines the orientation in space of the electron or atomic (39) _____.

   A (40) _____ is comprised of atomic (41) _____s that have the same n
value. With a (42) _____, (43) _____s divide the atomic (44) _____s
into different types, that have the same ℓ value and shape, which are then given letter designa-
tions (s, p, d, f,..). Within the (45) _____there is always an odd number of
(46) _____s that have different orientations. In addition, experimental evidence indicates a maximum of two electrons can occupy any (47) _____ requiring a fourth quantum number, m_s, to be defined to give the spin state of the electron. The (48) _____

stipulates that any two electrons that occupy the same (49) _____ must have opposite spins. (50) _____, or magnetic resonance imaging (MRI) measures the transition between spin states for the ^1H nuclei in an external magnetic field and relates it to molecular structure around the hydrogen.

Describing the Electrons on Atoms:

(51) _____s are used to describe the placement of electrons into the (52) _____s within the (53) _____ s and (54) _____s on the atom. The condensed version of electron configuration indicates the n value of the (55) _____ with a number followed by the letter indicating the (56) _____ type. The number of electrons in the (57) _____ is indicated using a superscript. (58) _____ says that electrons can pair only after each (59) _____ in a (60) _____ is occupied by a single electron which guarantees that any (61) _____ will contain the maximum number of unpaired spins possible. The specific orientation and spin of electrons are shown using arrows in the orbital box diagrams version of the (62) _____, The expanded version of the (63) _____ indicates the differences in orientation, but not spin.

Filled (64) _____s represent (65) _____ and electrons in unfilled (66) _____s are (67) _____. The (68) _____ replaces the (69) _____ with the symbol of the noble gas preceding the element in the periodic table, in the (70) _____. For the elements in groups 3A-8A in the fourth or higher periods, the filled d (71) _____ must also be included with the noble gas symbol.

The position of the element in the periodic table can also be correlated to the (72) _____ part of the configuration. The number of (73) _____s equals the group number for an element. Elements in groups 1A and 2A are the (74) _____ with while the (75) _____ appear in groups 3A-7A. Elements in the "B" groups, the (76) _____, generally have two (77) _____ in an s (78) _____ and the remaining electrons in the d (79) _____. Exceptions to this rule occur in groups 6B and 1B, which appear in the middle and near the end of the (80) _____. In these groups, only one (81) _____ is in the s (82) _____ so that the d (83) _____ can become half-filled (6B) or completely filled (1B).

Properties Related to Configurations:

When metals in the B groups form cations, electrons are lost from the s before the d (84) _____. Metals in the (85) _____ form cations by losing all the valence electrons to form a configuration that is (86) _____ with the noble gas preceding the element. Metals in the (87) _____ groups lose electrons from the p and then the s (88) _____s to form cations. Nonmetals in the (89) _____ add electrons to the p subshell to reach a configuration that is (90) _____ with the noble gas in at the end of the period when forming monatomic anions. The (91) _____ is a useful notation which shows the (92) _____ on an atom or ion as dots around the atomic symbol.

The (93) _____, a measure of size of neutral atoms, decreases going across a period, but increases going down a group. (94) _____ for cations are much smaller than the neutral atom, while monatomic anions have much larger (95) _____ than (96) _____.

Atoms that have only paired electrons are (97) _____, while those that have unpaired electrons are (98) _____ and will be attracted to a magnetic field. Some elements in the Fe, Ni and Co groups that are (99) _____ but act as permanent magnets are called (100) _____ elements.

The energy needed to remove an electron from a neutral gaseous atom, called the (101) _____ is always endothermic and is tabled with the notation of "first", "second", "third" indicating the order of removal. Removing a (102) _____ requires an extremely high (103) _____ , compared to removing a (104) _____ . The first (105) _____ of atoms varies in a regular fashion as it increases going across a period (row), but decreases going down a group in the periodic table. In contrast, the (106) _____ is a measure of the attraction of the nucleus of a neutral atom for an additional electron and does not have the same trends as the (107) _____. Very exothermic values are observed for the (108) _____ occur with the addition of an electron to a nonmetal atom in the (109) _____ .

- PRACTICE TESTS -

After completing your study of the chapter and the homework problems, the following questions can be used to test yourself on how well you have achieved the chapter objectives.

1. Tell whether the following statements would be always **True or False**:
    A) The effective nuclear attraction is nearly constant going across a row (or period) in the periodic table.
    B) Scandium forms a +2 ion that is isoelectronic with a noble gas.
    C) The atomic radius decreases going across a row (period) and down a group (family) in the periodic table.
    D) Group 7A elements have a single unpaired electron.
    E) Emission lines appear when an electron is removed from an atom.
    F) Electrons enter the fourth principal energy level once the third level is filled.
    G) The quantum number m_ℓ determines the shape of the orbital.
    H) The spectrum of Mg has an emission line at 266.8 nm which is a higher frequency than radiation with $\lambda = 315$ nm.
    I) The first ionization energy of Mg is greater than that of Na, but the second ionization energy of Mg is lower than that for Na.
    J) The radius of O^{-2} ion would be less than that of Be.
    K) The minimum frequency of light needed to induce an electrical current from a metal is the photoelectric effect and also gives the ionization energy for the metal.

2. Briefly answer the following:
    A) Explain why sodium vapor lamps do not produce a continuous spectrum of light but can still be very high intensity lamps.
    B) Explain what part(s) of the atom respond to visible light.
    C) Explain what must be true for a magnetic field to affect:
        a) an atom   b) a nucleus
    D) What is the connection between the principal quantum number and the types of subshells and the number of orbitals in shell?
    E) What is the difference between core and valence electrons on an atom?
    F) What is the difference in ions formed for the metals and nonmetals in the p-block elements and how is this related to the electron configuration for the atoms?
    G) Which of the two ions Ga^{+4} and Mn^{+4} is not a stable ion? Explain why not.
    H) The first ionization energy for oxygen is 1313.9 kJ/mol and the second ionization energy is 3381.1 kJ/mol.
        (a) Why is the second ionization energy much greater than the first?
        (b) How would the first ionization energy of sulfur compare to that for oxygen?

3. Green light has a wavelength of 550 nm.
    A) What is the energy of this photon of green light?
    B) Which region(s) would have a higher frequency than the photon: infrared light, radio waves or ultraviolet light?

4. The energy needed to produce the photoelectric effect in Cr(s) is 7.00×10^{-19} joules
    A) What is the threshold frequency for the effect?
    B) Which wavelength(s) of light could produce an electrical current from Cr(s): 220 nm, 300 nm, or 450 nm?

5. Suppose an excited electron on H drops to its ground state through a series of transitions:
    (a) $n = 7 \rightarrow n = 5$ (b) $n = 5 \rightarrow n = 4$ (c) $n = 4 \rightarrow n = 1$
    A) How many photons will the H atom produce in the process?
    B) Which of the transitions has the highest energy?
    C) In what regions will the emission lines appear?

6. For the orbitals that have the quantum numbers: (a) $\ell = 0$ (b) $\ell = 2$ (c) $\ell = 1$
    A) Sketch the proper orbital shape.
    B) In which n level would (a) - (c) first appear together?
    C) How many electrons could occupy (a), (b) and (c), respectively?

7. For the following sets of quantum numbers:
    A) Give the proper label to the orbital
    B) Tell for which element the orbital is first occupied, in a ground state configuration.
    (a) $n = 3,\ \ell = 0$ (b) $n = 4,\ \ell = 2$ (c) $n = 5,\ \ell = 1$

8. For each of the configurations, pictured below, tell whether it represents a possible or incorrect ground state configuration. If incorrect, explain what correction is needed.

A) 1s $\boxed{\downarrow\uparrow}$ 2s $\boxed{\downarrow\uparrow}$ 2p $\boxed{\downarrow}\boxed{\uparrow}\boxed{\downarrow}$

B) 1s $\boxed{\downarrow\uparrow}$ 2s $\boxed{\downarrow}$ 2p $\boxed{\downarrow\uparrow}\boxed{\ }\boxed{\ }$

C) 1s $\boxed{\downarrow\uparrow}$ 2s $\boxed{\downarrow\uparrow}$ 2p $\boxed{\uparrow}\boxed{\ }\boxed{\uparrow}$

D) 1s $\boxed{\downarrow\uparrow}$ 2s $\boxed{\downarrow\uparrow}$ 2p $\boxed{\downarrow\uparrow}\boxed{\uparrow}\boxed{\uparrow}$

9. A) Which of the atoms Sc, P, Si, or O, would have:
            (a) equal numbers of unpaired electrons?
            (b) the smallest noble gas core?
            (c) a +2 ion and give electron configuration of each +2 ion?

    B) Of the ions Cl^-, Ca^{+2}, Mn^{+2}, Br^- which would:
            (a) have the largest radius?
            (b) be paramagnetic?
            (c) be isoelectronic ions?

10. For a Ni atom that has 28 electrons:
    A)   Complete the configuration for the atom: [Ar] 4s $\boxed{\ }$ 3d $\boxed{\ }\boxed{\ }\boxed{\ }\boxed{\ }\boxed{\ }$ 4p $\boxed{\ }\boxed{\ }\boxed{\ }$
    B)   Considering the completed configuration, give the total number of electrons that:
    (a) are unpaired   (b) would have $\ell = 1$ (c) would have $m_s = +1/2$
    C)   Give the likely electron configuration for the +2 ion.

11. For the atoms (A) - (E) in the positions indicated in the figure on the right, give the appropriate:

(a) Lewis dot symbol for the element

(b) The electron configuration using the noble gas notation

(c) The electron configuration of the <u>valence electrons ONLY</u> using the orbital box method

1A	2A												3A	4A	5A	6A	7A	8A
H	He																	
	A																	Ne
		3B	4B	5B	6B	7B	<—	8B	—>	1B	2B			**D**				Ar
			B															Kr
													C					Xe
	La																**E**	Rn

12. Considering only groups of elements shown in the figure above for Problem 11, which group in the periodic table:

A) has the largest number of unpaired electrons for a ground state configuration?

B) contains the elements with the most exothermic electron affinity?

C) has a single electron in $\ell = 0$ and 10 electrons in a $\ell = 2$ subshells for a ground state configuration?

D) contains the element with the lowest value for a first ionization energy?

E) has 3 unpaired electrons with the $\ell = 1$ and the same spin quantum number?

F) contains the atom with the smallest atomic radius?

13. An ion has a +4 charge but has 2 electrons in the n= 1, 8 electrons in the n = 2 and 10 electrons in the n = 3 principal quantum levels.

A) Determine the atomic number and noble gas core of the ion

B) For the electron configuration of the ion, what is the total number of electrons in the:

(a) s subshell(s)?   (b) p subshell(s)?

(c) d subshell(s)?   (d) f subshell(s)?

C) What is the value for n and letter designation of for type of subshell from which electrons have been removed to form the ion?

14. Complete the following:

A) Arrange the following atoms or ions in order of increasing size:

(a) B, O, Li   (b) F, F $^-$, O^{-2}

B) Match the following radii with the ions: F $^-$, O^{-2} , Ne, Na^{+1} , Mg^{+2}

(a) 140 pm   (b) 65 pm   (c) 131 pm   (d) 95 pm   (e) 136 pm

C) Give the symbol of the atom or ion that in the FOURTH row of the periodic table that would meet the following criteria:

(a) has the lowest atomic number of elements that can form a +3 ion with no unpaired electrons.

(b) has the lowest atomic number of elements that can form a +2 ion with 5 unpaired electrons.

(c) has the lowest ionization energy with 3 unpaired electrons.

(d) forms a -2 ion with no unpaired electrons.

(e) is paramagnetic with the largest number of unpaired spins.

(f) has the highest atomic number among the ferromagnetic elements.

15. When an excited electron on a K atom falls from the 4d → 4s, a photon of $\lambda = 698$ nm is emitted. When the electron falls from 4d → 4p subshell, a photon with $\lambda = 365$ nm is emitted.

A) What will be the λ emitted when the electron falls from 4p → 4s on the K atom?

B) In what region of the electromagnetic spectrum will each photon appear?

C) Write the electron configuration for K when the electron is in the first excited state.

Chapter 8

Covalent Bonding

This chapter introduces a very common bond found in molecular compounds, the covalent bond which involves sharing of electrons between atoms in the molecule. Covalent bonding is the predominant bond in organic compounds and produces cyclic compounds as well as straight chain compounds introduced earlier in Chapter 3. Multiple covalent bonding, also common in organic compounds, further divides the hydrocarbons into the classes of alkenes, alkynes and aromatics. Learning the technique for drawing the appropriate Lewis structure for a molecule is a major goal of this chapter. The Lewis structure shows the distribution of valence electrons into covalent bonds and non-bonding pairs of electrons and gives the structural formula of the molecule. Knowledge of the structure will allow a number of properties of the molecule to be predicted and the common features of different molecules will also become apparent.

8.1 Recognizing the different types of covalent bonding

8.2 Use Lewis structures to represent covalent bonds in molecules and polyatomic ions.

Covalent bonds share valence electrons between the two atomic nuclei, with an overlap or merging, of the unfilled atomic orbitals of the two atoms. Sharing of valence electrons through bonding allows most nonmetal atoms, with the exception of H and B, to obtain enough electrons to achieve the eight electrons, or **octet**, of a noble gas configuration, a primary driving force for bonding. Not all valence shell electrons on the atom need to be shared to achieve the octet, however. In drawing the bonding pattern of the molecule, shared bonding electrons are shown as a dash between the atoms, while those not being shared, called lone pair electrons (or non-bonding electrons) are shown are they appear in the Lewis dot symbol. A **single covalent bond** is when two electrons are shared between two atoms, but sharing four electrons produces a **double bond**, as in O_2, and sharing six electrons results in a **triple bond** as in N_2.

Example: Lewis Dot Symbols and Bonds:

• single covalent bond :	• single covalent bond plus lone pairs:	• double bond plus lone pairs:	• triple bond plus lone pairs:
H-H	H-F̈	Ö=Ö	IN≡NI

Lewis structures for molecules are a convenient, effective way to show the distribution of valence electrons in a molecule. The atomic symbol is used to represent the nucleus and core electrons and only the valence electrons are drawn in as bonds or lone pairs. When complete, the total number of electrons depicted in the Lewis structure must always equal the sum of valence electrons that came from each the starting elements, minus the charge on the molecule. The five guidelines for writing Lewis structures are detailed on pages 320-321 of your text and you need to know them very well. Picking the correct central atom from the molecular formula is one of the most critical steps leading to a correct structure. The central atom has to form the <u>most bonds</u> to atoms in the molecule, so keep in mind that:

- Hydrogen, H and fluorine, F, form one bond and <u>neither</u> can be a central atom.
- As a Group 3 A element, boron, B, is the only nonmetal that is satisfied with less than an octet of electrons, needing only 3 bonds for a stable molecule.
- For groups 4A-7A, an octet of electrons is always the minimum needed and can not be exceeded for C,N,O or F atoms. You can then start with the premise that the number of bonds formed to these atoms is 8 - group number, i.e. C forms four bonds, N, three bonds, O two bonds, as it will apply most of the time.
- No multiple bonds are formed with the halogens, group 7A atoms

Also, remember to start with a symmetrical distribution of the other atoms around the central atom(s), since molecules are rarely linear. If this distribution violates the octet rule for C, N or O, then there is more than one central atom in the molecule. It is also a good idea is to start the molecules assuming only single covalent bonds and lone pairs are needed to complete the structure. Then it is very easy to determine if either multiple bonding or extra lone pairs appear in the molecule, using the guidelines 4 and 5.

8.3 Describe multiple bonds in alkenes and alkynes

8.4 Recognize molecules that can have cis-trans isomerism

Multiple covalent bonds can occur in Lewis structures that contain C, N and O, but the most variation occurs with C in the hydrocarbon family of compounds. Hydrocarbons can be divided into five groups (see table below) starting with alkanes, that have different bonding characteristics. Cyclic hydrocarbons and those with multiple bonds have a lower ratio of H/C in the molecule as compared to alkanes, since each additional C-C bonds replaces two C-H bonds. Note that the two groups of hydrocarbons with only single covalent bonds are called **saturated hydrocarbons**, while the three groups with multiple bonding are called **unsaturated hydrocarbons**.

Key features for Classes of Hydrocarbons:

Alkanes	Cycloalkanes	Alkenes	Alkynes	Aromatics
C_3H_8	C_3H_6	C_3H_8	C_3H_4	C_6H_6
propane	cyclopropane	propene	propyne	benzene
$CH_3 \cdot CH_2 \cdot CH_3$	$\begin{array}{c} CH_2 \\ /\ \backslash \\ H_2C - CH_2 \end{array}$	$CH_3 \cdot CH{=}CH_2$	$CH_3 \cdot C{\equiv}CH$	⬡
<--- saturated hydrocarbons--->		<---------- unsaturated hydrocarbons ------------>		
single C - C bonds only		At least one double bond	At least one triple bond	Alternating single and double bonds
highest H/C ratio	-->			lowest H/C ratio

The names of the hydrocarbons are based on the alkane name and still indicate the number of carbons in the longest continuous chain. Cyclic compounds have the prefix "cyclo-" before the alkane name, whereas the names for the non-cyclic hydrocarbons replace the "ane" suffix with "ene" and "yne" to indicate when a double or triple bond is present in the molecule (as shown in the table above).

Only alkenes can produce **cis-trans isomerism** as constitutional isomers. The major requirement is that two non-H groups be bonded, one each to a different C in the double bond. The double bond prevents free rotation around the carbon so that being on one side of the double bond versus the other produces molecules with distinctively different shapes. A prefix *cis-*, separated by a hyphen from the name, means the two groups are on the same side (either top or bottom) and including *trans-* indicates that the groups are across from each other.

cis- difluoroethylene:
$$\underset{H}{\overset{F}{\diagdown}}C=C\underset{H}{\overset{F}{\diagup}}$$
trans--difluoroethylene:
$$\underset{H}{\overset{F}{\diagdown}}C=C\underset{F}{\overset{H}{\diagup}}$$

8.5 Predict bond lengths from the periodic trends in atomic radii

8.6 Relate bond energy to bond length

8.7 Use bond enthalpies to calculate the enthalpy of a reaction

8.8 Predict bond polarity from electronegativity trends

Bonds have two very important properties - the **bond length** and bond energy- that are summarized in Table 8.1 and 8.2, respectively. The values given for the lengths and energies are average values. Bond length is determined by the atomic radii size and the number of electrons involved in the bond. Combining larger atoms produce longer bond lengths, while bonds get shorter as the shared electrons increase from single to double to triple bonds between the same two atoms.

In Chapter 6, the connection between ΔH for a reaction and the bond energies (B.E.) was introduced. With knowledge of the Lewis structures, the number and type of bonds broken and bonds formed can be determined and the bond energy (or bond enthalpy) can be used to calculate the enthalpy difference due to bonding changes.

ΔH reaction = (Total B.E. for bonds broken) - (Total B.E. for bonds formed)

The difference between the energy absorbed to break the bonds, and energy released when the new bonds are formed, explains the occurrence of endothermic and exothermic reactions, as shown in the diagram below.

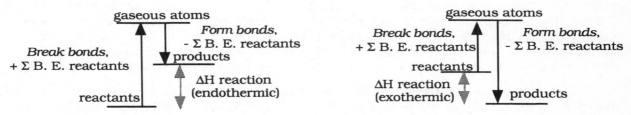

Electronegativity measures the ability of a nucleus to attract the shared electrons in a bond and can be used to predict bond polarity and type. Electronegativity increases going across a row and decreases going down a column in the periodic table, as shown in Figure 8.6 in the text.   When elements bond, and the electronegativity values of bonding atoms are the same, the bonding electrons are equally shared, producing a **nonpolar covalent bond**. However, when bonding is between atoms of significantly different electronegativity, the sharing is unequal and one atom has a partial positive charge, $\delta+$, and the other a partial negative change, $\delta-$. This partial separation of charge, which gets larger as the difference in the electronegativity increases, results in a **polar covalent bond**. If the difference gets large enough, no sharing occurs and the most electronegative atom in the bond gets both electrons, producing the full charges of an ionic bond. These characteristics can be summarized as:

Nonpolar Covalent Bond	**Polar Covalent Bond**	**Ionic Bond**
• *Equal sharing*	• *Unequal sharing*	• *No sharing of bonded pair*
• *No difference in electronegativity*	• *Some difference in electronegativity*	• *Very large difference in electronegativity*
• *No charge separation*	• *Partial charges*	• *Full charges, making ions*
I - I	$(\delta+)$ C = O $(\delta-)$	Na^+ Cl^-

8.9 Use formal charges to compare Lewis structures

8.10 Use resonance structures to model multiple bonding in molecules and polyatomic ions

In some molecules, more than one plausible Lewis structure can be drawn for a molecule and each one obeys the guidelines. The method of **formal charges** is then used to determine which structure is the most likely. Formal charges are not real charges, but a measure of the effective loss or gain an atom would undergo if it truly shared the bonded electrons in the structure. The value of the formal charge for each atom in the Lewis structure is calculated as:

Formal charge = [group no. atom] – $\frac{1}{2}$[bonded electrons to atom] – [electrons in lone pairs]

As a check of the calculations, remember that the sum of the formal charges assigned must equal the real charge on the molecule or ion. The three principles given on in the text (p. 342) are used to indicate which of the possible structures is the most likely. If two or more structures have equivalent distributions of formal charges they are equally likely. Such is the case with resonance structures. **Resonance structures** have exactly the same number and placement of single covalent bonds, but different placements of double bond(s) and lone pair(s). To best represent the actual structure, an average or composite structure with the "moving" electrons delocalized over several atoms is a more accurate picture, but the separate Lewis structures indicate which atoms are involved in the delocalization. A common molecule that exhibits resonance and is represented as a composite is the aromatic hydrocarbon, benzene, C_6H_6.

8.11 Explain why there are exceptions to the octet rule

8.12 Describe bonding and constitutional isomerism in aromatic compounds

The nonmetal elements typically bond to fill the valence octet. Occasionally the nonmetal atom will have less than eight electrons, such as in the molecule NO (a major neurotransmitter molecule in the human brain). Having fewer than an octet for one of the Group 3A-7A atoms produces a **free radical**. The free radicals are very reactive, short-lived molecules that combine quickly with other molecules, atoms, or themselves to produce stable molecules, a characteristic that is useful, but also detrimental, in many biochemistry reaction systems. **Antioxidants** are molecules that can react with free radicals and are utilized in biological systems to prevent inadvertent damage or control the number of free radicals.

It is possible for nonmetal elements in the third and higher row (period) to accept more than 8 electrons when they are central atoms. The expansion of the octet can occur because unfilled d-orbitals are available in the same n shell as the valence electrons in the s and p orbitals.

The structure of the six carbon, benzene ring of alternating single and double bonds, produces three constitutional isomers when two like groups are attached to the ring. The three constitutional isomers are called *ortho-* (for adjacent substitution), *meta-* (for having one carbon atom between the two carbons with the groups) and *para-* (for parallel substitution, when they are directly across from each other). Each constitutional isomer has a very different shape and properties that distinguish it from the other isomer forms.

Resonance forms: Resonance hybrid: Constitutional isomers: meta-   ortho-   para-

As with the cis and trans isomers, the name for the type of isomer will appear in front of the molecule name, separated by a hyphen.

Chapter Review - Key Terms

In order to make all the sentences below TRUE, insert the appropriate word or phrase from the list of key terms which best fits the context of the sentence.

NOTE: Any phrase or word from the list may used more than once.

alkenes	electronegativity	resonance hybrid
alkynes	formal charge	resonance structures
antioxidant	free radicals	saturated fats
aromatic compounds	Lewis structure	saturated hydrocarbons
bond length	lone pair electrons	single covalent bond
bonding electrons	multiple covalent bonds	*trans* isomer
cis isomer	nonpolar covalent bond	triple bond
cis-trans isomerism	octet rule	unsaturated fats
covalent bond	polar covalent bond	unsaturated hydrocarbons
double bond		

Concerning Covalent Bonding:
  The attractive force that causes one or more pairs of electrons to be shared between two atoms in a molecular compound is a (1) _____. The pair(s) of (2) _____ appear in the same valence shells of both atoms, while (3) _____ only exist in the valence shell of one of the atoms. The driving force that determines the number of (4) _____ for main body elements is to obtain enough electrons, through sharing, to achieve a noble gas configuration, which is summarized as the (5) _____. Many molecules contain nonmetal atoms that share more than one pair of electrons to produce (6) _____ which can be one of two types. Sharing of two pairs (4) electrons results in a (7) _____, while the sharing of 3 pairs (6) electrons between two atoms produces a (8) _____. (9) _____ produces bonds that have a higher bond energy than a (10) _____. A (11) _____ is used to show the distribution of the valence electrons, originally from each of the atoms in the molecule, in either (12) _____ and (13) _____ in the molecule.

  Carbon, oxygen, and nitrogen atoms generally obey the (14) _____ in the molecular structures written as a (15) _____. Molecules that have an odd number of electrons may contain exceptions to the (16) _____. Such molecules, called (17) _____, are very reactive molecules. Nonmetals in the third row or beyond, when central atoms in a (18) _____, may exceed the (19) _____ and have more either 10 or 12 electrons in the structure. Although common for other nonmetals, elements in group 7A do not share more than one pair of electrons to produce (20) _____ .

  Alkanes are hydrocarbon molecules that have only (21) _____ between the carbon atoms and represent the class of (22) _____. A second class in the hydrocarbon family includes molecules with (23) _____, called (24) _____, where H's have been replaced by an additional (25) _____(s) between the carbon atoms. Three types of molecules make up this class: (1) (26) _____ are hydrocarbons that contain at least one (27) _____ between carbons, whereas (2) (28) _____ have six electrons shared between two carbons in a (29) _____ in the molecule. and (3) (30) _____ have the characteristic structure of alternating (31) _____ and (32) _____ within a cyclic hydrocarbon molecule with six carbon atoms.

Different Characteristics of Bonding:
  The replacement of a (33) _____ by the (34) _____ in (35) _____, prevents free rotation about the C-C bond and produces (36) _____. A (37) _____ occurs when two H's have been replaced on the *same* side of the (38) _____ by two other atoms. A (39) _____, in contrast, occurs when the H's replaced are on *opposite* sides of the (40) _____. (41) _____ produces molecules with very different shapes which has very important implications for biological processes, such as in vision. Another important application is in the structure of fats. (42) _____ have only (43) _____ in the three carbon chains that are part of the fat molecule and consequently, the molecules pack well together due to the free rotation about the carbons. In (44) _____, at least one chain contains a (45) _____ which prevents effective packing, so that a liquid (oil) instead of a solid state is produced at room temperature. Most animal fats are (46) _____, while plants produce (47) _____. The isomer form for most natural (48) _____ is the (49) _____, but when (50) _____s appear, the chains are more linear and produce a overall molecular shape similar to that of (51) _____. (52) _____ in oils from plants can be made into

(53) _____ by adding H atoms to the (54) _____ which are used in many foods.

The (55) _____ of an atom is a measure of the attraction of an atom's nucleus for shared electrons in a (56) _____. Equal sharing occurs between atoms of the same (57) _____ and results in a (58) _____. A difference in the (59) _____ can produce either a (60) _____, in which unequal sharing occurs, or if very large, an ionic bond is produced with no sharing of the pair of electrons. The value of (61) _____ for main group elements increases going across a row (or period) in the periodic table, but decreases going down a group in the table.

The (62) _____ is influenced by the number of electrons being shared and the atomic radii of the atoms. A (63) _____ produces the shortest (64) _____ between the two atoms, and a (65) _____ produces a greater separation, but the longest (66) _____ for any two atoms always occurs in the (67) _____.

Properties from Lewis Structures:

The (68) _____ of an atom is very useful in determining which of several possible(69) _____s of a molecule is the most likely. To calculate the an atom's (70) _____ , one must subtract the sum of the (71) _____ and one-half of the (72) _____, based on the (73) _____ for the molecule, from the original number of valence electrons for the atom.

Molecules that have electrons are "delocalized" and not restricted to the same two atoms within the molecule then can exhibit (74) _____. which share the same number and placement of (75) _____ between atoms, but differ in the placement of (76) _____(s) or (77) _____(s) and (78) _____. A single, or "composite" structure, in which electrons in the moving bond(s) appear delocalized over several atoms, drawn in with a dotted line, is the (79) _____ for the molecule.

A molecule that has an odd number of valence electrons is a (80) _____. Typically the molecule contains (81) _____s and (82) _____, but one atom in the molecule does not have its minimum requirement of an (83) _____, so the molecule is unstable and very reactive. Molecules that can prevent the oxidation of other molecule by a (84) _____ are called (85) _____, Vitamin C and D are (86) _____ s, but each acts in different capacities in human biochemistry.

- PRACTICE TESTS -

After completing your study of the chapter and the homework problems, the following questions can be used to test yourself on how well you have achieved the chapter objectives.

1. Tell whether the following statements would be always **True or False**:

   A) Bonds that involve sharing of electrons are known as covalent bonds.

   B) Hydrogen is always found as a terminal atom in a molecule.

   C) Only the valence electrons, equal to the sum of the group numbers of atoms, are used in Lewis structures in the neutral molecules.

   D) Multiple bonds are formed when there are not enough electrons to make complete octets for all atoms with single covalent bonds and lone pairs.

   E) Nonmetals, except those in period (row) 1 and 2 will have more than eight valence electrons around them in a Lewis structure.

   F) Free radicals have one unpaired electron on the molecule that result in a very reactive material that often forms a dimer.

   G) A Se-O bond is longer than a S-O bond, but shorter than a Se-Se bond.

H) Resonance structures are a way to represent various electron arrangements in a given molecule that has lone pair electrons.

I) Bonding between different atoms produces polar covalent bonding since electrons will not be equally shared.

J) Nonzero formal charges of the same sign on adjacent atoms in a molecule or ion results in a much less stable and likely molecule.

K) Alkanes can have cis and trans type isomers, but alkynes and aromatic structures do not.

L) Electronegativity is the chief determining factor for bond polarity and is a measure of the attraction a nucleus has for bonded electrons.

2. For the Lewis structures (a) - (d) in the box on the right:
   A) Which are incorrect structures?
   B) State what is wrong with each incorrect structure and then show the correct structure for the molecular formula of the molecule.

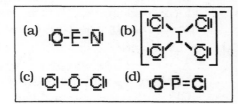

3. For the following compound names:
   (a) methylcyclobutane   (b) para-dibromobenzene   (c) 2- methylpropene
   (d) 1-chloro-3-methyl-1-hexyne   (e) cis-dichloropropene

   A) Draw the condensed or structural formula
   B) Give the correct molecular formula
   C) Tell whether the compound is saturated or unsaturated

4. Which of the following can exhibit cis-trans isomerism? Show the two isomer forms for each your choices.
   (a) $CH_3-CH=CH-CH_3$ (b) $CH_3-C\equiv C-CH_3$ (c) $HO-CH=CH-OH$ (d) $H_3C \cdot CH_2 \cdot CH=CH-CH_3$

5. For the pairs of bonds (a) - (e), indicated below:
   (a) C - O , N - O   (b) H - O , S - O   (c) P - N, N- Cl   (d) Si - F, O- F   (e) P - I, I - Cl
   A) Which bond <u>in the pair</u> would:
       (1) have the shortest bond length?   (2) be the most polar?
   B) Considering <u>all the bonds given</u> in (a) - (e), in which would:
       (1) O have a partial positive charge?   (2) N have a partial negative charge?
       (3) H have a partial positive charge?   (4) Cl have a partial negative charge?
       (5) I have a partial positive charge?   (6) the bond be nonpolar?

6. Given the molecules: (a) RbCl (b) NCl_3 (c) PH_3 (d) BaSe (e) IBr (f) SiF_4 (g) NaN_3 (h) KNO_3
   A) Which compounds would contain only ionic bond(s)?
   B) Which compounds would contain both covalent and ionic bonds?
   C) Which molecules would contain only covalent bonds?
   D) for the molecules that contain only covalent bonds:
       (1) Give the total number of valence electrons for the molecule
       (2) Draw out a plausible Lewis structure for each molecule

7. Suppose the molecule in the box below could be made (lone pairs electrons not shown in Lewis structure).

   A) In the molecule, which:
       (a) atom would have the greatest partial negative charge?
       (b) single covalent bond is the shortest and which the longest ?
       (c) bond is the most polar bond and which the least polar ?
       (d) single covalent bond has the highest bond energy?

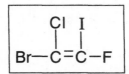

   B) Could the molecule have cis and trans isomers? Explain your choice.

8. For the following molecules or ions: CS_2, NH_3, $HCCl_3$, SF_6, BrO_2^-, SO_2
    A) Give the following information for each molecule or ion:
       (a) The central atom of each molecule?   (b) Total valence electrons for each?
       (c) Number of single covalent bonds?   (d) Number of lone pairs total?
    B) In which molecule(s) would there be:
       (1) atom(s) with expanded octet?   (2) multiple covalent bonding?
       (3) all atoms at a formal charge of zero?   (4) resonance structures possible?

9. Formamide, (a) $HCONH_2$ and (b) hydrazoic acid, HN_3 have the atom arrangements below:
    A) Complete the lewis structure for each
    B) Draw a resonance structure for each
    C) Calculate the formal charges for each resonance
    form, indicate which form is the most likely and if a
    resonance hybrid could exist.

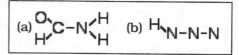

10. For the reactions: (a) $HCN(g) + 2 H_2(g) \rightarrow CH_3NH_2(g)$ (b) $CO_2(g) + H_2O(g) \rightarrow H_2CO_3(g)$
    A) Write the Lewis Structures for all the molecules involved.
    B) Make a list of the bonds broken and the bonds formed.
    C) Using bond energies, estimate the ΔH for the reaction.
    D) Tell whether the reaction is endothermic or exothermic.

11. Propionic acid, $CH_3CH_2CO_2H$, has the atom arrangement shown in the box below:
    A)   Complete the structure by filling in any missing bonds
    or lone pair electrons.
    B)   There are two types of C-O bonds with different lengths
    in the molecule. Explain how this could be true and tell
    which bond would be the shorter type.
    C)   When the acid molecule loses H^+ to become propi-
    onate ion, $CH_3CH_2CO_2^-$, the C-O bonds become equal in length. Explain why the
    equalization occurs.
    D)   Why would a resonance structure with a C=C in the molecule not be possible?

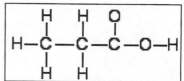

12. For the following molecules that contain only N-O bonds, indicate:
       NO_2, N_2O, NO_3^-, NO^+, N_2O_3
    A) the total number of valence electrons for each molecule.
    B) which does not have resonance structure(s) possible.
    C) in which structures all the atoms have formal charges of zero.
    D which would be a free radical.
    E) in which the formal charge on O is -2.
    F) which would contain the shortest bond between N and O.

13. The molecules below are related organic compounds:
    Phenol, C_6H_5OH; resorcinol, $C_6H_4(OH)_2$; Toluene $C_6H_5CH_3$; Anisole, $C_6H_5OCH_3$
    A) What structural similarity do they share?
    B) Excluding the C-H bonds, which has the most polar bonds in the structure?
    C) Which could have constitutional isomers and draw out the possible isomers.

14. An element X forms a compound with F. Tell whether X could be Si, P, N, C, Br or Xe for
    the compound having the characteristics below in (A)- (D) and prove your choice(s).
    A) XF_4 with no lone pairs on X   B) XF_4^{-1} with two lone pairs on X
    C) XF_4 with two lone pairs on X   D) XF_4^{-1} with one lone pair on X

15. A compound XO_2 reacts quickly with water to form $XO(g)$ and the acid molecule, HXO_3.
    When 0.523 g of XO_2 is added to 50.0 mL of water, 62.10 mL of 0.1225 M NaOH is needed
    to completely neutralize the acid formed.
    A) Write the balanced equation for the reaction of XO_2 with water
    B) Identify the element X and then draw a plausible Lewis structure of HXO_3. Prove
    your choice for X.

Chapter 9

Molecular Structures

This chapter continues the development of molecular bonding properties started in Chapter 8. The VSEPR Theory of bonding, that allows for prediction of molecular geometry, is developed and its ties to Lewis electron structures are shown. Valence bond theory that explains how atoms use their atomic orbitals to form the specific geometries and produce multiple bonds is also described. The type and strength of noncovalent interactions between molecules determined by the shape and polarity of molecules is then correlated to the physical properties of molecules, such as boiling points. The importance of noncovalent interactions in biological molecules is illustrated in the examples of cellular membranes and deoxyribonucleic acid (DNA). In addition, isomers that are distinguished by the ability to rotate light, chiral isomers, also common in biological systems, are introduced. Lastly, this chapter describes modern instrumentation in spectroscopy used by chemists to determine molecular shapes.

9.1 Recognize the various ways that the shapes of molecules are represented by models and on the printed page.

9.2 Predict shapes of molecules and polyatomic ions by using the VSEPR model.

9. 3 Determine the orbital hybridization of a central atom and the associated molecular geometry

Lewis structures of molecules are an efficient way to show the distribution of electrons into bonding and lone pairs. The three dimensional shapes of molecules must be represented using a different method, however, such as the ball and stick or the more realistic, space-filling models. Ball and stick models are often preferred since the important features, the bond angles and shape around the central atom, are apparent. **Valence shell electron repulsion theory** (VSEPR) explains the shapes observed for molecules. Its basic premise is that electron pairs will arrange themselves in a geometric pattern that minimizes the repulsion between the regions of electron density. Actual measured bond angles and the theoretical values from the model are in excellent agreement. The VSEPR theory produces five basic **electron pair geometries** with different bond angles and arrangements. These are described in detail in Figures 9.4-9.6 in the text, but a brief summary of the key features are:

Molecular arrangement: *(A = central atom, X = terminal atom, E = lone pair on central atom)*

	Electron Pair Geometry	Bond angle	Molecular Geometry
AX_2	linear	180°	1 shape possible
$AX_3 \rightarrow AX_2E_1$	trigonal planar	120°	2 shapes possible
$AX_4 \rightarrow AX_2E_2$	tetrahedral	109.5°	3 shapes possible
$AX_5 \rightarrow AX_2E_3$	trigonal bypyramid	90°, 120°	4 shapes possible
$AX_6 \rightarrow AX_4E_2$	octahedral	90°	2 shapes possible

Linear, trigonal planar and tetrahedral are very common and central atoms that have octets are limited to these types of electron geometries. Central atoms with expanded octets show either a trigonal bypyramid or octahedral geometry. Multiple bonds between atoms produce the same geometry as when a single bond occurs between the atoms, because the electrons still have the same alignment. The guidelines for predicting the proper geometry and shape around a central atom, once the Lewis structure is known, are summarized on pages 366-367 of the text.

Replacing bonds with lone pair electrons changes the **molecular geometry** or shape for a molecule, but not the basic electron geometry. Since lone pairs require more space that bonded electrons and also produce greater repulsive forces than bonded electrons, a small decrease in bond angles from that predicted by the VSEPR theory for the electron geometry also occurs.

In the **valence bond theory**, bonds result from overlap of half-filled atomic orbitals to produce the shared electron pairs. The actual geometries of the central atoms indicate that a change must occur in the atomic orbitals in order to bond. For example, for carbon to produce four equivalent bonds, such as in the molecule CH_4, the electrons in the 2s and 2p atomic orbitals must mix together into new orbitals. These new orbitals are known as hybrid orbitals and have an "average" of the energy, shape and orientation of the atomic orbitals mixed together. The **hybridization**, or mixing of orbitals, allows for more bonds to be formed and better overlap of atoms. All the hybrid orbitals produced are equivalent in energy and produce an arrangement similar to that predicted by the VSEPR theory, which can be summarized as follows:

Electron Pair Geometry:				
Linear,	*trigonal planar*	*tetrahedral*	*trigonal bypyramid*	*octahedral*

Hybridization	sp	sp^2	sp^3	sp^3d	sp^3d^2
No. Hybrids:	2	3	4	5	6

The actual hybridization the central atom has undergone to bond can be determined only from the Lewis structure of the molecule, much like the shape. Other key points you'll need to remember about hybridization are:

(1)   The atomic orbitals mixed together must come from the same n level. This is why the atoms C, N, and F are limited to an octet, since only 2s and 2p orbitals can be mixed. Only atoms in the third row (n=3) of the periodic table or beyond (n>3) can produce all five types.

(2)   Only lone pair or bonding electrons can occupy an hybrid orbital.

(3)   The number of hybrid orbitals always equals the number of atomic orbitals mixed together. The number of atomic orbitals of a particular type used to create the hybrid orbitals is indicated by a superscript in the name of the hybrid.

(4)   Hybrid orbitals are a consequence of bonding and do not occur on nonbonded atoms.

9.4 Describe covalent bonding between two atoms in terms of sigma or pi bonds, or both.

9.5 Use molecular structure and electronegativities to predict the polarities of molecules.

Two possibilities of two for overlap of orbitals to form a bond results in two different types of bonds between atoms. The types differ the location of the electron density for the bond. These bonds are known as **sigma bonds**, σ-bond, with head-to-head overlap and **pi-bonds**, π-bond, with a sideways overlap. The electron density in the sigma bond falls on the line between the nuclei of the bonding atoms. The sideways, parallel overlap of the pi-bond produces an electron density above and below, or in front and back, of the sigma bond. Key differences between the two types are:

sigma, σ-bond
- is always formed first
- is stronger than a pi-bond
- head-to-head overlap on internuclear axis
- formed from hybridized orbitals

pi, π-bond
- is the second and/or third bond in a double or triple bond
- sideways overlap of p orbitals
- formed only with non-hybridized p orbitals.

Whether a bond is sigma or pi does not affect its polarity, but the arrangement of the polar bonds does determine whether the molecule will be polar or nonpolar. **Polar molecules** have **dipole moments** which is a permanent separation of partially positive and negative charges within the molecule. **Nonpolar molecules** do not have a dipole moment and tend to have a symmetrical arrangement of terminal atoms around the central atom. The determination of whether a molecule will be polar or nonpolar involves identifying the polar bonds in the molecule, the molecular geometry of the molecule, and whether the polar bonds are arranged to produce equal but opposite forces within the molecule. The guidelines provided in the text on page 381 will be very useful in this determination. You should also keep in mind that lone pairs on the central atom will produce polar molecules for most geometries. The exceptions are the AX_2E_3, AX_3E_2 and AX_4E_2 geometries when the terminal (X) atoms are the same. Molecules with many central atoms, such as long chain alcohols or acids, can have both nonpolar and polar regions within the molecule.

9.6 Describe the different types of noncovalent interactions and use them to explain melting points and boiling points.

Chemical bonds are intramolecular forces within a molecule. However there are also forces between neighboring molecules, called **intermolecular forces** that are much weaker than chemical bonds and cause molecules to act as groups rather than as single particles. **Noncovalent interactions** produce the molecular forces that cause this interaction. Three important types of noncovalent interactions are: London forces, dipole-dipole attractions and hydrogen bonding. A summary of the key characteristics and energies needed to sever the intermolecular forces are:

London Forces (requires 0.05 - 40 kJ/mol)
- present in all molecules
- attraction results from induced temporary dipoles
- strength depends on polarization (size and shape of molecule)

Dipole-Dipole Attractions (requires 5 - 25 kJ/mol)
- occurs between polar regions or polar molecules only
- the bond polarity and arrangement of polar bonds determines the strength of the dipole moment

Hydrogen bonding (requires 10 - 40 kJ/mol)
- need a H bonded to O, N or F in molecule which can be attracted to a lone pair on a O, N or F on a neighboring molecule.

When hydrogen bonding is present, it is the most influential in affecting the forces between molecules and physical properties, such as boiling or melting points, of the molecule. The strength of London forces versus dipole-dipole attractions depends strongly on the size and shape of the molecule. The **polarization** of a molecule, the ability to induce temporary dipoles, increases with molar mass, as the number of electrons total and size of molecules increases. The shape dramatically influences the contact area in multicentric molecules. For example, a straight chain hydrocarbon will have a higher boiling point than a branched hydrocarbon of similar mass, because of stronger London forces.

For molecules that have comparable polarization, (molecular masses and shapes are similar) the boiling points and melting points follow this sequence:

Hydrogen bonding > dipole-dipole attractions> London forces

9.7 Identify the major components in the structure of DNA.

9.8 Define and describe the nature of chiral molecules and enantiomers.

Noncovalent interactions influence the shape of biologically important molecules such as **deoxyribonucleic acid (DNA)** molecules. Each DNA molecule is a **polymer** made up of hundreds of **nucleotides.** Each nucleotide has three components: a sugar (deoxyribose), an nitrogen-containing base, and a phosphate ions. The units polymerize with the deoxyribose units linked through phosphate groups to form a helical backbone, and the nitrogen-containing bases, bonded to the deoxyribose units, are folded inside a two strand helix. Only four nitrogen-containing bases appear in DNA: Adenine, Thymine, Cytosine, and Guanine. Hydrogen bonding that occurs between these nitrogen-containing bases holds the strands together to form a double helix.

A significant point to remember is that because of their structure, only adenine will form hydrogen bonds with thymine (A···T) and guanine with cytosine (G···C), producing complementary base pairs. The sequence of base pairs on one strand then forces a complementary sequence in the second strand in the DNA molecule. The storage of chemical information in the base sequencing is responsible for the genetic code for the DNA.

Biologically important molecules also have another feature related to structure. If a molecule contains a carbon atom that is bonded to four different groups, an asymmetric carbon is produced, called a **chiral** carbon. The molecule will have two isomers for each chiral carbon that are non-superimposable mirror images of each other, called **enantiomers** or chiral isomers. An enantiomer is an optically active molecule can rotate plane polarized light in one of two opposite directions, producing either a right (D) or left (L) "handedness". A carbon with less than four different groups bonded to it is an **achiral** carbon, which will not produce optical or chiral isomers.

9.9 Describe the basis of infrared spectroscopy (Tools of Chemistry) and UV-visible spectroscopy and (Tools of Chemistry) and how they are used to determine molecular structures.

Two instrumental techniques used to study molecular structure which use electromagnetic radiation to interact with molecules are infrared (IR) and ultraviolet-visible (UV-visible) spectroscopy. The modes of interaction between the two types of spectroscopy is different.

The energy of infrared radiation matches the energy of allowed motions of atoms in covalent bonds, such as stretching and bending, and varies with the bond type. The spectrum of absorbed IR wavelengths can be used as a fingerprint to identify the specific bonding and bond angles in the molecule.

Absorption of UV-visible radiation results from electron transitions to excited states within the bonding orbitals. The maximum energy of UV-visible absorbed is strongly dependent on the molecular structure and can be used to differentiate between isomers and identify groups within the molecule Colorless organic molecules can absorb UV radiation but not visible, since a molecules must appear colored to absorb in the visible region. Only organic molecules that have a conjugated system of alternating double and single bonds, extending over many atoms in the molecule, will be colored.

Chapter Review - Key Terms

In order to make all the sentences below TRUE, insert the appropriate word or phrase from the list of key terms which best fits the context of the sentence.

NOTE: Any phrase or word from the list may used more than once.

LIST:

achiral	hydrophilic	polymer
asymmetric	hydrophobic	replication
axial positions	induced dipole	sigma bond, σ bond
bond angle	intermolecular forces	sp hybrid orbital
chiral	lipid bilayer	sp^2 hybrid orbital
complementary base pairs	London forces	sp^3 hybrid orbital
deoxyribonucleic acid (DNA)	molecular geometry	sp^{3}d hybrid orbital
dipole-dipole attraction	noncovalent interactions	sp^{3}d^2 hybrid orbital
dipole moment	nonpolar molecule	spectroscopy (Tools of Chemistry)
electron-pair geometry	nucleotide	valence bond model
enantiomers	phopspholipid	valence-shell electron-pair repulsion (VSEPR) model
equatorial positions	pi bond, π bond	
hybridized	polar molecule	
hydrogen bond	polarization	

Geometry and Shapes of Molecules:

The application of the (1) _____ to a Lewis structure of a molecule is used to predict the (2) _____ of the molecule which accounts for the (3) _____s between the central atom and other atoms. The number of bonds and lone pairs on the central atom is used to determine the correct (4) _____. The (5) _____ gives the actual shape for the molecule and changes with the number of lone pairs. When five atoms are bonded to a central atom, the (6) _____ is a trigonal bypyramid and both (7) _____ and (8) _____ positions are possible for the terminal atoms. The (9) _____ positions, with (10) _____ of 90° are directly above and below the triangular plane that contains the central atom and atoms in the (11) _____ positions. When lone pairs replace the atoms bonded to the central atom in this geometry, they always occupy (12) _____ positions in the molecule, to minimize repulsion. The actual structure of a molecule can be determined through the use of (13) _____, methods that use electromagnetic radiation, from the infrared, ultraviolet or visible region, to study molecule shapes.

The (14) _____ says that covalent bonds result from overlap of two, half-filled electron orbitals. Head-to-head overlap results in a (15) _____ where the electron density lies right on the internuclear axis. Sideways overlap of parallel p orbitals results in a (16) _____, which cannot be formed by itself, but only in conjunction with a (17) _____. Double bonds consist of a (18) _____ and a (19) _____, while triple bonds have two (20) _____s and one (21) _____.

In order to maximize the number of bonds that can be formed by an atom, electron orbitals from the same n level can be (22) _____ or mixed together to form orbitals for bonding. The number of orbitals (23) _____ depends on the bonding situation and is determined after the Lewis structure is known. The mixing produces orbitals that can be used for (24) _____ bonds or orbitals of lone pairs on the central atom. (25) _____ cannot be formed with (26) _____ orbitals. Mixing an s atomic orbital with one p atomic orbital produces two (27) _____ which can then be used to form two (28) _____, 180° apart. Mixing an s with two p atomic orbitals from the same n level produces three (29) _____ , with trigonal planar electron pair geometry, while mixing the s with all three p atomic orbitals produces the tetrahedral geometry typical of four (30) _____. Expanded octets on a central atom requires that unfilled d orbitals be available to form five (31) _____ or six, (32) _____. Structures made using (33) _____ orbitals have the same predicted (34) _____ and (35) _____ for the molecule as those predicted with the (36) _____.

Polarity and Noncovalent Interactions:

In an electrical field, (37) _____s align their partially positive or negative ends of the contain permanent dipole moments, of the field whereas (38) _____s are unaffected by the electrical field. A (39) _____ has one or more polar bonds and has a (40) _____ that does not allow for the polar bonds to cancel. A (41) _____ either has a (42) _____ with a symmetrical distribution of bonds with the same polarity or no polar bonds.

There are three basic categories of (43) _____ that produce (44) _____ which act as attraction forces between all types of molecules that determine characteristic physical properties of molecules such as boiling and melting points. (45) _____, which can be the weakest of the three, appears in all substances independent of the type of bonding. The (46) _____ are a result of nonpermanent (47) _____s formed in the electron distribution of the molecule caused by neighboring molecules. The strength of the nonpermanent (48) _____ depends on the (49) _____ properties of the molecule. Generally, the (50) _____ of a molecule increases with the total number of electrons in the molecule, so that the attractions due to (51) _____ in higher mass molecules are greater. Another type of interaction is caused by the attraction of the opposite partial charges in adjacent molecules called (52) _____ that can occur only between (53) _____ which have permanent (54) _____s. The third type of (55) _____, (56) _____, occurs in only molecules that have H atoms bonded to either a O, N or F atom.

The strength of either type of (57) _____, can be related to the temperature needed to boil or melt substances. For molecular substances with similar (58) _____ , determined by the molar masses and total number of electrons in the molecules, the ones with (59) _____ only will have much lower boiling points than those with (60) _____. If the molecule has H bonded to either a O, N or F atom, the (61) _____ produced creates very specific arrangements and stronger interactions and these molecules will have even higher boiling points than that of (62) _____ with similar mass. Additionally, the (63) _____ which determines the shape of the molecule plays a very important role in the strength of any of the (64) _____ since shapes that allow for greater contact area increase the (65) _____ that can exist between molecules.

In biological cells, (66) _____ play an important role for the transport, interactions and organization of the molecules within the cell. Cellular membranes are composed of (67) _____ which are dual region molecules that have a polar "head" which produces (68) _____ an a nonpolar "tail" which interact with (69) _____. The molecules self-organize into a (70) _____ which is two molecules thick. The nonpolar tails create a (71) _____ (water-hating) region inside the membrane and the polar heads two (72) _____ (water-loving) layers on the inner and outer edges of the membrane.

Superimposable Isomers and Structure of Biological Substances:

The primary molecule for encoding genetic information in the cell, (73) _____, also relies on (74) _____ to produce self organization. It is a (75) _____ of thousands of repeating units called (76) _____s which bonded together through a backbone of alternating sugar and phosphate molecules. The most critical information contained in DNA is in the order of the (77) _____s, which is set by the allowed (78) _____ between the four nitrogen bases which produces only two kinds of (79) _____. The (80) _____ are a direct result of the (81) _____ which allows guanine to (82) _____ only with cytosine and adenine with thymine.

Molecules that contain (83) _____ carbons are also biologically important and have isomers that are non-superimposable mirror images called (84) _____. The physical property of the molecule affected by the (85) _____ carbon is the ability to rotate planes of polarized light in opposite directions. The two types of "handedness" produce the (86) _____ are given the notation, right (D) or left (L), to indicate their rotation of polarized light. Such a molecule requires a very specific (87) _____, one that has a hybridized carbon atom with (88) _____ and four different groups bonded to it producing an (89) _____ molecule, indicating it has no plane of symmetry. Alternatively, if a rotation of the mirror image of the molecule produces a superimposable mirror image, then the molecule is a (90) _____ molecule and has only (91) _____ carbons.

- PRACTICE TESTS -

After completing your study of the chapter and the homework problems, the following questions can be used to test yourself on how well you have achieved the chapter objectives.

1. Tell whether the following statements would be always **True or False**:

A)   The VSEPR theory predicts that there are five basic electron geometries for molecules that differ in bond angles and molecular geometry.

B)   Lone pairs will distort a electron geometry because they require less space than bonding electron pairs.

C)   Both square planar and tetrahedral structures are possible geometries for distributing four pairs of electrons on a central atom.

D)   Lone pairs in axial positions will result in less repulsion than those in equatorial positions in the trigonal bypyramid geometry.

E)   Multiple bonded atoms have the same effect as single bonds in the VSEPR geometries.

F)   The molecules NCl_3 and BCl_3 would have the same bond angles.

G) H_2O and CO_2 are both polar molecules, but H_2O has a higher boiling point because of hydrogen bonding.

H) CH_2Cl_2 and $CHCl_3$ would have the same molecular geometry, but different intermolecular forces.

I) NCl_3 and CCl_4 would have the same electron geometry, but different molecular geometry.

J)   A hybrid orbitals results from mixing half filled atomic orbitals on a atom from the same n level.

K)   Sigma bonds are the same as single bonds.

L)   Pi-bonds are typically formed with non-hybridized p orbitals on an atom.

M)   The number of hybrid orbitals formed by an atom will equal the total number of bonds and lone pairs shown for the atom in the Lewis structure.

N)   It is possible for a molecule to have both hydrophilic and hydrophobic character.

O)   The molecule with the formula $CH_3CBrICl$ contains a chiral carbon atom.

P)   Square planar molecular geometry always results in nonpolar molecule.

Q)   Infrared spectroscopy gives information about molecular structure because the energy of a chemical bond matches the energy of IR radiation.

R)   The nitrogen base units in DNA are covalently bonded to phosphate units but also produce noncovalent interactions which holds the two strands together in the double helix.

2. Briefly answer the following:

A)   What is meant by the term noncovalent interactions and how do they compare to covalent bonds?

B)   What is the difference between dipole-dipole attractions and London forces?

C)   Explain why the resonance structures for a molecule could all have the same molecular geometry or shape, but constitutional isomers could have very different shapes.

D)   Explain why the boiling points of methanol, CH_3OH and CH_4 differ by 230° (65°C versus -164°C) while the boiling point of 1-decanol $C_{10}H_{20}OH$, is only 55°C higher than that for decane, $C_{10}H_{22}$ (229°C versus 174°C).

E)   Changes in the number of intermolecular interactions is responsible for the fact that ice is less dense than liquid water. What is the intermolecular force and why does the number change when ice melts?

3. Acetic acid CH_3CO_2H mixes easily with water to form solutions.
   A) What type of noncovalent interactions do you expect to be the strongest interactions between water and acetic acid molecules?
   B) Is acetic acid hydrophobic or hydrophilic?
   C) Carboxylic acids have the general formula, $CH_3(CH_2)_xCO_2H$,
      (1) Would you expect the molecule to show more, less or about the same solubility in water as x increases? Explain.
      (2) How would the solubility in fat be affected as x increases?

4. The structures of the two molecules are shown below, (using the shorthand notation for the molecules with any hydrogens bonded to C's not shown.)
   benzene,   1, 3 dichlorobenzene,
   C_6H_6 $C_6H_4Cl_2$
   b.p. = 80.1 °C   b.p. = 173 °C
   A) What noncovalent interactions do you expect to occur in pure benzene?
   B) What noncovalent interactions do you expect to occur in pure 1, 3 dichlorobenzene?
   C) Explain the major reason(s) why the boiling point of 1, 3 dichlorobenzene is greater.

5. For the molecule: $SeBr_4$
   A) Draw the proper Lewis structure for $SeBr_4$
   B) How many sigma bonds are in the molecule?
   C) How many pi-bonds are in the molecule?
   D) What's the electron geometry and molecular geometry of the molecule around Se?
   E) What would be the two types of Br-Se-Br bond angles in the molecule?
   F) What type of bond is the Se-Br bond: polar covalent, nonpolar covalent or ionic?

6.   A) Of the molecules: Br_2, HBr, N_2, HCl
         (a) Which would have the strongest London forces?
         (b) Which would have the strongest dipole-dipole attractions?

      B) Of the following pairs of molecules, which would have the stronger London forces?
         (a) CH_3Cl or CH_3Br (b) CH_3CH_2Cl or $CH_3CH_2CH_2CH_2Cl$
         (c) butane or 2-methylpropane

      C) The forces of attraction between molecules of H_2 are:
         (a) hydrogen bonds   (b) dipole-dipole attractions   (c) covalent bonds
         (d) London Dispersion forces   (e) none of the above

7. Which of the following molecules would have:
              (a) NF_3 (b) N_2O (c) XeF_2 (d) PCl_5 (e) OF_2
   A)   a permanent dipole moment?
   B)   a molecular geometry that is the same as its electron pair geometry?
   C)   three lone pairs on the central atom?
   D)   a molecular geometry named angular (refer to Table 9.1 as needed)?
   E)   both pi and sigma bonds to the central atom?

8. For the molecules: (a) FNO   (b) IF_3 (c) SiF_4 (d) SO_2
   A)   Draw the Lewis structure for each molecule.
   B)   Tell the hybridization for N in FNO, I in IF_3, Si in SiF_4 and S in SO_2
   C)   Redraw the structures, representing: (1) the actual shape or bond angles, and
        (2) all polar bonds shown as arrows($\longmapsto$) pointing from the (δ+) to (δ-) charges.
   D)   Tell whether each molecule is polar or nonpolar.
   E)   What is the dominant type of noncovalent interaction (intermolecular force) for each: London forces, dipole-dipole attraction, or hydrogen-bonding?
   F)   Which, if any, of the molecules could exhibit resonance? Does the shape around the central atom change in the resonance structures?

9. For the molecules: (a) BCl_3 (b) XeF_4 (c) Cl_2SO (d) HCOF   (e) SF_4

Tell which of the molecules, if any, would :

A) have bond angles of 90° and be a nonpolar molecule?

B) have bond angles of about 120° and be a nonpolar molecule?

C) have bond angles of about 109°?

D) have a central atom that has mixed three different types of atomic orbitals from n = 3 to create hybrid orbitals?

E)   show H-bonding forces as intermolecular forces?

F) be expected to have the strongest London forces in the group?

G) have a planar shape but also polar molecule?

10. Tell which of the molecules (a) -(e), if any, would :

(a) NI_3 (b) $AsBr_5$ (c) $CHCl_3$ (d) BrF_5 (e) N_2H_2

A) have a central atom with sp^3 hybridization?

B) have a central atom with sp^3d^2 hybridization?

C) have a central atom with sp^3d hybridization?

D) have a central atom with sp^2 hybridization?

E) have pyramidal shapes, but differ in bond angles and type of pyramid?

F) have an atom with both pi and sigma bonds?

G) have London forces only as intermolecular forces?

H) show H-bonding forces as intermolecular forces?

11. For the five compounds below:

(a) $CH_3CH_2CH_3$ (b) CH_3OCH_3 (c)CH_3Cl (d) CH_3CHO (e) CH_3CN

A)   Tell whether each molecule is polar or nonpolar.

B)   What is the dominant type of noncovalent interaction (intermolecular force) for each: London forces, dipole-dipole attraction, or hydrogen-bonding?

C)   Tell the hybridization and approximate bond angle for each of the central (underlined) atoms in the molecules

D)   Match the molecular formulas above with the ball and stick representations of the molecules shown below.

(1) _____   (2) _____   (3) _____   (4) _____   (5) _____

E)   Tell to which molecules the two space-filled models below correspond to.

(1) _____   (2) _____

F)   What do you think are the main reasons CH_3CN has the highest boiling point at 82°C of the five and $CH_3CH_2CH_3$ at - 42 °C, is the lowest?

12. Melamine, $C_3N_3(NH_2)_3$, a cyclic compound, is a common plastic molecule made by combining three H_2NCN molecules.

A) Complete the Lewis structures for the H_2NCN and melamine in the figures given, adding in any missing pi-bonds and lone pairs, so that all the atoms have only zero formal charges in the molecule.

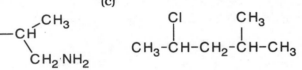

Melamine:

B)   Although the formal charges don't change, tell whether the hybridization changes for either C or N when the melamine molecule is formed from H_2NCN.

C)   The shape of the melamine molecules and the noncovalent interactions are important for the properties of melamine. What is the shape of the molecule and what are the strongest intermolecular forces in melamine?

13.   Identify the hybridization and electron geometry for the carbon marked with the asterisk in the molecules below and tell whether it is a chiral or achiral carbon.

(a)   (b)   (c)   (d)

$CH_3-CH-CH_3$ $CH_3\cdot CH_2\cdot CH\cdot CH_3$ $CH_3-CH-C-OH$ CH_3-C-OH

14.   Indicate which of the following molecules can have enantiomers. For those that can, indicate which carbon(s) will produce the enantiomers, by circling the chiral carbon(s).

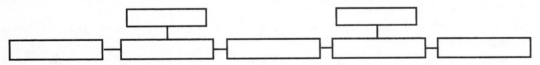

(a)   (b)   (c)

15. Concerning DNA molecules:

A)   In the diagram below, representing one strand of the DNA molecule, identify each box as being either a: <u>sugar</u> (deoxyribose), <u>phosphate</u> or <u>nitrogen base</u>.

B)   Write the complementary sequence of bases for a strand of DNA that has the base sequence: - A - G - C- A - T- C- T-

C)   How are the deoxyribose and phosphate units linked together in DNA? What function do these parts serve in the molecule?

D)   What forces hold the strands together?

E)   For the complementary base pair, adenine and thymine, shown in the box on the right in approximately the same position as they would appear in the DNA strands, indicate which groups are producing the <u>two</u> hydrogen bonding interaction between the bases in DNA, by showing dashes between the H and the lone pair it is attracted to.

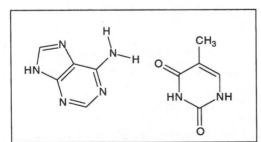

Chapter 10

Gases and the Atmosphere

All gases have a number of properties in common. In addition to discussing these properties and the theories that explain them, this chapter discusses characteristics of the earth's atmosphere and chemical reactions that occur between gases.

10.1 Explain the properties of gases.

10.2 Describe the components of the atmosphere.

Properties of gases include compressibility, pressure, expansion to fill available space, and complete mixing. The quantities that are measured with respect to gases are pressure (P), temperature (T), volume (V), and amount (number of moles, n). **Pressure** of a gas is the force with which the molecules hit the walls of the container. Units of pressure include Pa, mm Hg, torr, atm, and bar. A **bar** is approximately equal to one atmosphere, a definition based on the force that the earth's atmosphere exerts on everything it contacts. Note that this relationship of one bar to one atmosphere is approximate. Conversion factors of 1 atm = 1.01325 bar = 101.325 kPa = 760 torr = 760 mm Hg should be used in precise calculations.

The earth's atmosphere consists mainly of N_2 and O_2, with much smaller amounts of Ar, CO_2, Ne, H_2, and other gases. The amount of H_2O in the atmosphere varies considerably. Concentrations of atmospheric gases are usually expressed in %, ppm, or ppb by volume, depending on the amount of the gas. The atmosphere can be described by its layers, with the **troposphere** being closest to the earth's surface, and the other layers, the **stratosphere**, mesosphere, and thermosphere, increasingly distant from the surface. We live in the troposphere, and the weather that so much affects our daily life occurs in this layer. About 99% of the mass of the atmosphere is included in these two layers.

10.3 State the fundamental concepts of the kinetic-molecular theory and use them to explain gas behavior.

The kinetic-molecular theory of gases explains the properties of gases. The basic concepts are:

1. The volume of the particles in a gas is much, much smaller than the volume of the entire sample.
2. The movement of gas particles is random in speed and direction.
3. Attraction and repulsion between gas particles is minimal except when collisions occur.
4. The collisions that occur between the particles are elastic, meaning that energy can be transferred, but it is conserved.

Another concept that helps to explain the properties of gases is that the average kinetic energy of gas particles is directly proportional to the absolute (Kelvin) temperature. It is also useful to remember that K.E. = $1/2(mv^2)$, which means that for two different kinds of particles at the same temperature, the larger one will move considerably more slowly than the smaller one. This last concept can be represented graphically with a slightly asymmetrical bell curve which becomes flatter as the temperature increases.

10.4 Solve problems using the appropriate gas laws.

Various laws are used to describe the relationships between various properties of gases. These include

- **Boyle's Law:** V and P vary inversely when n and T are constant, or PV = k
- **Charles's Law:** V and T vary directly when n and P are constant, or V/T = k. Note that this law necessitated the development of the absolute temperature scale, where the lowest temperature possible is 0 K, which is -273°C.

- **Avogadro's Law:** V and n vary directly when T and P are constant, or V/n = k. This law explained Gay-Lussac's earlier law of combining volumes, stating that gases combined chemically in ratios of small whole numbers.

These laws can be combined into the **Ideal Gas Law**, PV = nRT, which allows the calculation of any variable if the other three are known. R, the **ideal gas constant**, is an experimentally determined proportionality constant. Its value varies according to the units in which it is expressed. A convenient set of values for use with the Ideal Gas Law is STP, **standard temperature and pressure**. At these conditions, one atmosphere(bar) and 0 °C, the volume of one mole of any gas is 22.4 L, commonly called the **standard molar volume**. If the amount of gas remains constant, a convenient form of the ideal gas law is $\frac{P_1V_1}{T_1} = \frac{P_2V_2}{T_2}$. Note that temperature in gas law calculations must always be expressed in Kelvins.

10.5 Calculate the quantities of gaseous reactants and products involved in chemical reactions.

10.6 Apply the ideal gas law to finding gas densities and molar masses.

Stoichiometry calculations involving gases can often be simplified by using the ideal gas law and the law of combining volumes. PV = nRT can be solved for the number of moles and this value used as needed in the problem. The ideal gas law can also be rearranged and the value of molar mass included to allow the value of density of a specific gas to be calculated at a given set of conditions. This formula is $d = \frac{PM}{RT}$, where d = density and M = molar mass of the gas. Another application of the ideal gas law is to determine the molar mass of a gas sample in which mass, pressure, volume, and temperature are known. This formula is $M = \frac{mRT}{PV}$, where M = molar mass of the gas and m = mass of the sample.

10.7 Perform calculations using partial pressures of gases in mixtures.

Dalton's law of partial pressures emphasizes that the properties of gases are independent of their identity. This law states that $P_{Total} = P_A + P_B + P_C + \cdots$, or the total pressure of a mixture of gases is the sum of the partial pressure of each gas comprising the mixture. The **partial pressure** of each gas is equal to the mole fraction of that gas in the mixture times the total pressure of the mixture. **Mole fraction** of gas X is defined as $\frac{n_x}{n_{total}}$. A common application of this law is the correction for the vapor pressure of water that must be made to the value of pressure when a gas is collected over water.

10.8 Describe the differences between real and ideal gases.

The ideal gas law is generally accurate for moderate temperatures and pressures, but at temperatures considerably lower than 0 °C and pressures considerably greater than one atmosphere, its predictions become less accurate. The two assumptions that become problematic under extreme conditions are that the volume of the gas particles is negligible in comparison to the total volume of the gas sample and that there is no interaction between the particles. The **van der Waals equation** includes these factors in a modification of the ideal gas law, where the correction factors a and b vary according to the identity of the gas being studied. This formula is $\left(P + \frac{n^2a}{V^2}\right)(V - nb) = nRT$.

10.9 Describe the main chemical reactions occurring in the atmosphere.

Unlike a mixture of ideal gases, gases in the atmosphere undergo chemical reactions because the system is exposed to light, allowing photochemical reactions to occur. One type of photochemical reaction is electronic excitation, in which NO_2, a naturally-occurring free radical, gains extra energy from a photon, increasing its reactivity. The other type of photochemical reaction is formation of free radicals by **photodissociation**, in which a chemical bond is broken to produce two particles that each have an unpaired electron. Since free radicals are extremely reactive, these processes have a significant effect on the composition of the atmosphere.

10.10 Explain the main features of stratospheric ozone depletion and the role of CFCs in it.

Ozone is an important atmospheric chemical. In the stratosphere, it protects the earth's surface from harmful ultraviolet light by absorbing up to 99% of this radiation. A normal cycling between O_3 and O_3 occurs, maintaining the necessary level of ozone. However, when ozone is destroyed by reaction with free radicals, this process becomes unbalanced and the protective ozone layer becomes depleted. **Chlorofluorocarbons**, CFCs, are major contributors to this process. They are particularly harmful because they produce chlorine free radicals which persist in the stratosphere.

10.11 Explain the main chemicals found in and the reactions producing industrial pollution and urban pollution.

Pollution in the troposphere is often seen as a more immediate concern because humans are directly exposed to these pollutants on a daily basis. **Primary pollutants**, which enter the atmosphere directly from the source include particulates, aerosols, and individual molecules, atoms, or ions. SO_2 is a major component of smog and acid rain; it is a product of combustion of fossil fuels. NO_x includes nitrogen oxides produced at the high temperatures of internal combustion engines. These compounds are also components of smog and acid rain. Hydrocarbons are also primary atmospheric pollutants. In the troposphere, ozone is a **secondary pollutant**, resulting from the reaction of NO_2 free radicals with oxygen. One form of urban air pollution is chemically reducing smog, formed from smoke containing SO_2, fly ash and other particulates, and organic compounds. The other form of urban air pollution is chemically oxidizing smog, usually called **photochemical smog**. This smog is formed when large concentrations of NO_x, O_3, and hydrocarbons are produced in areas where it is very sunny, allowing photodissociation reactions to produce additional pollutants. This type of smog is sometimes worsened by geographical features which trap a warm polluted air mass over a large city.

Chapter Review - Key Terms
In order to make all the sentences below TRUE, insert the appropriate word or phrase from the list of key terms which best fits the context of the sentence.
NOTE: Any phrase or word from the list may used more than once.

LIST:		
absolute temperature scale	ideal gas constant	photochemical smog
aerosols	ideal gas law	photodissociation
air pollutant	Kelvin temperature scale	pressure
Avogadro's Law	law of combining volumes	primary pollutant
bar	milliliters of mercury	secondary pollutant
barometer	mole fraction	smog
Boyle's Law	Newton (N)	standard atmosphere
Charles's Law	NOx	standard molar volume
chlorofluorocarbons	ozone hole	standard temperature and
combined gas law	partial pressure	pressure (STP)
Dalton's law of partial pressures	particulates	stratosphere
electronically excited state	Pascal (Pa)	torr
ideal gas	photochemical reactions	troposphere
		van der Waals equation

Describing Behavior of Gases:

The kinetic molecular theory can account for the observed behavior of all gases. Particles in a gas state move freely at very high speeds, producing many collisions with the walls of their container which results in a force per unit area called (1) _____ of the gas. The SI unit for (2) _____ is the (3) _____, which is equal to a force of one (4) _____ per square meter. Atmospheric gases produce a (5) _____ that can be measured with a mercury (6) _____, made from an evacuated tube, open on one end, placed in a pool of mercury. The height of the column of liquid in the tube is equal to the atmospheric (7) _____ and is measured in (8) _____, (symbol = mmHg) or (9) _____. The (10) _____, represented by the symbol *atm*, occurs when the height in the (11) _____ is 760 mm Hg and normal pressures are near this value. Although the (12) _____ is very convenient, it is not an SI unit and is actually equal to 101,325 (13) _____s. Another SI unit, however, that is nearly equal to the (14) _____ is the (15) _____, which equals 100,000 (16) _____s, so that 1.00 (17) _____ = 1.01 atm.

When representing the kinetic energy, or motion, of any particle, particularly gases, by its temperature in a calculation, the (18) _____ must be used instead of the Celsius scale. The zero on the (19) _____ corresponds particles having zero motion, which is -273.15°C. The degree Celsius and degree Kelvin represent the same amount of energy, so that the any temperature in °C can be converted to the (20) _____ by only adjusting for the zero and adding 273 to the T(°C).

Three laws were separately determined that relate the (21) _____ of a gas to its volume, temperature and number of gas particles, as the moles of gas. (22) _____ says that at a constant temperature and fixed number of moles, the (23) _____ of a gas multiplied by its volume is a constant, so the (24) _____ is always increased when the is volume decreased under these conditions. (25) _____ says that volume divided by its temperature, on the (26) _____, is a constant, so that increasing the temperature of fixed amount of gas in a closed container will increase the (27) _____ of the gas. The third law, (28) _____, says that the (29) _____ of a gas will increase when the number of moles gas particles increases, in a closed container at a constant temperature. Combining the three laws together produces the (30) _____, PV/nT = R = constant, where the R is called the (31) _____. The value of the (32) _____ varies with the units used, but has the value of 0.0821 L-atm-K^{-1}-mol^{-1}, when atm are the (33) _____ units used. Under the conditions of (34) _____, 0°C and 1 atm, the volume of 1.0 mole of any gas is 22.41 L, which is called the (35) _____ of a gas at STP. The (36) _____ quantitatively predicts the behavior of gases at pressures of a few atmosphere and temperatures well above the boiling points of the substance.

Modifications of the Ideal Gas law and Comparing Two gases:

For a fixed amount of a single gas, undergoing a change in conditions of either P,V or T, the (37) _____ is the most useful modification of the (38) _____ to predict the new value of either P,V, or T for the fixed amount of gas. To define the density of a single gas, in grams/L, the (39) _____ can be modified to calculate the density by substituting the mass divided by molar mass for n, the number of moles of the gas. The same substitution will also allow the (40) _____ to be used to solve for the molar mass of an unknown gas, if the mass, P T and V of the gas are known.

Gases in a mixture will still act as (41) _____ and exert the same pressure as if alone in the container, but the (42) _____ of the gas is called a (43) _____, to indicate it is only part of the total. According to (44) _____, the total (45) _____ of the mixture is the sum of all the (46) _____s of each gas in the mixture. The ratio of the (47) _____ of a gas to the total pressure of the mixture is also equal to the ratio of moles of that gas to the total moles in the mixture, called the (48) _____ of the gas in the mixture.

In a different view, two gases, A and B, involved in the same reaction, can be viewed as having the same pressure and temperature, so that the application of the (49) _____ to both gases produces the equality: $n_A/n_B = V_A/V_B$. It then follows that the molar ratio given by the coefficients in the balanced chemical reaction of the two gases, will equal the volume ratio in which they will react (or be produced), which is the behavior first described by Guy-Lussac and stated as the (50) _____.

When at high (51) _____s or temperatures near the boiling point of the substance, many gases change from being an (52) _____ to a real gas and show deviations from the (53) _____. The two factors responsible for this change, the volume occupied by the gas particles themselves and the attractive (or repulsive) forces that can act between them, are taken into account in the (54) _____, which is then a much better predictor of the value of P,V or T of a gas under high pressures or low temperatures.

Reaction of Gases in the Earth's Atmosphere:

The two innermost layers that form part of the Earth's atmosphere are the (55) _____, which extends from the Earth's surface to about 10 km altitude and the (56) _____, which extends from the top of the troposphere to about 50 km altitude. The major gas particles in the (57) _____ are O_2, O_3 (ozone) and O while the (58) _____ is mainly composed of N_2 and O_2 molecules, with a host of other trace gases. Because of the high flux of light photons through the layers from the sun, reactions between gas molecules in either layer can take place by the absorption of the photons, which are called (59) _____. A common type of (60) _____ is the absorption of a photon which results in the breaking apart of a covalent bond in the molecule, called (61) _____, which produces two free radicals as the end product. The absorption of the photon may also produce an (62) _____, which can cause a free radical and a new molecule to be formed as products form the original molecule.

Ozone in the (63) _____ is produced by a series of reactions involving the (64) _____ of O_2 molecules. Ozone, in turn, absorbs a second type of harmful ultra-violet radiation, which is cancer-causing radiation to humans, to reproduce O_2 molecules. The balance between the production and consumption of ozone in the (65) _____ has been upset, in recent years, by the introduction of manmade (66) _____ into the atmosphere. (67) _____ of the C-Cl bond in the (68) _____ occurs when the molecule rises to the (69) _____ and produces atomic Cl, which is highly reactive. Cl atoms then interfere with the O_2/O_3 cycle, resulting in a greater destruction of ozone than production. This series of reactions, occurring under very cold arctic winters conditions, has produced an (70) _____ in the (71) _____ over Antarctica that can be directly correlated to the Cl atom concentration.

Substances that degrades air quality of the inermost layer of the stmosphere, the (72) _____ , are called (73) _____s which are one of two types, either a (74) _____, introduced directly into the atmopshere or a (75) _____ formed from by a series of atmospheric reactions. A (76) _____ can be introduced from natural processes, such as volcanoes, or human activities, such as burning of fossil fuels for electricity, heating or transportation. Extremely small solid particles, such as soot or fly ash, are called (77) _____ are considered a health hazard, but which also act as reaction surface for the production of (78) _____s from other types of (79) _____s. An (80) _____, a colloidal suspension of (81) _____in fine water vapor droplets, often seen as a "haze" in the atmosphere, is also a facilitator for the chemical conversion of (82) _____s. The term (83) _____, a shortened form of "smoke combined with fog", is an example of such an (84) _____.

Gaseous free radicals NO_2 and NO from N_2 and O_2 , called the (85) _____ compounds, primarily result as a byproduct of the high temperature combustion processes. The combination of sunlight, (86) _____ and gaseous hydrocarbons together produces an (87) _____, called (88) _____ that has major adverse health effects on plants, animals and humans. Reactions that occur within the (89) _____ produce ozone, then classified as a (90) _____, which is one of the major causes of the adverse effects. The (91) _____s that produce ozone also require some time to occur, so that wind currents often cause the pollution from the ozone to occur far away from the source.

- PRACTICE TESTS -

After completing your study of the chapter and the homework problems, the following questions can be used to test yourself on how well you have achieved the chapter objectives.

1. Tell whether the following statements would be always **True or False**:

    A)   A gas exerts a pressure that is always directly proportional to number of particles present

    B).   Standard pressure is 100 mmHg or 1.0 bar.

    C).   The ideal gas law only applies to gases at STP.

    D).   A volume of 5.0L of any gas will contain the one-half the number of particles as 2.5 L of any other gas.

    E).   As pressure increases, the density of the gas also increases.

    F).   Every gas in a mixture exerts a part of the pressure and has a mole fraction equal to the ratio of its partial pressure over the total.

    G). SO_2 molecules move slowly than CO_2 molecules at the same temperature.

    H).   Air pollution can apply to the stratosphere as well as the troposphere.

    I).   Aerosols involve particles larger than those classified as particulates.

    J).   Photochemical smog is typical of large cities where large numbers of transportation vehicles emitted pollutants from their exhaust.

    K).   Free radicals are very reactive species and participiate in many atmospheric reactions.

    L).   Ozone can be detrimental or beneficial to humans depending on which layer of the atmosphere in which it is produced.

    M).   Aerosols can obtain pollutants by both adsorption and absorption mechanisms.

    N).   Collecting gases over water always results in a greater pressure because the water reacts with the gas, producing more products.

2. Select the best answer for the following questions (A) - (F).

    A) If the temperature of the oil is raised from 68°C to 136°C, infthe apparatus pictured below, which of the following statements about the change in volume of the gas trapped in the capillary tube would be correct, the volume of the gas is:

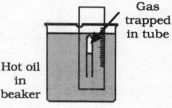
Gas trapped in tube

Hot oil in beaker

    (a) reduced by 20%   (b) increased by 14%

    (c) increased by 20%   (d) reduced by 14%

    (e) cut in half.   (f) doubled.

2. B) Suppose that the average molecular speed of argon gas molecules trapped in a closed container doubled. Which of following statements describes how the increase could have been accomplished?

    (a) The argon gas was heated until T(°C) was doubled.

    (b) One-half of the argon gas leaked out of the container, the P changed, but T constant

    (c) The argon gas was heated until the P doubled.

    (d) More argon gas was added until the P doubled, but the T remain unchanged.

    (e) None of the above would double the speed.

  C) If 0.50 mole He and 1.0 mol H_2 gases are at the same T and P, they have equal:

    (a) weights

    (b) average molecular speeds

    (c) kinetic energies

    (d) volumes

    (e) densities

  D)   What is the partial pressure of CH_4 in a 10.0 L flask that contains 0.50 mol CH_4, 0.20 mol N_2, and 4.30 mol O_2 at 25°C if the total pressure was 2.0 atm?

    (a) 0.667atm   (b) 0.25 atm   (c) 0.20 atm   (d) 0.512 atm   (e) 0.605 atm

  E)   Which of the gases below will diffuse more slowly than oxygen, O_2, at the same T?

    (a) F_2 (b) Ne   (c) NO_2 (d) CH_4 (e) NO

  F) Will the volume of a gas increase, decrease, or stay the same when each of the following occurs?

    (a) The pressure is increased from 1 atm to 2 atm, while the temperature is increased from 100°C to 200°C.

    (b) One mole of gaseous XeF_6 completely dissociates into Xe(g) and F_2(g) at a constant temperature and the pressure changes from 1 atm to 2 atm.

3. Briefly answer the following:

  A)   What is the significance of -273.15 °C?

  B)   Concerning ozone in the stratosphere:

    (a) What effect do chloroflurocarbons on the natural cycle of O_3/O_2?

    (b) What are the two active form(s) of the chlorine species?

    (c) What conditions result in the destruction of O_3 in the first weeks of spring, when sunlight returns to the South (and North) Poles?

  C)   What is(are) the basic difference(s) between primary and secondary pollutants?

  D)   What evidence is there that the layers of the atmosphere can mix?

  E)   What are the factor(s) that make ozone production in the troposphere a regional problem, instead of a local problem?

  F)   What are the three necessary ingredients to produce photochemical smog?

  G)   What are the three major secondary pollutants formed in photochemical smog that cause the health problems and from which primary pollutant(s) are they produced?

  H)   Which is chemically reducing - smog or photochemical smog- and what is the difference in the composition that makes one reducing and the other not?

4. A sample of laughing gas, N_2O, is contained in a closed 250 mL flask at a temperature of 40 °C and has a pressure of 113.0 kPa.

A) What is the mass of the N_2O sample in the flask?

B) What is the density of N_2O, in grams per liter, under these conditions?

C) Which of the following factors could be changed without affecting the density of this sample of N_2O: P, V or T? Explain the reason(s) for your choice.

5. Consider two flasks of equal volume: Flask A, is filled with 5.00 g Ar(g) at 400 K and Flask B, is filled with 10.0 g of CS_2(g) at 200 K.

A) Which flask contains the greatest number of moles of particles?

B) Which flask contains the gas with greatest pressure?

C) Which flask has the particles with the least kinetic energy?

C) At what T(°C) would the gas in Flask A have the same pressure as in Flask B?

6. Acetaldehyde is a common liquid that vaporizes readily. If a pressure of 331 mmHg is observed in a 125 mL flask at 0°C when the density of the acetaldehyde vapor is 0.855 g/L, what's the molar mass of acetaldehyde?

7. A sample of sulfur hexafluoride gas occupies a volume of 5.10 L at 198°C. Assuming the pressure is constant, what temperature in °C is needed to reduce the volume to 2.50 L?

8. "Strike anywhere" matches contain the compound tetraphosphorus trisulfide, which burns in O_2 (g) to form tetraphosphorus decaoxide and sulfur dioxide gas.

A) Write the balanced chemical reaction for the burning of the compound in O_2(g)

B) How many milliliters of sulfur dioxide measured at 725 torr and 32 °C can be produced from the burning of 0.800 g of tetraphosphorus trisulfide?

9. If all the propane, C_3H_8, in a 30.0L tank at 20°C and pressure of 300 atm is burned, how many 15.0 L tanks of O_2 gas, where the O_2(g) is also at 20°C and 300 atm, would be required to completely burn (or react) the propane?

10. A sample of a liquid hydrocarbon which is known to consist of molecules containing only 5 carbons is vaporized in a 0.204L flask by immersion in a water bath at 101°C. The barometric pressure is 767 torr and the gas in the flask condenses to 0.482 g of liquid when cooled. What is the molecular formula of the hydrocarbon?

11. The volcano reaction occurs when ammonium dichromate decomposes according to the reaction: $(NH_4)_2Cr_2O_7$(s) → N_2(g) + 4 H_2O(g) + Cr_2O_3(s).

If 20.3g of $(NH_4)_2Cr_2O_7$(s) were completely decomposed and all the product gases are trapped in a 5.00L flask at 20°C, what is the:

A) partial pressure of each gas in the flask?

B) total pressure in the flask, in atm, after the reaction is complete?

12. Acetylene, C_2H_2, is produced in the lab by reacting calcium carbide, CaC_2 with water: CaC_2(s) + H_2O(l) → C_2H_2(g) + $Ca(OH)_2$.

1.5 g of calcium carbide was added to excess water and the C_2H_2 produced was collected over water. The volume collected at 23°C was 528 mL when the total gas pressure was 738 mmHg. If at 23°C, the vapor pressure of water is 21 mmHg,

A) How many grams of C_2H_2 were in the sample collected?

B) What is the % yield of the reaction?

13. NO_2 has an odor threshold of 10 mg/m^3 and acute inhalation level of 200 ppm.

A) Convert the odor threshold to its ppm value, in μL/L, assuming STP conditions.

B) What is the partial pressure of NO_2 in air at a concentration of 200 ppm?

14. A gaseous mixture contains 5.78 g CH_4, 2.15 g Ne and 6.80 g SO_2 , all as gases.
     A) What is the total pressure exerted by the mixture inside a 75.0L flask at 85°C ?
     B) Which gas is exerting the greatest pressure in the flask and what is its value?
     C) Which gas has the greatest density in the flask and what is its value?

15. If a 60.0 L tank is filled with $N_2(g)$ at 170 atm at 21°C:
     A) What weight of $N_2(g)$ is stored in the tank?
     B) How does the density of the N_2 in the tank compare to the density of N_2 at STP?
     C) Is the N_2 most likely to be acting as real or ideal gas in tank?

16. 1.9 moles of gas are in a flask at 21°C and 697 mmHg. The flask was opened and
    10.5g of gas X was added to the flask. The pressure changed to 795 mmHg and the
    temperature of the mixture was 26°C in the flask.
     A) How many moles of gas total are in the flask after the addition of gas X?
     B) How many moles of gas X were added to the flask?
     C) What is the identity of gas X: O_2, Cl_2 or NO_2?

Chapter 11

Liquids. Solids, Materials

In chapter 10, kinetic-molecular theory as it applied to gases was considered. In this chapter, the ideas of particles in motion are used to explain the properties of liquids and solids.

11.1 Explain the properties of surface tension, capillary action, vapor pressure, and boiling point, and describe how these properties are influenced by intermolecular forces.

Liquids are a "condensed phase." This means that the particles are closer together than in a gas; however, the particles have enough mobility to be fluid. **Viscosity** is the resistance of a liquid to flow. The particles of a viscous liquid have strong intermolecular forces. Because liquids do not expand to fill their container, they have **surface tension**, which is the energy required to expand the surface. Surface tension reflects the different interactions between the particles in a bulk sample of a liquid versus particles at the surface which interact with another phase. **Capillary action**, which allows movement of liquids through solids, is also a result of intermolecular forces within the liquid and between the liquid and solid phases. The **volatility**, or tendency to vaporize, of a liquid, is determined by the strength of intermolecular forces. The vapor pressure gives an indication of the volatility of a liquid. The **normal boiling point** of a liquid is the temperature at which the vapor pressure is equal to one atmosphere. Liquids with low boiling points because of weak intermolecular forces are volatile.

11.2 Calculate the energy associated with vaporization and fusion.

11.3 Describe the changes of phases that occur between solids, liquids, and gases.

11.4 Use phase diagrams to predict what happens when temperatures and pressures are changed for a sample of matter.

11.5 Understand critical temperature and critical pressure.

Phase changes occur when a sample of matter loses or gains energy. Liquid-vapor transitions are vaporization and condensation; liquid-solid transitions are melting and crystallization. Solid-vapor transitions occur directly with some substances; these are called sublimation and deposition. In all of these transitions, energy is required to go from a condensed phase to a disordered phase, so the processes of vaporization, melting, and sublimation are endothermic. Conversely, energy is released when the particles slow down and become closer together, so that condensation, crystallization, and deposition are exothermic. The energy associated with these transitions can be calculated using the appropriate value of $\Delta H°$/mol and allowing for the size of the sample. These transitions can be represented graphically by heating curves and phase diagrams. **Heating curves** show an upward trend when one phase is being heated and a vertical portion when the energy added is being used to cause a phase change. **Phase diagrams** illustrate the relationship of P and T to the three phases and the six possible transitions among them. Important points on a phase diagram are the triple point and the critical point, which identifies the critical temperature and pressure.

11.6 Explain the unusual properties of water.

Water is such a common substance it is often used as an example to illustrate the properties described above. However, because the water molecule is small, very polar, and capable of forming hydrogen bonds in three dimensions, these properties are often not quite as predicted. Unlike other substances, solid water (ice) is less dense than its liquid (water), and therefore floats. Water has an unusually large value of heat capacity, heat of fusion, and heat of vaporization. All of these properties involve transfer of energy, making water a strong influence in weather and temperature patterns on the earth. Water has a very large surface tension and thermal conductivity.

11.7 Do calculations based on knowledge of simple unit cells and the dimensions of atoms and ions that occupy positions in those cells.

11.8 Differentiate among the major types of solids.

The study of solids illustrates the influence of interactions between particles on the macroscopic properties of matter. The particles of a solid are very close together and have very little freedom of movement. Solids can be classified as ionic, metallic, molecular, network, or amorphous. Ionic solids are characterized by definite crystalline structures held together by very strong attractions between oppositely charged particles. Metallic solids are characterized by orderly arrangements of atoms resulting from definite arrangements of atoms with freedom of movement of valence electrons among the nuclei. The structure of molecular solids is determined by London forces, dipole-dipole interactions and hydrogen bonding. Network solids are held together by two- or three-dimensional covalent bonds involving large numbers of particles in definite arrangements. **Amorphous solids** also involve covalent bonds, but there is no pattern to the arrangement. The arrangement of atoms in definite patterns can be described in terms of the crystal lattice. The **unit cell** is used to describe the smallest repeating unit in a crystal, and can be classified in one of seven categories. The most common is the cubic unit cell, which can be simple, body-centered, or face-centered. Another way to describe arrangement of particles in crystalline solids is by the packing arrangement, which can be either hexagonal or cubic. The packing arrangement affects the density of the material.

11.9 Explain the basis of materials science.

Materials science is the study of the relationships between the structure of matter used to construct various devices and the physical and chemical properties of those materials. Three major classes of materials are metals, polymers, and ceramics. **Composite** materials include more than one of these types in an attempt to maximize desirable properties and overcome limitations of each individual material.

11.10 Explain metallic bonding and how it results in the properties of metals and semiconductors.

11.11 Describe the phenomenon of superconductivity.

Metals are characterized by high electrical and thermal conductivity, ductility, malleability, luster, and insolubility in water and other solvents. These properties can be explained by a model of **metallic bonding** in which the solid consists of metal cations surrounded by a mobile "sea" of electrons. The electrons are visualized as moving freely in an energy band composed of a series of orbitals with similar energies. This model includes a valence band, where the electrons are normally found, and a **conduction band** of higher energy orbitals. In a metal these two bands overlap and the electrons move freely. in a **semiconductor**, there is a relatively narrow gap between the bands and under some conditions the electrons can move into the conduction band. In an **insulator**, the band gap is so large that electrons cannot move into the conduction band. When a solid is cooled to a critical temperature at which moving electrons encounter no resistance, it becomes a **superconductor**. Materials which are superconductors at readily achievable temperatures would be useful for many practical applications.

Silicon in very pure form is a semiconductor whose properties can be changed by adding small amounts of impurities. This doping process produces n-type or p-type conductors which have an excess or deficiency of electrons, respectively. When these types of semiconductors are joined at a **p-n junction** a gate is formed that allows flow of current in only one direction. This combination of properties has allowed the development of computer chips which can perform very complex functions.

11.12 Explain the bonding in network solids and how it results in their properties.

11.13 Explain how the lack of regular structure in amorphous solids affects their properties.

The many covalent bonds in **network solids** result in the formation of extremely large molecules. This results in extremely strong materials which have high melting points and that are generally unreactive except in extreme conditions. Diamonds, graphite and various silicates are the most common examples of network solids. Examples of amorphous solids are cement, ceramics, and glass. **Cement** is a mixture of various metal oxides that is worked in its hydrolyzed form and then allowed to slowly dry into an irregular network of crystals joined by water molecules. **Ceramics** are materials that are shaped while in a hydrolyzed form and then fired at extremely high temperatures. **Glass** is an amorphous solid consisting of oxides of metalloids or non-metals that are melted and then hardened.

Chapter Review - Key Terms

In order to make all the sentences below TRUE, insert the appropriate word or phrase from the list of key terms which best fits the context of the sentence.

NOTE: Any phrase or word from the list may used more than once.

LIST:		
amorphous solids	cubic unit cell	p-type semiconductor
boiling	deposition	phase diagram
boiling point	doping	semiconductor
capillary action	energy band	solar cell
cement	equilibrium vapor pressure	sublimation
ceramics	evaporation	superconductor
closest packing	glass	supercritical fluid
composites	heating curve	surface tension
concrete	hexagonal close packing	triple point
condensation	insulator	unit cell
conduction band	materials science	valence band
conductor	meniscus	vapor pressure
critical pressure	metallic bonding	vaporization
critical temperature (Tc)	n-type semiconductor	viscosity
crystal lattice	network solids	volatility
crystalline solids	normal boiling point	x-ray crystallography
crystallization	optical fiber	zone refining
cubic close packing	p-n junction	

Concerning Liquid Properties, Phase Changes and Phase Diagrams:

Liquids have the ability to flow, but the particles are very close together and an experience many intermolecular interactions. The shape of the molecules and its intermolecular forces determine its resistance to flow measured as the (1) _____ of the liquid. Another physical property of the liquid, the (2) _____, which is the tendency of a drop of liquid to expand (or flatten) its surface area, also depends strongly on the intermolecular forces. If the forces between the molecules are strong, the liquid has a high (3) _____ and it will not "wet" or spread out on the another surface easily. When a glass tube is placed in the liquid and the liquid rises into the tube, this is the result of (4) _____, which is due to the attractive forces between the interaction of the glass surface with molecules on the liquid surface. A competition then results between the strength of the two types of attractive forces which can produce a curved surface called a (5) _____ in the level of the liquid in the tube. The meniscus will curve up if the (6) _____ is greater than forces causing the (7) _____ and down if the reverse is true.

Converting a liquid substance to a gas is the change of state known as (8) _____. Molecules at the surface of the liquid have fewer forces holding them than

molecules in the bulk and can escape into the gas phase in a process known as
(9) _____, a term used to indicate the temperature is below the (10) _____ of
the liquid. The pressure exerted by the gas molecules that escape the liquid surface due to
(11) _____ is called the (12) _____, to indicate the liquid must be present for
this pressure to exist. To describe the general tendency of the liquid to vaporize, the term
(13) _____ is used, which does not have a specific value associated with it. Liquids
with weak intermolecular forces have a high (14) _____ and will exert higher
(15) _____s than liquids with stronger forces at the same temperature.

The opposite of (16) _____ is (17) _____ where gas molecules lose
excess kinetic energy and become liquid again. When the liquid is in a closed container, the
competition between the two processes results in an dynamic equilibrium and a constant
value for the (18) _____ called the (19) _____ which changes value only when
the temperature of the liquid changes. The surface area of the liquid will affect the rate of
(20) _____ of the liquid, but not the (21) _____ observed for the liquid. When
the (22) _____ of the liquid equals the atmospheric pressure, (23) _____ of
the liquid occurs, in which the bubbles of vapor are seen rising from all parts of the liquid.
Therefore, the (24) _____ for a liquid varies with the atmospheric pressure, even
though the (25) _____, set by the temperature, is not changing. The temperature at
which the liquid boils when the atmospheric pressure is 1.0 atm is called the
(26) _____ of the liquid.

Solids can also exert a (27) _____ through a process known as
(28) _____, where surface molecules obtain enough kinetic energy to escape the forces
that hold them in the solids state to become gaseous. The energy required
for (29) _____ is equal to the sum of the heats of vaporization and fusion for a substance. The gas molecules return to the solid state, in the process known as
(30) _____ , which can be used to produce thin films of substances on other materials. (31) _____ differs from (32) _____, which occurs when a substance
changes from the liquid to the solid at the melting point, in that the molecule converts from the
gaseous state.

The thermal energy needed to produce these changes in state can be directly measured
from the (33) _____ of the substance, which is a plot of temperature versus heat
added to the substance. Alternatively, the (34) _____ for a substance indicates the
phase(s) present at any given temperature and pressure for the substance. The lines shown on
the (35) _____ are the pressure and temperatures at which any two of the three states
of matter (solid, liquid or gas(vapor)) are in equilibrium with each other. The point where the
three lines meet is called the (36) _____ and indicates the single combination of P and
T when all three phases will be present. The point were the liquid-gas(vapor) line ends corresponds to the (37) _____ and (38) _____ for the substance. When at or above
the (39) _____, the substance cannot be converted to the liquid state, no matter how
high a pressure applied. Although invisible like a gas, the substance can flow like a liquid
above the (40) _____, and so it is called a (41) _____.

Concerning Arrangements of Atoms in Solids and Types of Materials:

The size and shape of molecules, as well as intermolecular forces determines the specific arrangement assumed by particles in a solid. Solids that have arrangements with long
range order, reflected in planar faces and sharp angles are called (42) _____ . In
contrast, (43) _____ lack the long range regular arrangement, but still have many

properties of solids. Each (44) _____ has a specific shape that reflects the internal regularity of the solid and the exact position of each particle is given by the (45) _____ that is made up of small, identical segments called (46) _____s which reproduce the whole crystal when repeated in three dimensions. The arrangement that maximizes the attractive forces of the particles and minimizes empty space in the solid is called the (47) _____ arrangement. Metals and ionic solids often have a (48) _____ arrangement that has a (49) _____ with sides of equal length which meet at 90° angles and each particle has 6 nearest neighbors. An alternate arrangement where each particle has 12 nearest neighbors is the (50) _____ arrangement. (51) _____ is a method that can be used to determine which of the (52) _____ arrangements appears in a (53) _____ and also determine the specific dimensions of the repeating (54) _____ for the crystal. Solids where covalent bonds hold atoms together instead of intermolecular forces are extremely hard materials and are called (55) _____.

(56) _____ is the study of the relationship between crystalline structure and the chemical and physical properties of solids. There are three main classes of materials, metals, polymers and (57) _____. Materials that are made by combining components of all three classes are called (58) _____. A diverse class of solids that are made from firing wet mixtures containing aluminosilicates (or clay) to produce Si-O-Si covalent bonds to link atoms in the material are (59) _____ . Although covalent bonds occur between the Si and O atoms, as in (60) _____ these materials are typically (61) _____ since they lack an repeating (62) _____indicative of long range order.

A (63) _____ is a class of solid that is made largely from silicates from sand (SiO_2), combined with carbonates, that also has to be heated to produce the final material, Unlike (64) _____ , a (65) _____ will be transparent to light and have a hard, shiny surface. The unique properties of a (66) _____ allow it to be spun into long fibers, called an (67) _____, that can be used transmit voice and data instead of electrical wires. Similarly two other classes of materials, (68) _____ and (69) _____ also are made largely from silicates as sand (SiO_2), and other calcium and iron containing materials, that do not require firing, but rather water, to achieve the final solid state by undergoing recrystallization and reaction to bond to surfaces and each other.

Concerning the Conduction Properties of Solids:

Metallic solids are atomic solids that behave as though there are metal cations existing in a sea of delocalized, mobile electrons. The term (70) _____ is used to describe the interaction that has occurred between the atoms in the solid to create the "sea" of electrons. A merging of electron states of the atoms in the solid produces two distinct (71) _____s, called the (72) _____ and the (73) _____. The (74) _____ is made from merging the electron orbitals that hold the valence electrons which is then only partially filled by the valence electrons. The (75) _____ is formed from merging higher, unfilled electron orbitals on the atom. In a metal, which is a good (76) _____ of electricity, the two bands overlap so that the electrons move easily into states in the (77) _____. The shiny luster associated with metals is due to absorption, and then emission, of visible radiation which is sufficient energy to excite the electrons from the (78) _____ into the (79) _____. In an (80) _____, a large energy gap appears between the two bands and the electrons cannot be excited easily into the (81) _____. In a (82) _____, an energy gap also exists but is small enough that the electrons can be excited into the (83) _____ with the application of an electrical field. All three types have varying resistance to electron flow, with metals having the least resistance, but there are a few solids that have zero resistance to electron flow below a certain temperature and these are called (84) _____ s.

Solid state electronic devices require (85) _____ s that are ultrapure materials for consistent electrical properties where the amount and type of impurity is strictly controlled. One method used to obtain ultrapure materials is (86) _____, which is based on the principle that impurities will be more soluble in a molten phase than in the crystallized solid. Deliberating adding impurities to a (87) _____ is called (88) _____ where the impurity atoms to create either negatively-charged electron conductors, called an (89) _____, or positively- charged conductor of "holes", a (90) _____. If the impurity atom that has a similar size but one more valence electron than the atoms that make up the bulk of the solid, a (91) _____ has been created. To make a (92) _____, the impurity should have one less valence electron. Putting the two types of (93) _____ together creates a device with a (94) _____ that allows an electrical current to flow in only one direction through the device. In a common type of (95) _____, doped silicon acts as a photoconductor, which means that sunlight absorbed on its surface is sufficient energy to cause electrons (or holes) to move into the (96) _____ to produce an electrical current.

- PRACTICE TESTS -

After completing your study of the chapter and the homework problems, the following questions can be used to test yourself on how well you have achieved the chapter objectives.

1. Tell whether the following statements would be always **True or False**:

    A).   Any crystalline material must be hard and have high melting point.

    B)   Most materials are more dense in the solid state than in the liquid state.

    C).   The triple point will be close to, but lower than, the melting point of a substance.

    D)   The heat of vaporization for all types of solids discussed in this chapter is the greatest for ionic solids because of the strong attractive forces between particles.

    E)   Hexagonal close packing produces fewer nearest neighbors for an atom than cubic close packed.

    F)   All crystalline solids have a crystal lattice that can be defined by a unit cell.

    G)   A face-centered cubic unit cell always contains 4 atoms.

    H)   The ratio of ions in the unit cell always matches the formula unit for the ionic compound.

    I)   The crystal lattice in ionic crystals is usually made from the anions with the cations in the holes.

    J)   Metals and network solids exhibit a high thermal and electrical conductivity.

    K)   Insulator materials have a narrow band gap between energy bands.

    L)   Nitrogen atoms could be a dopant for silicon to make n-type semiconductors.

    M)   Glasses and ceramics are examples of amorphous solids.

    N)   Particles in a liquid state have less kinetic energy than those in the gas state

    O)   When a gas condenses, energy is released to the environment.

    P)   Crystallization is the opposite of fusion.

    Q)   The boiling point of a given liquid is always constant.

2.   A)   Explain why SO_2 is a gas while SeO_2 is a solid at room temperature.

    B)   A chemist has to be careful about correcting measured boiling points to 1 atm, but seldom worries about this for melting points. Why not?

C)   (a) What is X-ray crystallography used for?   (b) What must be the relationship between the wavelength of incident radiation and spacing of the atoms in the crystal? (c) Why are X-rays the best for this?   (d) Why can't method be used to determine spacing of layers in amorphous crystals?

D)   (a) How does the size of the energy gap between the valence and conduction band determine whether a substance will be a insulator, metal or semiconductor at room temperature? (b) Does doping change the size of the energy gap?

E)   Compare the properties of cement and ceramics in terms of composition and what is needed to harden the material and produce the network of covalent bonds in the final material.

3. The first semiconductor were based on germanium.
    A) To what group in periodic table does geranium belong?

    B) What atoms could be used to dope germanium to obtain n-type and p-type semiconductors? Explain the reason(s) for your choice.

4. Which of the following would have larger $\Delta Hvap$? Give the reason(s) for choice.
    A) $Ag(s)$ or $I_2(s)$ B) NaI_3 or NI_3 C) CCl_4 or CS_2 D) H_2S or H_2O

5. A liquid in a closed container always establishes the same equilibrium vapor pressure, but a liquid in an open container may completely evaporate before it establishes an equilibrium vapor pressure while at the same temperature. Why is this true?

6. Consider the forces that can produce a meniscus.
    A) What two forces produce the meniscus in a liquid placed in a glass tube?
    B) What situation produces a meniscus that curves upward? Which downward?
    C) Why is no meniscus observed when a plastic graduated cylinder is used to measure water instead of glass?

7. As hot coffee cools, the cooler liquid at the surface sinks to the bottom of the cup and the hotter coffee rises to the top. (a) Why does this occur? (b) How could you construct an experiment to observe this phenomena occurring and measure temperatures necessary for the change to occur (i.e. how cool the surface must get, etc.)

8. Arrange the enthalpies below in order of increasing energy involved, putting any negative values first, with the most exothermic listed as the lowest.
    $\Delta H_{vaporization}$, ΔH_{fusion}, $\Delta H_{condensation}$, $\Delta H_{crystallization}$, $\Delta H_{deposition}$, $\Delta H_{sublimation}$

9. CsCl has a body centered cubic unit cell. Why couldn't $CaCl_2$ have same crystal structure (unit cells)?

10. Tell what type of solid (molecular, ionic, network covalent, metallic) the compounds (a) - (g) would be, based on properties given. Also, name the forces in solid and, if molecular, tell whether forces are H-bonding, dipole-dipole attractions, or London forces.

Compound:	Observed Properties:
(a) $SnCl_4$	soft crystals, m.p. -30.2°C, liquid is electrically nonconducting
(b) $B(s)$	very hard crystals, m.p. 2250°C, solid is semiconductor
(c) $Ga(s)$	shiny, m.p. 29.8°C, b.p. 2260°C, electrical conductor as solid
(d) $TiBr4$	soft orange-yellow crystals, b.p. 230°C, liquid poor conductor
(e) S_8	yellow brittle solid, m.p. 113°C, b.p. 445°C, poor conductor
(f) BaH_2	gray crystals, m.p = 675°C, b.p. = 1400°C liquid conducting
(g) AsH_3	colorless gas, b.p. - 62°C, nonconductive as liquid

11. Given the critical temperatures for the following, discuss effect of structure, forces and mass on Tc to the explain order observed.

	CO_2	C_3H_8	NH_3	CH_3COCH_3	CH_3CH_2OH
T_c	31°C	97°C	132°C	236°C	243°C

12. Match the properties (A) - (F) with one of the six types of solids given in the table below.

  (a) Ionic (b) Network Covalent (c) Metallic

  (d) Molecular (H-bonding) (e) Molecular (dipolar) (f) Molecular (London forces)

  _____ A) very low to moderate melting points, dependent on polarization, sublime easily, very poor conductors

  _____ B) low to moderately high melting points, soluble in solvents with hydrogen-bonding

  _____ C) hardness varies from soft to hard, melting points also vary low to high, have luster as solids, ductile, excellent conductors electricity.

  _____ D) hard, brittle, moderate to high melting points, good electrical conductors in liquid state

  _____ E) low to moderate melting points, have dipole moments, but generally nonconductors of electricity, soluble polar solvents such as water

  _____ F) very hard, sublime or melt at very high T, nonconductors of electricity in solid and liquid

13. For the phase diagram given on the right for substance X, indicate at which point(s):

  A) there will be an equilibrium between:

    (a) deposition and sublimation: ___

    (b) vaporization and condensation: ___

    (c) crystallization and melting: ___

  B) represent the P and T region where X would be:

    (a) solid   ___

    (b) liquid   ___

    (c) gas   ___

  C) there would be three phases present: ____

  D) represent the beginning of the supercritical fluid region: ___

  E) would be the normal boiling point of X : ___

  F) Referring to the heating curve given for X in the second diagram (P = 1 atm), tell <u>at what point on the phase diagram</u> X would be when in the sections marked by the number (1)- (5) on the heating curve:

    (1) _____ (2) _____ (3) _____ (4) _____ (5) _____

14.   Suppose 12.6 g of I_2(s), is placed in a 50.0 mL sealed flask at 25°C which has the following properties: ΔH_{vap}= 41.8 kJ/mol (normal b.p.= 184°C), ΔH_{fus} = 15.52 kJ/mol (m.p. = 114°C), Triple point = 114°C (v.p.= 91 mmHg) , T_c = 512°C, P_c = 116 atm; v.p. = 0.466 mmHg at 25°C, molar heat capacity solid = 54.44 J/mol-°C.

  A) Soon after sealing, a violet vapor appears, characteristic of I_2(g)

    (a) What is the appropriate names for the change(s) in state occurring in the flask?

    (b) What mass of I_2 has vaporized in the flask to achieve the equilibrium vapor pressure at 25°C?

14.   B) If the flask is heated to 114°C:
         (a) How many phases are present in the vial at 114°C and what are they?
         (b) What mass of I_2 would now have vaporized? Will the vapor appear darker?
         (c) Compare the amount of heat needed to heat the I_2(s) from 25°C to 114°C to that
             needed to melt I_2(s) at 114°C.
       C) If the container temperature is increased to 512°C, will the I_2 be acting as a
       supercritical fluid in the flask? Prove your answer.

15. Silver crystallizes in face centered cubic unit cell and has a density = 10.5 g/cm^3
       A) Calculate the length of an edge of unit cell in nanometers.
       B) Calculate the radii of Ag atoms.
       C) What type of forces exist between Ag atoms in crystal?

16. Above 1000°C, Fe(s) changes from body center cubic, with edge length = 0.28654 nm to
     face centered cubic, edge length = 0.363 nm
       A) What is the mass contained in each type of unit cell?
       B) What is density in g/cm^3 of each crystalline form of Fe(s)?
       C) Does radius of the Fe atoms change when the unit cells changes? Prove your
       answer.

17.   A) Magnesium crystallizes in a hexagonal close packed unit cell. Draw a diagram of the
         first three layers of the crystal structure.

         B) For the 2 dimensional pattern of points (representing atoms) in the diagram on the
         right:
            (a) Draw a unit cell that would contain
                ONE atom.
            (b) Draw a unit cell that would contain
                TWO atoms.

18. Ammonium iodide crystals have cubic structure like NaCl, with the radius of NH_4^+ = 148
    nm and the radius I$^-$ = 220 nm
       A) What is the length of the edge of unit cell?
       B) How many I$^-$ ions are nearest neighbors to NH_4^+ in structure?
       C) Does the unit cell produce the empirical formula? Prove your
       answer.
       D) Considering the diagram for a single face of the unit cell in the
       crystal on the right, determine whether the grey circles are the
       iodine ions or the ammonium ions.

Chapter 12

Fuels, Organic Chemicals and Polymers

This chapter presents an introduction to organic chemistry from the viewpoint of petroleum products and polymers.

12.1 Describe petroleum refining and methods used to improve the gasoline fraction.

Fossil fuels, including coal, natural gas, and petroleum, are the most common fuels in use today. Crude oil is a complex mixture of various hydrocarbons that must be refined into useful products. This refining is done by fractional distillation, in which simpler mixtures of hydrocarbons are collected according to their range of boiling points. Each fraction has specific uses, such as the lowest boiling fraction, which is used for fuel or for reactants in plastic manufacturing, and the highest boiling fraction, which is used for lubricating oils. A particular fraction may be further refined to enhance desirable properties. Molecules in the kerosene fraction (C_{12} to C_{16}) undergo **catalytic cracking** to produce molecules in the gasoline fraction (C_5 to C_{12}). The smaller molecules can undergo **catalytic reforming** to make branched hydrocarbons that burn more efficiently in internal combustion engines. The octane number and efficiency of combustion of gasoline can be further improved by adding small amounts of alcohols to the mixture.

12.2 Identify the major components of natural gas.

12.3 Identify processes used to obtain organic chemicals from coal, and name some of their products.

Natural gas consists mainly of C_1 to C_4 hydrocarbons. Coal is used primarily for production of electricity, although some is pyrolyzed by heating at high temperatures without oxygen to produce coal tar and other products. Fractional distillation of coal tar produces aromatic compounds that are used as starting materials for many products.

12.4 Convert equivalent energy units.

12.5 Relate atmospheric CO_2 concentration to the greenhouse effect and to global warming.

Energy conversions can be used to make comparisons between various fuels. The most common quantities are fuel value, described as J/g of fuel, and energy density, described as J/L of fuel. These values can be calculated from values of ΔH^o for the fuel and the molar mass or density of the fuel. Since no combustion process is completely efficient, this factor must also be considered in comparing fuels.

Combustion of fossil fuels releases CO_2 into the atmosphere. Because CO_2 absorbs infrared radiation, it acts as a "blanket" in the upper troposphere and traps heat close to the earth's surface. Before excessive combustion of fossil fuels increased this concentration, that energy was radiated back into outer space. Although methane, water, and ozone also contribute to this greenhouse effect, carbon dioxide is the only gas whose concentration has been drastically affected by human activity. This greenhouse effect has contributed to an overall trend toward global warming in the past 100 years.

12.6 Identify major organic chemicals of industrial and economic importance.

12.7 Name and draw the structures of three different functional groups produced by the oxidation of alcohols.

12.8 Name and give examples of the uses of some important alcohols.

12.9 Identify or write the structures of the functional groups in alcohols, aldehydes, ketones, carboxylic acids, esters, and amines.

Organic chemicals occur naturally or can be manufactured from fossil fuels. Small, unsaturated hydrocarbons and aromatic compounds are used as feedstocks for the petro-chemical industry. Important organic compounds considered here include alcohols, carboxylic acids, esters, and polymers. Alcohols contain the -OH (hydroxyl) functional group. The location of the -OH group on the carbon chain and its ability to form hydrogen bonds have a profound effect on the physical properties of alcohols. Primary alcohols can be oxidized to produce aldehydes, which can be further oxidized to carboxylic acids. The functional group in **aldehydes** is a carbon at one end of the carbon chain with a double bond to oxygen. Secondary alcohols can be oxidized to produce ketones, but tertiary alcohols cannot be oxidized. The functional group in **ketones** is a carbon atom that is not on the end of the carbon chain with a double bond to oxygen. Many large, naturally occurring molecules contain the -OH group. Two important examples of alcohols are methanol, used as a starting material for the manufacture of formaldehyde; and ethanol, used as a beverage and as a fuel.

12.10 List some properties of carboxylic acids, and write equations for the formation of esters from carboxylic acids and alcohols.

Carboxylic acids contain the -COOH functional group, where the carbon atom has a double bond to one oxygen and a single bond to the -OH group. These acids react with bases to produce salts; the molecules are polar and form hydrogen bonds with water. Diacids contain two -COOH groups; some of these are very important raw materials in the production of polymers. Carboxylic acids react with alcohols to produce **esters** in a condensation reaction. Small ester molecules are often used as flavor and odor components in foods. Other common esters include aspirin and triglycerides, which are components of fats and oils. Triglycerides containing saturated fatty acids are (solid) fats; those containing unsaturated fatty acids are (liquid) oils. Oils can be hydrogenated to convert them into solids.

12.11 Explain the formation of polymers by addition or condensation polymerization; give examples of synthetic polymers formed by each type of reaction.

12.12 Draw the structures of the repeating units in some common types of synthetic polymers and identify the polymers that form them.

12.13 Identify the types of plastics most successfully being recycled.

Polymers are large molecules formed by combining many simpler molecules. They can be characterized by their properties as **thermoplastics**, which soften and flow and can be easily reshaped upon heating, or **thermosetting plastics**, which form a rigid structure that cannot be altered upon reheating. A classification based on the chemical reactions of polymerization is addition polymers and condensation polymers. **Addition polymers** are formed from monomers that contain one or more double bonds which are caused to form free radicals by an initiator. These free radicals attack other double bonds, forming new covalent bonds and additional free radicals. The reaction continues as long as there are unsaturated monomers present. Polystyrene, polyvinyl chloride, and natural and synthetic rubbers are all examples of addition polymers. **Condensation polymers** are formed when monomers containing two or more functional groups react to join together, eliminating a small molecule such as water or HCl. Nylon and polyester are examples of condensation polymers. Note the diagrams of monomers and polymers in your text. Addition polymers are readily recycled.

12.14 Illustrate the basics of protein structures and how peptide linkages hold amino acids together in proteins.

12.15 Differentiate among primary, secondary, and tertiary structures of proteins.

12.16 Identify polysaccharides, their sources, the different ways they are linked, and the different uses resulting from these linkages.

Biopolymers include protein and carbohydrate molecules. Protein molecules are condensation polymers formed when a-amino acids are joined by peptide linkages. These bonds are formed between the -COOH group of one amino acid and the -NH$_2$ group of the next amino acid, resulting in a peptide backbone with R groups extending outward. The 20 amino acids in human biochemistry differ in their R groups, which determine the properties of the overall protein molecule. The **primary structure** of a protein is the identity and sequence of the amino acids that are joined by the covalent peptide linkages. The **secondary structure** of a protein is usually described as a-helix or b-pleated sheet and is determined by hydrogen bonding. The **tertiary structure** of a protein is the overall shape of the molecule, often described as globular or fibrous. Tertiary structure is determined by covalent bonds or noncovalent interactions between R groups along the peptide change. Tertiary structure is a major factor in determining the function of the protein. Polysaccharides are formed when monosaccharides are joined by glycosidic linkages in a condensation reaction. Starches are formed by plants for energy storage, and glycogen by plants for the same purpose. Cellulose is formed by plants for structure. Humans cannot digest cellulose because the shape of the glycosidic linkages in cellulose is different from those in starch and glycogen.

Chapter Review - Key Terms
In order to make all the sentences below TRUE, insert the appropriate word or phrase from the list of key terms which best fits the context of the sentence.
NOTE: Any phrase or word from the list may used more than once.

LIST:

addition polymer	ester	petroleum fractions
aldehyde	global warming	polyamides
alpha carbon	glycosidic linkage	polyester
amide	greenhouse effect	polymer
amide linkage	hydrolysis	polypeptide
amine	ketone	polyunsaturated acid
amino acid	macromolecule	primary structure
carboxylic acid	monomer	saponification
catalytic cracking	monounsaturated acid	secondary structure
catalytic reforming	octane number	tertiary structure
catalyst	partial hydrogenation	thermoplastics
condensation polymer	peptide linkage	thermosetting plastics
copolymer		

Concerning Fuels and Carbon Dioxide Emissions:

Nearly all fuels are carbon-based compounds such as fossil fuels which are complex mixtures of thousands of different organic compounds. Petroleum, a liquid fossil fuel, can be separated into several other types of mixtures, based on boiling point range, known as (1) _____, through fractional distillation. The composition of the fractions can be further refined through processes using a substance called a (2) _____, which increases the rate of chemical reaction and participates in the reaction, but is not permanently converted like the reactants. One such process is (3) _____, which uses a (4) _____ at high temperature and pressure to break long chain hydrocarbons into short chain molecules, increasing their volatility. Another is (5) _____, which converts straight chain hydrocarbons to branched hydrocarbons or to aromatic hydrocarbons with the use of a different (6) _____. One goal of these conversions is to produce gasoline with properties that allow it to be burned in the internal combustion engine. The burning properties of the gasoline can be classified by its (7) _____, which indicates how smoothly the gasoline will burn in the engine. Mixtures with only straight chains have a very low (8) _____, which produces "knocking" in the engine. Increasing the proportion of branched or aromatic compounds in the gasoline mixture will increase the (9) _____.

The industrial revolution fostered the wide use of coal and petroleum as fuel sources and one of the effects has been a rise in the carbon dioxide level in the troposphere. Carbon dioxide contributes to the (10) _____ and it is the only one of the four gases that human activities have any great effect on. The four atmospheric gases that produce the (11) _____ have a vital role in maintaining the temperature in the troposphere. There is, however, strong scientific evidence to connect increased carbon dioxide emission with (12) _____, a potential increase in worldwide atmospheric temperature. Many of the consequences of (13) _____ may not beneficial, such as melting polar ice caps, high sea levels or changes in rain patterns that would dramatically affect agriculture worldwide.

Functional Groups for Carbon Compounds:

Millions of compounds can be classified as organic compounds. Many of these compounds contain functional groups, with O or N bonded to carbon. In an (14) _____, the C is both bonded to an O in a C= O and at least one other H. In a (15) _____, the C in the C= O is also bonded to an OH group, often written as -COOH in the structure. A (16) _____ group occurs when the C in the C= O has no H's bonded to it. The -OH group in a primary alcohol can be oxidized to produce an (17) _____ group and then further oxidized to an (18) _____. Secondary alcohols can only be oxidized to produce molecules with (19) _____ groups.

A fatty acid is a long chain hydrocarbon with a (20) _____ group on one end of the molecule. Fatty acids that have one C = C in the chain, are called (21) _____s. If the chain contains two or more double bonds, the fatty acid is then a (22) _____. A fat (triglyceride) that is an oil at room temperature contains a high percentage of chains that are either (23) _____ or (24) _____ where the double bonds produce shapes with weaker London forces. Such fats can be converted to semisolids (called vegetable shortening or margarine) by (25) _____, a reaction that adds H_2 to some of the C= C bonds in the molecule, converting them to single bonds.

A condensation reaction of an alcohol with a (26) _____ produces an (27) _____ and water as the products. The reverse of this reaction occurs when an (28) _____ is reacted with water, or strong acid or base, called a (29) _____ reaction and an alcohol and a (30) _____ are produced. Triglycerides result when three fatty acids react with glycerol to produce three (31) _____ groups which link the chains to a single glycerol molecule. When triglycerides are reacted with aqueous NaOH, a (32) _____ reaction takes place which is also called (33) _____ where the sodium salts of the fatty acids produced are the main ingredient of soap.

When a nitrogen is bonded to a C as a -NH_2 group, the functional group is called an (34) _____. Like alcohols, an (35) _____ can also be reacted with (36) _____ to produce water and another molecule, where the (37) _____ has lost an OH group and the (38) _____ a H to form an (39) _____. The specific bonding pattern, where the C in the C= O group is bonded to an NH group, linking two formerly separate molecules is called an (40) _____.

Concerning Polymers and Reactions that Produce Them:

When hundreds of small molecules, called (41) _____s, are bonded together with covalent bonds, a (42) _____ is formed. Since the molecule is extremely large, a more descriptive term is to call it a (43) _____. If the molecule produced does not have an analog in the natural world, it is called a synthetic (44) _____ or plastic. The two basic types of plastics are (45) _____, that can undergo reversible changes from solid to liquid which soften and flow when reheated, and (46) _____ that will first soften and flow, but then harden to a rigid structure that cannot be remelted.

There are two major chemical reaction mechanisms by which a (47) _____ can be formed, which require a very different structural characteristics in the (48) _____. In one mechanism, the reaction starts with two (49) _____ s, each containing a C= C, adding directly to each other, changing the pi-bonds into new covalent bonds. A (50) _____ formed by this mechanism is called an (51) _____ and has either branched or largely single chains. In the second type of mechanism, the (52) _____ s contain two functional groups, one on either end of the molecule, such as two alcohol, (53) _____, (54) _____ groups or a combination of these groups. The different type of (55) _____ , called a (56) _____ is then formed when the functional groups react, producing a small molecule, such as water, and link the molecules together with covalent bonds. When the chemical structure of the two (57) _____ s reacting to form the single chain is different, a (58) _____ is formed, which can be either type, an (59) _____ or a (60) _____ since either mechanism can be used.

The (61) _____ s formed from the second mechanism are further classified by the type of linkage that results. When (62) _____ and alcohol groups combine to form a (63) _____, water is lost, and an (64) _____ linkage occurs and the (65) _____ is called a (66) _____, where Dacron is an example of this type. When monomers with (67) _____ and (68) _____ groups react, by the same mechanism, water is also produced, but an (69) _____ occurs between the units and a (70) _____, such as nylon or polypeptides has been formed. Many natural polymers, particularly those that have a biological role, are a formed as (71) _____ s with either type of linkage.

Concerning the Structure of Biological Macromolecules:

One of the largest classes of these are proteins which are (72) _____, based on the type of linkage formed. The (73) _____ s for proteins are molecules called (74) _____ s which have both (75) _____ and (76) _____ groups in a single molecule. In the important (77) _____ s common in human biochemistry are distinguished by having the (78) _____ group bonded to the first carbon next to the (79) _____ group, called the (80) _____.

In a protein molecule the linkage is referred to as a (81) _____ since the order of appearance of the (82) _____ and the nitrogen are always the same in the chain, which is unique to this type of (83) _____. The general name for the type of (84) _____ then changes to a (85) _____, to reflect the special linkage. Another unique feature is that the (86) _____ s that form the protein do not have the same exact structure as do most (87) _____ s, so that the side groups on the chain are not the same nor do they repeat the same manner. If only fifty or so (88) _____ s have been linked to form a single chain, the molecule is called a (89) _____ rather than a protein, since the number of (90) _____ s joined together is small in compared to the thousands bonded together in a typical protein molecule.

Each protein has a distinctive shape, which influences its function in a biological system for which can be classified as having three contributing parts. The first part, the interactions of portions of the chains, is called the (91) _____ of the protein which is determined by the specific sequence of (92) _____ s in the protein chain. The position of the "R" or side groups on the (93) _____ bonded together can result in either intramolecular H-bonding (between adjacent R groups) or intermolecular H-bonding between separate chains or portions of chains. The second part, the (94) _____ , describing how the chains interact, is usually of two types, an α-helix, due to intramolecular H-bonding, or a β-pleated sheet, due to intermolecular H-bonding. The final 3-dimensional arrangement of the protein, called its (95) _____.

Polysaccharides, a major class of biomolecules, are formed with D-glucose molecules as the (96) _____s. When the D-glucose units are polymerized, a water molecule is lost, so a (97) _____ is formed and the specific type of linkage used is called a (98) _____ which occurs between OH groups on the 1 and 4 carbons in different glucose units. If the OH groups on the 1 and 4 carbons are in the "*trans*" position in the final (99) _____, cellulose is produced. If the OH groups are in the "*cis*" position, either a (100) _____ called starch (in plants) or glycogen (in animals) is formed which serves as the an chemical energy storage molecule. These polysaccharides can act as storage molecules for glucose, since like (101) _____s or (102) _____s, they can undergo (103) _____ reactions to reproduce the original (104) _____ structures.

- PRACTICE TESTS -

After completing your study of the chapter and the homework problems, the following questions can be used to test yourself on how well you have achieved the chapter objectives.

1. Tell whether the following statements would be always **True or False**:

   A). Coal mixtures contain more aromatic compounds than crude oil.

   B) Fractional distillation separates each component of petroleum.

   C) Octane numbers cannot be higher than 100, nor lower than zero.

   D) The function of catalytic reforming is to convert long hydrocarbons to shorter hydrocarbon chains and that of catalytic cracking is to form branched molecules.

   E) Octane enhancers are aldehydes and aromatics that can added to gasoline to increase octane number.

   F) Oxygenated gasolines are used to help CO emissions from the exhaust.

   G) Evaporation rates of gasoline blends can be reduced by lowering the amount of aromatic hydrocarbons present in the fuel.

   H) The primary reaction for esters is hydrolysis.

   I) $CH_3CH_2CH_2OH$ is a tertiary alcohol and CH_3CH_2OH is a secondary alcohol.

   J) All alcohols produce an aldehyde when oxidized, but only primary alcohols can be made into carboxylic acids..

   K) Methanol is called "wood alcohol" because the primary source of it is cellulose-based plants, whereas ethanol can be made from starch-based plant material.

   L) An 80% pure ethyl alcohol solution is also 80 proof.

   M) A certain plastic material was found to be flawed after production. The material was then reheated and recast which indicates it was a thermosetting plastic.

   N) A polymer made from an acetic acid, CH_3COOH, and methanol, CH_3OH, is a condensation polymer.

   O) Low density polyethylene means the polymer chains contain fewer carbons.

   P) Condensation polymers involve a free radical in the initiation and formation steps of the polymers.

   Q) Synthetic rubber is an example of a copolymer that has been formed by a condensation mechanism.

1.
R)   Alpha- amino acids will have the amine group on the carbon adjacent to the carboxylic acid group.

S)   Proteins have many function in the human body and result from a condensation reaction that produces a peptide linkage.

T)   Soaps are produced by saponification of esters of fatty acids called triglycerides.

U)   Proteins and polysaccharides can all undergo a hydrolysis reaction to reproduce their respective monomers, amino acids and D-glucose molecules.

V)   Fats have a higher percent of polyunsaturated triglycerides than oils.

2.
A)   Explain why ethanol, although obtained from plant material, is not a fossil fuel.

B)   Why is ethylene an important product in refining petroleum??

C)   Considering the text discussion and data below, what are two benefits of burning oxygenated gasoline versus just octane enhanced (unleaded) gasoline?

Additive: Benzene   Ethanol   Methanol   Methyl-T-Butyl ether (MTBE)
Octane rating: 99   115   123   123

D)   When the following functional groups react what is the product for:
    (a) an acid and a alcohol react (b) a dicarboxylic acid and diamine react.

E)   How could two chemists, each staring with a compound having the formula C_3H_8O, get two different classes of compounds upon initial oxidation?

F)   How do the boiling points of acids compare to that of alcohols of the nearly the same molecular mass and what is the primary reason for the difference?

G)   (a) Why is it easier to recycle thermoplastics, instead of thermosetting plastics?
    (b) If you were going to make a bench that needed to be outside in all weathers, which type of plastic would you choose to make it out of and why?

H)   What effect on the polymer properties such as hardness, density, rigidity, do the following have: (a) branching (b) cross-linking (c) the R (or side) groups on chain

I)   What characteristic of cellulose makes it impossible for humans to digest?

J)   Considering proteins: (a) What noncovalent forces are responsible for the secondary structure of proteins?
    (b) When cross-linking through covalent bonds occurs in proteins, what type of linkage and groups are typically involved?

3. A) Match the following fractions typically obtained from crude oil with the boiling point range:

(a) Volatile gases   _____   A) boiling points < 20°C
    (b) asphalt residue   _____   B) boiling points 20- 180°C
    (c) kerosene   _____   C) boiling points 180-270°C
    (d) lubricating oils   _____   D) boiling points 250-370°C
    (e) gasoline (or naptha) _____   E) boiling points 340-470°C
    (f) fuel oils   _____   F) boiling points. ≥ 470°C

B) When passing by a refinery, you will see that flames typically appear at the top of some the refinery stacks. What is being burned at the top of the stack? Why is it necessary to burn these gases?

C) Jet fuel falls between gasoline and kerosene in its composition. Give the approximate range of for the number of carbons in jet fuel and the approximate boiling range.

4. Arrange the following in order of increasing boiling point:

(a) CH_3OH (b) CH_3CH_3 (c) CH_3CO_2H (d) CH_3CH_2OH

5. Classify each of the following as either an:

(a) amine
(b) alcohol
(c) carboxylic acid
(d) amino acid
(e) ester
(f) ketone
(g) aldehyde

A) $CH_3 \cdot CH_2 \cdot \overset{\overset{\displaystyle O}{||}}{C}-OH$

B) $H_2N-CH_2 \cdot CH_3$

C) $H_2N-CH_2 \cdot \overset{\overset{\displaystyle O}{||}}{C}-OH$

D) (cyclohexadienone-type ring with =O)

E) (benzene ring)$-\overset{\overset{\displaystyle O}{||}}{C}-CH_3$

F) (benzene ring with two CH_3 and OH)

G) $H_2N-CH_2 CH_2 CH_2 \cdot \overset{\overset{\displaystyle O}{||}}{C}-OH$

H) $CH_3 \cdot CH_2 \cdot \overset{\overset{\displaystyle O}{||}}{C}-O-CH_3$

I) $CH_3 \cdot CH_2 \cdot CH_2 \cdot \overset{\overset{\displaystyle O}{||}}{C}-H$

6. For the following alcohols:

A) Indicate which are primary, secondary and tertiary alcohols

B) For those which are primary alcohols show what would be the structures of two products possible from the oxidation of the alcohol.

C) For the secondary alcohols, indicate what the condensed molecular structure would be for the oxidation product.

(a) $CH_3 \cdot OH$

(b) $Cl-$(benzene ring)$-CH_2-OH$

(c) (benzene ring)$-\underset{\underset{\displaystyle OH}{|}}{\overset{\overset{\displaystyle CH_3}{|}}{CH}}$

(d) $CH_3-\underset{\underset{\displaystyle OH}{|}}{CH}-CH_2 \cdot \overset{\overset{\displaystyle O}{||}}{C}-NH_2$

(e) $CH_3-\underset{\underset{\displaystyle CH_3}{|}}{\overset{\overset{\displaystyle OH}{|}}{C}}-CH_2 \cdot CH_3$

7. Chewing gum is mostly polyvinylacetate. The structure of the <u>monomer</u> is shown below:

A) What's the name of the monomer and what functional group does it contain?

B) Draw out a portion of the polymer chain for polyvinylacetate, showing at least two repeating units of the chain

C) Is polyvinylacetate an addition or condensation polymer?

D) What type of noncovalent interactions can occur between the side groups on the chain?

$CH_3-\overset{\overset{\displaystyle O}{||}}{C}-O-CH=CH_2$

8. PETE, polyethyleneteraphthlate, is a common plastic used for rigid clear containers like ketchup or soda bottles and is a polymer made by reacting with teraphthalic acid with ethylene glycol. The structures of each monomer is shown on the right:

A) Draw out a section of the polymer chain formed, showing 2 of the repeating units

B) Is the polymer formed a condensation or addition polymer?

C) Since PETE is a recycled plastic do you think it is a thermoplastic polymer or thermosetting plastic? Give the reasons for your choice.

Ethylene glycol

$HO-CH_2 \cdot CH_2 \cdot OH$

$HO-\overset{\overset{\displaystyle O}{||}}{C}-$(benzene ring)$-\overset{\overset{\displaystyle O}{||}}{C}-OH$

Teraphthalic acid

9. Saran wrap is a copolymer made from $CH_2=CCl_2$ (vinylidene chloride) and $CH_2=CHCl$ (vinyl chloride) which is an addition polymer.

A) Draw out a section of the polymer chain formed, showing 4 of the repeating units, assuming a 50-50 copolymer is made.

B) Saran wrap clings tightly to a glass surface. What forces occuring between the polymer chains and the glass that are responsible for this effect? (*Refer to glasses in Chapter 11, if needed*)

C) In a microwave, heating a bowl sealed with Saran wrap causes the plastic to expand when heated (trapping the steam leaving the food) and then contract when the bowl cools. Is Saran wrap behaving like a thermoplastic or thermosetting plastic? If it were not this type of plastic, how would be its behavior in the microwave heating cycle change?

10. For the molecules below, which:

(a) $H_2C{=}CH\text{-}Cl$ (b) $H_2N-\text{⟨benzene⟩}-NH_2$ (c) $HO\text{-}C(=O)-\text{⟨benzene⟩}-C(=O)-OH$ (d) $\text{⟨benzene⟩}-CH{=}CH_2$ (e) $H_2C{=}CH\text{-}OH$ (f) $HS-CH_2-\overset{NH_2}{\underset{}{CH}}-\overset{O}{\overset{\|}{C}}-OH$

  A) molecule(s) could form addition polymers with themselves?
  B) molecule(s) could form condensation polymers with themselves?
  C) two molecules could be combined to form a copolymer with an amide linkage?
  D) molecule(s) could form a polymer with a peptide linkage?
  E) molecule(s) would form polymers that would exhibit H-bonding between the chains, not just on the ends?
  F) molecule(s) would form a polymer called polystyrene?
  G) molecule(s) could form a polymer with disulfide linkages between chains?

11. For a mixture that is 10% CH_3OH and 90% gasoline by volume, with the gasoline component assumed to be isooctane, C_8H_{18} (density equal to 0.692 g/mL).
  A) Will the mixture meet the criteria of being at least 2.7% by mass O for an oxygenated fuel? (density CH_3OH = 0.791 g/mL)
  B) What volume of CO_2 (g), in liters, measured at STP, would be produced per gallon? Is this more or less than what would be produced if the fuel were pure C_8H_{18}? (1 gal = 4 qt = 3.785 L)
  C) How many liters of O_2 must be consumed for the combustion of one gallon of the mixture? Does this change if the gallon were pure isoctane? Prove your answer.

12. Dextran is a branched polysaccharide found in yeast and bacteria that has an cis-type linkage between the D-glucose units. In an experiment, the molecular weight of dextran extract was found to be 68,500 g/mol
  A) How many glucose units are in this polysaccharide?

  B) Using ⟋‾⟍ for cis or ⟍‾⟍ for trans linking of the glucose rings, which of the diagrams (a)- (c) below would best represent the basic repeating pattern in dextran?

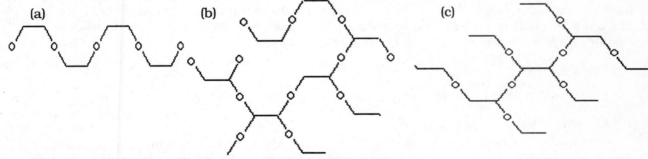

(a)   (b)   (c)

13. For the peptide with the structure: Ala-Ser-Gly-Tyrosine
  A) Draw out the structure of the peptide
  B) Indicate how many peptide linkages there are in the molecule and circle each one.
  C) What would be the name of this peptide?
  D) Could this peptide be called a protein? Explain your answer.
  E) Indicate which groups on the peptide backbone would be likely to produce hydrogen-bonding between the chains (for intermolecular bonding).

Chapter 13

Chemical Kinetics: Rates of Reaction

In this chapter the rate of chemical reactions is considered along with the pathways by which the particles involved in the reaction are rearranged into different compounds.

13.1 Define reaction rate and calculate average rates.

The speed of a chemical reaction in which all the substances are in the same phase (homogeneous) is determined by:

- the structure and bonding in the reactants and products
- the concentrations
- the temperature of the reaction system
- catalysts, if present

If the reaction system involves substances in more than one phase (heterogeneous), the reaction rate is also affected by the characteristics of the surfaces at the interface between the phases. The **reaction rate** is defined as the change in concentration of any substance in the reaction with respect to time, using units of M/t. The relative rates for substances in the reaction system are determined by stoichiometry. The rate is negative for reactants, and positive for products. Since reaction rates vary as a reaction proceeds, the **instantaneous reaction rate** at a given time is often more useful.

13.2 Describe the effect reactant concentrations have on reaction rate, and determine rate laws and rate constants from initial rates.

13.3 Determine reaction orders from a rate law, and use the integrated rate law method to obtain orders and rate constants.

13.4 Calculate concentration from time, time to reach a certain concentration, and half-life for a first-order reaction.

The **rate law** describes the relationship between reaction rate and concentration(s) using a proportionality constant, k, called the **rate constant**. A large value of k means a reaction is fast. Values of k are unique to a specific reaction at a certain temperature. Rate laws must be determined experimentally by comparing reaction rates as concentrations are varied systematically. Rate laws are usually expressed as "rate = k $[A]^m[B]^n$," where m and n are the **order of reaction** with respect to reactants A and B, respectively. Zero order means the rate is unchanged as the concentration varies; first order means the rate varies directly with concentration; second order means the rate varies as the square of the concentration. The **overall reaction order** of the reaction equals the sum of the orders of each reactant. Another way to express rate laws is in the integrated form:

- first order: $\ln[A]_t = -kt + \ln[A]_0$

- second order: $\frac{1}{[A_t]} = kt + \frac{1}{[A_0]}$

- zero order: $[A]_t = -kt + [A]_0$

These forms can be used to identify the order of reactions based on plots of time and concentration data. They can also used to make predictions based on time or concentration. An important concept is half-life, $t_{1/2}$, the amount of time required for one-half of a reactant to be consumed. For a first-order reaction $t_{1/2} = \frac{0.693}{k}$. Note that a large value of k means a short-half-life. It is important to remember that rate laws and order of reactions can only be determined by experiment. The orders of reaction in the rate law are not determined by stoichiometry of the balanced reaction.

13.5 **Define and give examples of unimolecular and bimolecular elementary reactions.**

13.6 **Show by using an energy profile what happens as two reactant molecules interact to form product molecules.**

Elementary reactions are used to describe the actual molecular interactions as a reaction occurs. In a **unimolecular reaction**, a single particle changes to produce a different particle or particles. In a bimolecular reaction, two particles interact and rearrange into different particles. The term **transition state** or **activated complex** is used to describe the high-energy particles that fleetingly exist during these elementary rearrangements. The energy required to reach this state is called the **activation energy**, E_a. A high activation energy is associated with a slow reaction. Energy diagrams illustrate the relationships between activation energy and the initial and final states of a reaction.

13.7 **Define activation energy and frequency factor, and use them to calculate rate constants and rates under different conditions of temperature and concentration.**

A more detailed consideration of the relationship between temperature and reaction rate uses the **Arrhenius equation** to connect activation energy, E_a, and Kelvin temperature to the value of k. Other considerations are the steric factor, which describes the orientation with which particles must collide, and the frequency factor, which describes the number of effective collisions which are likely to occur. Determinations of k at various temperatures allows calculation of the value of E_a.

13.8 **Derive rate laws for unimolecular and bimolecular elementary reactions.**

13.9 **Define reaction mechanism and identify rate-limiting steps and intermediates.**

13.10 **Given several reaction mechanisms, decide which is (are) in agreement with experimentally determined stoichiometry and rate law.**

Rate laws can be used to determine the probability of a sequence of elementary reactions suggested as a possible mechanism for a reaction. The molecularity of the slowest, or **rate-limiting**, step must coincide with the order of the overall rate law. An **intermediate** is produced in an early step of the mechanism and consumed in a later step. When all of the steps are combined and species found on both sides eliminated, the result should be the same as the original reaction.

13.11 **Explain how a catalyst can speed up a reaction; draw energy profiles for catalyzed and uncatalyzed reaction mechanisms.**

13.12 **Define the terms enzyme, substrate, and inhibitor, and identify similarities and differences between enzyme-catalyzed reactions and uncatalyzed reactions.**

Catalysts speed up chemical reactions by lowering the activation energy. Catalysts actually change during early steps of the reaction, but are always regenerated during a later step. The energy diagram for a catalyzed reaction differs from that for the uncatalyzed reaction only by having a lower hump for activation energy.

Enzymes are protein molecules which act as biological catalysts, greatly increasing the efficiency of chemical reactions in living things. Enzymes lower the activation energy of their specific reaction by use of an active site, a specific location on the molecule where the substrate is held in place and changed in some way to allow the desired reaction to occur. After the reaction, the enzyme returns to its original state. Enzyme function can be disrupted by **denaturation**, in which changes in secondary or tertiary structure of the enzyme affect the active site. Enzymes can also be inhibited when some molecule other than the substrate blocks access to the active site.

13. 13 Describe several important industrial processes that depend on catalysts.

Industrial catalysts are usually **heterogeneous catalysts**, meaning that the catalyst is in a different phase than the reactants. These catalysts often are transition metals. Common examples of industrial catalysts are the rhodium(III) iodide used in the production of acetic acid, the platinum-based catalytic converters used in automobiles, and the platinum or rhodium catalysts used to produce methanol from methane.

Chapter Review - Key Terms

In order to make all the sentences below TRUE, insert the appropriate word or phrase from the list of key terms which best fits the context of the sentence.

NOTE: Any phrase or word from the list may used more than once.

LIST:

activated complex	frequency factor	overall reaction order
activation energy (Ea)	half-life	rate
active site	heterogeneous catalyst	rate constant
Arrhenius equation	heterogeneous reaction	rate law
average reaction rate	homogeneous catalyst	rate-limiting step
bimolecular reaction	homogeneous reaction	reaction intermediate
chemical kinetics	induced fit	reaction mechanism
cofactor	inhibitor	reaction rate
denaturation	initial rate	steric factor
elementary reaction	instantaneous reaction rate	substrate
enzyme	intermediate	transition state
enzyme-substrate complex	order of reaction	unimolecular reaction

---

Describing Rates of Reaction:

The study of the speed or (1) _____ of reaction, the time and energy needed for reactions to occur, the influence of temperature and catalysts, as well as the step-wise progress of the reaction, are all part of (2) _____. The conversion of reactants to products always take place with certain velocity or (3) _____ of reaction. When reactants and products are in the same phase, the reaction is a (4) _____, whereas if any of the reactants and products are in different phases, the reaction is a (5) _____. Surface area of the reactants are very important in a (6) _____ as the (7) _____ of reaction increases in direct proportion to the surface area. Adding solid catalyst (or one of a different phase) called a (8) _____, to the reactants in a (9) _____ system, can also be used to provide a reaction surface which can speed up the (10)_____.

A graph of concentration versus time always shows that the (11) _____ changes over time, starting high and decreasing to lower values as the time since mixing the reactants increases. The (12) _____ calculated by taking the difference in concentration of a reactant at two different times divided by the difference between the two times, is the (13) _____ for that time interval. In contrast, the tangent to the curve at a particular time gives the (14) _____ at that point in time. The (15) _____ of reaction, is the (16) _____ at the time of mixing, when t= 0.

The (17) _____ is the equation that describes the rate of reaction and its dependence on the concentration of reactants. Two other factors must appear in the (18) _____, which must be experimentally determined: (1) the exponents (or power) for each reactant concentration giving the (19) _____ for that reactant and (2) the

(20) _____, with the symbol k, which is the measured proportionality constant between the measured rate and concentration of reactant(s) in the reaction. The (21) _____ is the temperature dependent term in the equation and is also specific to the reaction. If the reaction has more than one reactant, the (22) _____ for each reactant can be added together to give the (23) _____. The (24) _____ for the reaction, which is the time needed for the concentration of a reactant to drop to one half of its initial value is a convenient way to gauge the speed of a reaction. The (25) _____ of a reaction may be calculated from the (26) _____, or vice versa, but the initial concentrations are also needed for zero and second order reactions.

Concerning Reaction Mechanisms and Energy Profiles:

There are two fundamentally different ways reactants can be transformed to products on the nanoscale level of single atoms, ions or molecules. When a single reactant particle is rearranged or has a bond broken within it to make products, a (27) _____ takes place. When two particles have to come together to exchange atoms or join together to form a larger particle, a (28) _____ has taken place. Most reactions occur as a series of steps, called the (29) _____, to produce the overall balanced reaction. An (30) _____, is then a single step in the (31) _____, which can be either a (32) _____ or (33) _____, that describes the exact nanoscale particle(s) undergoing change in that step. The (34) _____ for each reactant is related to its stoichiometry in the slow step of the (35) _____, called the (36) _____ of the reaction which is often very different than the coefficient in the balanced chemical reaction. Consequently, the (37) _____ for reactant can not be predicted but must be measured in the laboratory. Furthermore, substances, called an (38) _____ or more specifically a (39) _____, can be produced in one step of the reaction mechanism, and then consumed in a later step and do not appear in the overall balanced equation, even though they have a role in the reaction. Similarly, substances that act as (40) _____ are always in the same phase as the reactants and will increase the (41) _____ of reaction, but also cancel out of the balanced overall reaction.

The make products from reactants, an energy state which is higher than that of either the reactants or the products, known as the (42) _____ or (43) _____, must be achieved which acts as an energy barrier for the reaction. The minimum energy needed to surmount the energy barrier or "hill" is called the (44) _____ for the reaction. The (45) _____ determines the value of the (46) _____ in the (47) _____, but not the concentrations or (48) _____. The equation that defines the exact relationship between the (49) _____, (50) _____ and temperature is called the (51) _____. The (52) _____ has two parts, the (53) _____, represented by A, and an exponential term, $e^{-Ea/RT}$, which is a measure the fraction of particles which have energy greater than or equal to the (54) _____. The (55) _____ results from, in part, the (56) _____ of a reactant which is a geometric constraint, related to the shape of the reactant particle, which results in only a fraction of collisions being successful. Catalysts increase the (57) _____ by altering the (58) _____ and lowering the (59) _____ for the conversion.

Concerning Enzymes as Catalysts:

Large, usually globular proteins, called (60) _____s function as highly efficient and molecular specific catalysts in biochemical reactions. Each (61) _____ acts on a specific molecule type called a (62) _____ for which it has a high affinity. A small portion of the protein, called the (63) _____, is where the (64) _____ molecule actually binds to the (65) _____. The (66) _____ formed from the binding acts as the (67) _____ for the reaction. As a consequence of binding, a change in shape in either the (68) _____ or (69) _____ occurs, called an (70) _____, which is critical to forming product. An (71) _____ may also require that a (72) _____, a second inorganic or organic molecule that is not a reactant or product, for the reaction to occur. If the (73) _____ concentration is limiting, the (74) _____ will be first order for the concentration of substrate. If the (75) _____ concentration is limiting, the reaction will appear to be a zeroth order reaction, independent of (76) _____ concentration, but will be proceeding at its maximum rate of conversion. An increase in temperature can adversely affect an (77) _____ -catalyzed reaction. If the temperature becomes high enough to cause the protein chain to undergo (78) _____, the normal globular shape transforms into a more fibrous form, and the reaction rate decreases dramatically, since the (79) _____ of the (80) _____ no longer has the specific shape needed to produce the (81) _____. An (82) _____ can also act to impede an (83) _____- catalyzed reaction, by binding to the (84) _____ of the (85) _____ and preventing the (86) _____ from binding.

- PRACTICE TESTS -

After completing your study of the chapter and the homework problems, the following questions can be used to test yourself on how well you have achieved the chapter objectives.

1.   Tell whether the following statements would be always **True or False**:

   A).   Reaction rates are independent of volume in which the reaction takes place.

   B)   Chemical kinetics is a study of reaction rates and reaction pathways and the energy profile for the reaction pathway is based on the kinetic energies of reactants.

   C)   The symbolism [ ] in the rate law means that the molarity of reactant named within the brackets must be used when using the rate law.

   D)   The rate of reaction of Al(s) reacting with HCl(aq) would vary with the surface area of the Al(s) and the concentration of HCl, but [Al] will not appear in the rate law.

   E)   The larger the rate constant value the faster a reaction will proceed.

   F)   The overall order of the reaction will equal the sum of the coefficients in the balanced, overall reaction.

   G)   If a reaction is first order in a reactant, doubling the concentration of the reactant will quadruple the rate of reaction.

   H)   Unimolecular reactions produce first order reactions and bimolecular ones, second order reactions.

   I)   The greater the activation energy, the easier it will be for the reaction to occur.

   J)   The rate of disappearance of a reactant can always be related to the appearance of product through the stoichiometry of the balanced reaction.

   K)   The slowest step in a reaction sequence is the rate limiting step and determines the orders of the reactants in the rate law.

   M)   The Ea is a measure of a reaction's energy barrier which catalysts, intermediates and inhibitors can all affect.

   N)   A reaction that is very exothermic has a low energy barrier and is likely to be a very fast reaction.

   O)   A reaction occurs and evidence is found to suggest that three steps are involved. This means some intermediate products must be formed.

2. A)   What is the purpose of the spark plug in an automobile engine?

    B)   Explain why reaction rates decline as molecularity increases.

    C)   Sometimes doubling a reactant's concentration increases the reaction rate and at other times it has no effect. Explain why these differences can occur.

    D)   Why is it necessary to trap intermediates to determine the mechanism of the reaction? Will catalysts give the same information?

    E)   Would it be likely that inhibitor and substrate molecules for an enzyme share some structural characteristics? Explain.

3. Which of the following best explains why a reaction rate changes with time?
    A) the rate constant changes
    B) concentration of reactant changes
    C) the order of reaction changes
    D) both the rate constant and the order changes
    E) both the rate constant and concentration of reactant(s) change
    F) both the concentration of reactant(s) and order change

4. If the rate = $k[A][B]^2$, which of the following produce the fastest and slowest initial rates?
    A)   0.50 mol A, 0.50 mol B in 2.0 L
    B)   0.50 mol A, 0.50 mol B in 1.0 L
    C)   2.00 mol A, 1.00 mol B in 1.0 L
    D)   1.00 mol A, 2.00 mol B in 1.0 L

5. The Haber reaction for ammonia is : $N_2(g) + 3 H_2(g) \rightarrow 2 NH_3(g)$. Without doing any experiments which of the following could you say must be TRUE.
    A)   The reaction rate equals $-\Delta[N_2]/\Delta t$.
    B)   The reaction is first order in N_2.
    C)   The rate of disappearance of N_2 is 3 times the rate of disappearance of H_2.
    D)   The rate of disappearance of H_2 is 1.5 times the rate of disappearance of NH_3.
    E)   The reaction mechanism has at least two steps.
    F)   The energy of activation is positive.

6. For a reaction occurring between A and D where the rate equation is: rate = $k[A]^2[D]$,
    A) Will the following increase, decrease or not affect the reaction rate?
            (a) cut the concentration of A in half, but keep D the same
            (b) double the concentration of A , but cut that for D in half.
            (c) double D, but decrease A to one-fourth of the original value
            (d) cut the concentration of A in half, but quadruple that for D
    B) Of those that affected the rate, which will have the greatest effect?

7. The reaction: $2 ICl(g) + H_2(g) \rightarrow I_2(g) + 2 HCl(g)$ is first order in each reactant.
    A) What does this make the overall order for the reaction?
    B) Suppose you measured the rate of reaction is 4.89×10^{-5} M/s when [ICl] = 0.100M and [H_2] = 0.300M, calculate the value of k for the reaction.
    C) At what concentration of ICl would the rate be 7.85×10^{-2} M/s when the H_2 concentration is twice that of ICl?
    D) How will the disappearance of ICl compare to that of H_2?
    E) Is it plausible that the rate-limiting step is bimolecular? Explain your reasoning.

8. For the data collected from the decomposition of C_4H_6, given in the table and graphed in the figure below:

A) Calculate the average reaction rate between:
   (a) the start of the reaction and 1000 seconds,
   (b) the start of the reaction and 3000 seconds
   (c) Draw the line for each rate on the graph. Which is the best fit to the actual rate of reaction?
   (d) Does the rate increase or decrease as the reaction proceeds

B) Estimate the instantaneous reaction rate at 1000 and at 3000 seconds.

C) Estimate the value of the rate constant for the reaction and indicate where or how you obtained the value it.

D) From which data, the table or plot, could you most easily determine the half-life of the reaction? Explain the reason(s) for your choice.

time(s)	$[C_4H_6]$
0	1.59
1000	0.92
2000	0.53
3000	0.31
4000	0.18
5000	0.10

9. The percentage of organic matter in the sample that was oxidized (i.e. converted to products) varied with time in the following manner:

Time (days)	1	5	10	20
% Oxidized	21	68	90	99

   A) Is the reaction following a first or second order rate law? Prove your answer.
   B) What is the value of k for the reaction and its units?

10. The decomposition of NOBr to NO and Br_2 is a second order reaction with k = 0.080 $M^{-1}s^{-1}$ If the initial concentration of NOBr was 0.600M in a 1.0 L flask,
   A) What will be the concentration of NOBr left after 5 minutes?
   B) How many moles of Br_2 would be produced by the reaction after 5 minutes?
   C) What is the thermal energy lost or gained by the reaction in (A) when the ΔH_f°= 82.2 kJ/mol for NOBr(g) and NO(g) has ΔH_f°= 90.4 kJ/mol. *(Review Chapter 6 as needed)*

11. The decomposition of cyclobutane, C_4H_8, at 750°C is first order. If 25% of a cyclobutane sample has decomposed in 80 seconds,
   A) What is the half-life, in minutes, of the reaction?
   B) What can you say about the molecularity of the rate-limiting step for this reaction?

12. Hydrogen peroxide, H_2O_2, undergoes the decomposition: $H_2O_2(aq) => 2 H_2O + O_2(g)$
   The rate of decomposition was measured for different initial concentrations at 40°C to obtain the following data:

$[H_2O_2]$	0.100	0.200	0.300
Initial rate (M/min)	1.93×10^{-4}	3.86×10^{-4}	5.79×10^{-4}

   A) What is the order of reaction for H_2O_2 in the rate law?
   B) Solutions of H_2O_2 are sold as 30% (by mass) H_2O_2. If the same decomposition occurs at 40°C, how long will it take for a 30% solution to become 10% by mass?
   C) The following is a possible 2-step mechanism for the decomposition in an aqueous solution to which a small amount of NaBr was added.

   Step I: $H_2O_2 + Br^- \rightarrow BrO^- + H_2O$ Step II: $H_2O_2 + BrO^- \rightarrow Br^- + H_2O + O_2$
      (a) Is the balanced overall reaction the same as the decomposition of H_2O_2?
      (b) What species is acting as a catalyst, if any, in this mechanism?
      (c) What species is acting as an intermediate, if any, in this mechanism?

13. The following data was collected for the reaction between NO(g) and O_2(g):

Experiment	[NO]	[O_2]	Initial rate (M/sec)
1	0.090	0.045	0.0711
2	0.045	0.045	0.0160
3	0.090	0.090	0.1280
4	0.045	0.022	?

   A) Determine the orders for each reactant,

   B) Write the rate law (equation), and

   C) Give the value of the rate constant for the reaction.

   D) What is the value of the rate in the fourth combination?

14. The decomposition: SO_2Cl_2(g) → Cl_2(g) + SO_2(g) is first order and has a rate constant of 2.83×10^{-3} min^{-1} at 600 K. If you start with a container that contains a partial pressure of SO_2 at 0.80 atm,

   A) what will be the percent SO_2 decomposed after 1.0 hour has elapsed?

   B) will the pressure due to the reaction gases be greater than, less than or the same as the initial pressure?

   C) what is the half-life of the reaction? Is this a relatively fast or slow reaction?

15. The reaction 2 NO_2(g) → 2 NO(g) + O_2 (g) has a measured rate equal to 0.498 M/s at 319°C and a rate of 1.81 M/s at 354 °C.

   A) What is the value of the activation energy for this reaction?

   B) What is the value of ΔH for the reaction? (*Consult Chapter 6 concepts as needed.*)

   C) Using the graph provided, sketch the energy profile of the reaction showing the relative position of reactants and products, the activated complex (or transition state) and show the size of ΔH and Ea - drawn to scale with respect to each other- for the reaction.

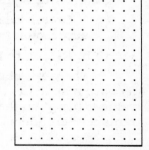

16. The decomposition of ethyl bromide is first order with an Ea = 226 kJ/mol. At 450°C, the half-life of the reaction is 78 seconds.

   A) What will the half-life be at 500°C?

   B) At what temperature would the half life be increased to 4.0 minutes?

17. The following data was collected for the hydrolysis reaction: *Ester* + H_2O --> products

Trial	[ester], M	[H_2O], M	T (°C)	k, M/s
1	0.100	0.200	15	0.051
2	0.100	0.200	25	0.101
3	0.100	0.200	35	0.184
4	0.100	0.200	45	0.332

   A) What is the activation energy for the reaction?

   B)   Estimate the value of A, the frequency factor from plot of ln k versus 1/T

   C) How many times faster will the reaction occur at 45°C than it does at 25°C?

   D) If this were an enzyme-catalyzed reaction, can you expect the rate constant to increase as the temperature is increased beyond 45°C. Explain your choice.

Chapter 14

Chemical Equilibrium

Many chemical reactions obviously proceed from reactants to stable products. Other reactions are reversible, with products readily converted back into reactants. In an equilibrium system, the rate of conversion of reactants into products is exactly equal to the rate of conversion of products into reactants.

14.1 Recognize a system at equilibrium and describe the properties of equilibrium systems.

14.2 Describe the dynamic nature of equilibrium and the changes in concentrations of reactants that occur as a system approaches equilibrium.

If an equilibrium mixture consists mostly of products, the system is described as product-favored; if mostly reactants, as reactant-favored. Important characteristics of chemical equilibrium are:

- *dynamic*: At equilibrium, chemical change is occurring at the nanoscale level, but since it occurs at the same rate in both directions, no net change can be observed. The symbol $\Leftrightarrow$ illustrates this.

- *independent of approach*: For a given reaction at a specific temperature, the equilibrium state is the same regardless of how that equilibrium was achieved.

- *catalysts*: The presence of a catalyst does not affect the position of the equilibrium; it only affects the time required to reach equilibrium.

14.3 Write equilibrium constant expressions, given balanced chemical equations.

14.4 Obtain equilibrium constant expressions for related reactions from the expression for one or more known reactions.

14.5 Calculate K_p from K_c, or K_c from K_p for the same equilibrium.

The value of the **equilibrium constant**, K_c, is the ratio of the rate constant for the forward reaction to the rate constant for the reverse reaction. The expression that defines this value is based on the balanced reaction as written. Written in terms of the molarities of reactants and products, the equlibrium constant expression is:

For the general reaction: $\quad aA + bB \mathrel{<=>} cC + dD \qquad K_c = \dfrac{[C]^c[D]^d}{[A]^a[B]^b}$.

However, these concentrations only refer to dilute solutions or to gases. The concentrations of solids and pure liquids are not included in equilibrium expressions. Note that a K_c expression refers to a reaction as written. The K_c expression for a reaction written as a combination of several other reactions includes the respective expressions for those reactions.

K_p expressions can be written for reactions using pressures of gases instead of concentrations. The form of the expression is the same, but $[A]^a$ is replaced by $P_A{}^a$, so that:

For the general reaction: $\quad aA + bB \mathrel{<=>} cC + dD \qquad K_p = \dfrac{P_C{}^c\, P_D{}^d}{P_A{}^a\, P_B{}^b}$

The values of K_p and K_c are not equal unless there are no gases in the equilibrium reaction, or the moles of gases are the same on both sides of the reaction arrow. The relationship between the two forms can be summarized by the equation:

$K_p = K_c (RT)^{\Delta n}$, where Δn = (no. moles gaseous products)- (no.moles gaseous reactants)

14.6 Calculate a value of K_c for an equilibrium system, given information about initial concentrations and equilibrium concentrations.

14.7 Make qualitative predictions about the extent of reaction based upon equilibrium constant values; that is, be able to predict whether a reaction is product-favored or reactant-favored based on the size of the equilibrium constant.

The value of K_c can be determined by measuring the concentrations of all species present at equilibrium. However, it is more often found by allowing a known amount of reactant to reach equilibrium, measuring the concentration of one species, calculating the concentration of the other species, and then calculating the value of K_c. The table used for calculating concentrations uses rows labeled "initial, change, and equilibrium." A large (>> 1) value of K_c indicates that the system is product-favored, while a small (<< 1) value indicates a reactant-favored system. A value close to 1 indicates that the system contains similar amounts of both products and reactants.

14.8 Calculate concentrations of reactants and products in an equilibrium system if K_c and initial concentrations are known.

14.9 Use the reaction quotient Q to predict in which direction a reaction will go to reach equilibrium.

One application of equilibrium constants is to determine the equilibrium concentration of one species if initial concentrations or equilibrium concentrations are given. A chart of "initial, change, and equilibrium" concentrations is usually used to organize the information needed.

Another application of equilibrium constants is in predicting the direction of a particular reaction mixture, given the concentrations at any particular time. A **reaction quotient**, Q, is determined using an expression mathematically identical to that of K_c. The value of Q is compared with K_c as follows:

- $Q = K_c$: The reaction as described is at equilibrium.

- $Q < K_c$: The reaction as described will continue in the forward direction until the amount of products has increased to reach equilibrium.

- $Q > K_c$: The reaction as described will proceed in the reverse direction until the amount of products has decreased to reach equilibrium.

14.10 Show by using Le Chatelier's principle how changes in concentration, pressure or volume, and temperature shift chemical equilibria.

Le Chatelier's Principle is used to describe changes that will occur if changes are imposed on a reaction system at equilibrium. According to this principle, if the conditions of a system at equilibrium are changed, the position of the equilibrium will shift in such a way that the change is counteracted. A shift in the forward direction means that the reaction goes toward the products or to the right. These three terms are used interchangeably, and the opposite terms apply in the other direction.

- When the concentration of a species is increased, the equilibrium shifts away from that species; likewise, if a concentration is decreased the position shifts toward that species. The value of K cannot change, however.

- When gases are present an increase in pressure or a decrease in volume favors the side of the reaction with fewer moles of gas. Note that adding an inert gas has no effect on the position of the equilibrium.

   Increasing pressure- system tries to contract- shift to side with least moles of gas

   Decreasing pressure - system tries to expand- shift to side with most moles of gas

- Another change that is often considered is temperature. Increasing the temperature of an endothermic reaction causes a shift in the forward direction. This can be thought of as adding heat, a reactant. The value of K will change when temperature is changed, but the sign of ΔH sets the direction of change, so that:

When + ΔH (endothermic):   T ↑, K ↑ and T↓, K ↓

When - ΔH (exothermic):   T ↑, K ↓ and T↓, K ↑

14.11 Use the change in enthalpy and the change in entropy qualitatively to predict whether products are favored over reactants.

Equilibrium principles can be explained by considering reactions at the nanoscale. In general, reaction systems are favored in the direction of lower enthalpy. Also, reactions with greater disorder or entropy are generally favored. At the nanoscale, then, a low enthalpy, highly disordered system is most favored. As temperature increases, the importance of disorder increases over the importance of enthalpy.

14.12 List the factors affecting chemical reactivity, and apply them to predicting optimal conditions for producing products.

Consider the following generalizations about equilibrium:

• K_c (or K_p) > 1 means a product-favored system.

•   Exothermic reactions are generally product-favored.

•   An increase in entropy is generally product-favored.

•   A reaction that is product-favored at low T is probably exothermic.

•   A reaction that is product-favored at high T probably has greater disorder in products than in reactants.

Consider the following generalizations about reaction rates:

•   Gas-phase reactions are faster than liquid or solid phase reactions.

•   Reactions are faster at high T than low T.

•   Reaction rates increase as reactant concentrations increase.

•   Reactions involving solids are faster when the solids are very small particles.

•   Reactions are faster when a catalyst is present.

Chemical processes can be designed based on consideration of the above principles. This sometimes requires a balance between two different trends. For example, in the production of ammonia from nitrogen and hydrogen, a higher temperature would seem to favor faster production by increasing the number of effective collisions between reactant molecules. However, the decomposition of ammonia is endothermic. Therefore an intermediate temperature is used. The principles of equilibrium and kinetics must both be considered in process design.

Chapter Review - Key Terms

In order to make all the sentences below TRUE, insert the appropriate word or phrase from the list of key terms which best fits the context of the sentence.

NOTE: Any phrase or word from the list may used more than once.

LIST:		
chemical equilibrium	equilibrium constant	product-favored
dynamic equilibrium	equilibrium constant expression	reactant-favored
entropy	Haber-Bosch process	reaction quotient
equilibrium concentration	Le Chatelier's Principle	shifting an equilibrium

Concerning Chemical Equilibrium:

A characteristic of all chemical reactions is the tendency to achieve a state known as (1) _____ where the concentrations of all reactants and products appear to remain constant over long periods of time. All chemical reactions have both a forward rate, which is the speed at which reactants convert to products and a reverse rate, where the opposite process occurs. The state of (2) _____ is not static, even though the concentrations appear constant, but rather is a (3) _____ where the rate of production of the products is balanced by an equal rate of disappearance of products. At the point of (4) _____, the concentration of reactants and products will be defined by a ratio called the (5) _____, with the symbol, K, which will have a finite, nonzero value. A (6) _____ reaction will mean that there are many more products than reactants at the point of chemical equilibrium and that the (7) _____ is greater than 1.0. A (8) _____ reaction will mean that there are many more reactants than products and the (9) _____ is less than 1.0.

The exact form of the (10) _____ as a the mathematical equation can be used to quantitatively define the (11) _____s of each recatant and product in the chemical reaction. The concentration (or pressure) of each product appears in the numerator of the (12) _____ while those for the reactants are in the denominator and each is raised to the power of its stoichiometric coefficient in the balanced reaction. Pure solids, pure liquids and a solvent (in a dilute solution) cannot appear as part of the (13) _____, since the value for moles per liter (the molar density) of these substances is constant even if volume changes of the system are made.

Shifting a Chemical Equilibrium:

The value of the (14) _____, Q, which has the same mathematical form of the (15) _____, but uses the actual concentrations instead of (16) _____s, is used to determine whether or not a certain combination of reactants and products will be at (17) _____. Only when the value of the (18) _____ is equal to the value of K, will the combination represent the state of (19) _____. Once (20) _____ has been achieved, changing the conditions such volume, pressure, temperature or the concentration of either reactant or product, which is termed (21) _____, may produce a (22) _____ that is not equal to K at that point. To compensate either the forward or reverse rate must dominate to reestablish (23) _____ and produce a (24) _____ equal to K. To summarize this behavior, (25) _____ says that when (26) _____ occurs, the system reacts to reestablish a new state of (27) _____ in a way that partially counteracts the change (or stress) applied. (28) _____ will apply to changes in conditions such as removing or adding of reactant or product, changing the volume (or pressure) of the system, or changing the temperature for a system at (29) _____. The value for the (30) _____ will stay the same in the first two types of changes, but changes in temperature changes the value of the (31) _____. If the reaction is endothermic, the value of the (32) _____ and temperature will change in the same direction, but if the reaction is exothermic, the changes in the (33) _____ and temperature will be in opposite directions.

The temperature changes in the (34) _____ result from the energy effect, but there is a second independent effect, called the (35) _____ effect, which also helps to set the value of the (36) _____. That disordered chemical systems, which have a larger number moles of products than reactants or a more diverse array of bonds in the products, tend to be (37) _____ is due to (38) _____. For any chemical system, both the energy and (39) _____ effects must be taken into account. At high temperature, the (40) _____ effect will generally be the more important in determining the actual value of the (41) _____.

A prime example of (42) _____ to maximize the yield of product is the (43) _____, in which $N_2(g)$ is reacted with $H_2(g)$ to produce $NH_3(g)$ which has had a dramatic impact on worldwide agriculture. The use of high pressures in the (44) _____, causes the chemical reaction to become more (45) _____, even though the (46) _____ has not changed. In addition, when the product, NH_3, is liquefied and removed from the reaction phase, this forces more product to be made by the gaseous system, by (47) _____. Catalysts are also used so that the temperature can be kept low, which maximizes the value of (48) _____ for the exothermic reaction and also improves the rate of reaction.

- PRACTICE TESTS -

After completing your study of the chapter and the homework problems, the following questions can be used to test yourself on how well you have achieved the chapter objectives.

1. Tell whether the following statements would be always **True or False**:

    A). Equilibrium is a state for chemical reactions that occurs only when the forward rate of reaction equals the rate for the reverse reaction.

    B) The rate constants for the forward and reverse rates must be equal for equilibrium to occur.

    C) When reactants are first combined, the system is not at equilibrium and the reaction quotient, Q, is greater than K.

    D) The equilibrium ratio of concentrations will be the same no matter what the starting concentrations of reactants and products are, if T is held constant.

    E) The values of K, the equilibrium constant, range between 0 and 1.0 because it is always a ratio, which must be a fraction.

    F) Pure solids, liquids and gases do not appear in the equilibrium constant expression because their molar concentration (molar density) is constant.

    G) Since catalysts and inhibitors change the rate of reaction, they also change the value of K for the equilibrium.

    H) An equilibrium constant of 10 indicates a product favored reaction and a value of 0.010 would be reactant-favored.

    I) According to Le Chatelier's Principle, heating a reaction with a negative ΔH value will favor product formation.

    J) For the reaction, $2\ SO_2(g) + O_2(g) <=> 2\ SO_3(g)$ decreasing the volume of the container will cause more products and K will get larger.

    K) K_p and K_c have the same values only when the number of moles of product gases are equal to moles of reactant gases.

    L) If you double all the coefficients in the balanced reaction, you need to double the value of K as well.

2. A) Why does an increase in temperature always favor the entropy effect, but not the energy effect on equilibrium?

    B) For the reaction $SO_2(g) + Cl_2(g) <=> SO_2Cl_2\ (g)$ which is endothermic.

    (a) Describe four changes that would shift the equilibrium to the LEFT (reactant side) and tell whether the value of K would increase, decrease or stay the same.

    (b) Based on the energy and entropy effect, what would you expect to be true about the value of K_p at 25°C? Explain the reason(s) for your choice.

3. The following two reactions happen in sequence in the atmosphere, where the first happens at very high temperatures inside the internal combustion engine, while the second occurs rapidly in the atmosphere at 15°C to produce the air pollutant, NO_2.

    (1) O_2 (g) + N_2 (g) <=> 2 NO (g) ΔH = + 180.8 kJ $K_{(1)}$ = 2.7 X 10^{-18}

    (2) 2 NO(g) + O_2 (g) <=> 2 NO_2(g) ΔH = -113 kJ $K_{(2)}$ = 6.0 X 10^{13}

    A) What is the overall reaction when the reactions are added together?

    B) Which is the correct way to calculate the value of K for the overall reaction?

        (a) $K_{(1)}$ + $K_{(2)}$ (b) $K_{(1)}$ /$K_{(2)}$ (c) $K_{(1)}$ × $K_{(2)}$ (d) $K_{(1)}$ - $K_{(2)}$

    C) Is the overall reaction exothermic or endothermic and what is the value of ΔH of reaction?

    D) Will the high pressure of the combustion chamber in an engine favor product formation in: (a) only the first reaction   (b) only the second reaction

                        (c) both reactions   (d) neither of the reactions

4. Suppose solid HgO, liquid Hg and O_2(g) were allowed to reach equilibrium through the following endothermic reaction: 2 HgO(s) <=> 2 Hg(l) + O_2(g). If the following changes were made to the equilibrium system, how would the amount of Hg(l) in the flask be affected- would it *increase, decrease or stay the same*- as compared to the amount in the flask BEFORE the change?

    A) Add some HgO(s), but not lose any O_2(g) in the process.

    B) Remove some O_2 gas from flask

    C) Increase the pressure on the equilibrium in the flask.

    D) Lower the temperature of the flask.

    E) Increase the size of the flask to become double the original volume.

5. Make a qualitative judgement of the effect on [CO] in an equilibrium system for the reactions (a) - (c) given below for the following changes. To indicate the direction of change, use: INC for increase, DEC for decrease, NC for no change and CT for can't tell when the change produces opposing effects in the concentration.

        (a) CO (g) + 2 H_2 (g) <=> CH_3OH(g) ΔH = - 90.7 kJ

        (b) C(s) + 2 H_2O (g) <=> CO(g) + H_2(g) ΔH = + 131 kJ

        (c) 2 C(s) + O_2 (g) <=> 2 CO(g) ΔH = - 221 kJ

    A) Increase P   B) Increase T   C) Decrease P and T

    D) Increase P and decrease T   E) Add a catalyst

6. For the reactions given in the table below, which of the reactions (a) -(e), if any, obey the following criteria:

    A) K_p is greater than K_c.

    B) There are fewer products than reactants at equilibrium?

    C) K_p is equal to K_c.

    D) There is only one molarity in the K_c expression.

    E) You could solve for "x" amount of change without using the quadratic equation only if the initial concentrations of reactants are the same.

Reaction	K_c at 25°C	ΔH(kJ)
(a) 2 CH_4(g) ↔ C_2H_6(g) + H_2(g)	9.5 X 10^{-13}	+ 64.9
(b) CH_3OH(g)+ H_2(g) ↔ CH_4(g) +H_2O(g)	3.6 X 10^{20}	- 115.4
(c) H_2(g) + Br_2(g) ↔ 2 HBr(g)	2.0 X 10^9	- 72.5
(d) Mg(OH)$_2$(s) ↔ MgO(s) + H_2O(g)	1.24 X 10^{-5}	+ 81.1
(e) 2H_2(g) + CO(g) ↔ CH_3OH(g)	3.76	- 90.7

    F) You could solve for "x" amount of change without using the quadratic equation for all possible initial concentrations of reactants.

    G) An increase in T(°C) will mean an increase the amount of products.

    H) A decrease in volume of the system will increase the amount of products.

7. The following reaction has $K_c = 3.59$ at 900°C: $CH_4(g) + 2\,H_2S(g) <=> CS_2(g) + 4\,H_2(g)$

    A)   For combinations (a) - (b) of reactant and products given ,

        (1) tell how Q is related to K_c .

        (2) If Q IS NOT EQUAL to K_c, tell in what direction the shift will occur, toward reactants or products?

Mixture	$[CH_4]$	$[H_2S]$	$[CS_2]$	$[H_2]$
(a)	1.00	2.00	1.00	4.00
(b)	1.00	1.00	1.00	1.00
(c)	1.24	1.30	1.21	2.63
(d)	1.24	1.52	1.15	1.73

    B)   Suppose one started by mixing 1.0 M $CH_4(g)$ with 2.0 M $H_2S(g)$ for the reaction. Indicate, qualitatively, what happens to the concentrations of each reactant and product over time in the 2 regions below by drawing a diagram, similar to that shown in Figure 14.6 in the text. Use curves to represent the concentration profile for each reactant and product of the system and label each curve with the substance you are representing by the profile.

    Region (1): Show the changes that occur from the start of the reaction as the reaction proceeds and then achieves equilibrium

    Region (2): Show the changes that occur if once equilibrium is established, some amount of H_2 is added to the system and a new equilibrium is reestablished.

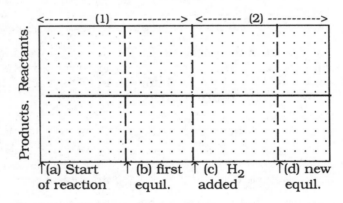

    C)   How does Q compare to K at the points (a) - (d) given by the arrows on the diagram?

8. The reaction $2\,A(g) + B(g) <=> 2\,C(g)$ has a $K_c = 20.9$ at 227°C and $K_c = 5.0 \times 10^6$ at 430°C.

    A) Is the combination of 0.015 mol A, 0.012 mol B and 2.60 mol C in a 2.0 L container an equilibrium mixture at either temperature? Prove your answer.

    B) Is the reaction endothermic or exothermic? Are the energy and entropy effects exerting opposite or similar forces on the equilibrium as the temperature is changing? Explain the reason(s) for your choices.

9. For the chemical reaction : $NO_2(g) + NO(g) <=> N_2O(g) + O_2(g)$ the Kp = 0.914 at 25°C. Suppose 2.00 moles each of $NO_2(g)$ and $NO(g)$ are placed in a 5.0 L container at 25°C and allowed to reach equilibrium.

    A) Write out the form of the Kp expression

    B) Prepare a table showing the initial pressures, change in pressures and the equilibrium pressures that result, letting x = the change in NO(g).

    C) What would be the equilibrium pressure of NO(g) in the container at equilibrium?

10. At 700 K, the following equilibria occurs: $CCl_4(g) <=> C(s) + 2\ Cl_2(g)$ When the initial pressure of CCl_4 is 1.0 atm, the total pressure at equilibrium is 1.35 atm at 700 K.

    A) Prepare a table showing the initial pressures, change in pressures and equilibrium pressures. Letting x = the change in CCl_4 pressure.

    B) Write out the Kp expression with the equilibrium pressures from the table in (A).

    C) What is the value of Kp for the reaction?

    D) Which of the following nanoscale pictures best represents the equilibrium mixture using ● for CCl_4, ◐ for C(s) and ○ for Cl_2 particles?

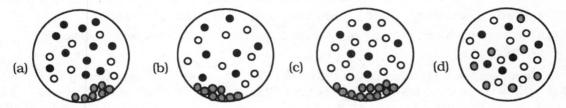

(a)   (b)   (c)   (d)

11. Initially 2.50 moles of NOCl was placed in a 1.50 L chamber at 400°C and after equilibrium was established, it was found that 28% of the NOCl had dissociated, with the reaction: $2\ NOCl(g) <=> 2\ NO(g) + Cl_2(g)$

    A) What is the value of Kc for the equilibrium at 400°C?

    B) What is the value of Kp for the NOCl equilibrium at 400°C?

    C) If in an equilibrium mixture at 400°C, the P_{NO} = 0.35 atm and that for P_{Cl2} = 76 torr (mmHg), what's the equilibrium partial pressure of NOCl in the same mixture?

12. Given the reaction: $Fe_2O_3(s) + 3\ H_2(g) <=> Fe(s) + 3\ H_2O(g)$ where Kp = 8.11 at 1000 K. If 100.0 g of Fe_2O_3 solid and 1.00 g of $H_2(g)$ were placed in 2.0L vessel and heated to 1000 K:

    A) What would be the partial pressure of $H_2O(g)$ in the flask at equilibrium?

    B) What would be the mass of $Fe_2O_3(s)$ that has reacted?

    C) Considering the amount of reactants and products at 1000K, does this reaction follow the prediction that reactions with K > 1.0 are always product-favored systems? Explain your reasoning.

13. The dissociation of N_2O_4, $N_2O_4(g) <=> 2\ NO_2(g)$ has a $K_c = 4.63\ X\ 10^{-3}$ at 25°C. Starting with 13.5 g of N_2O_4 in a 5.0L vessel, what would be the concentrations of N_2O_4 and NO_2 in the vessel when equilibrium established at 25°C?

14. Cyanogen, C_2N_2 , is a very poisonous gas made from the decomposition of hydrogen cyanide, HCN by the reaction is $2\ HCN(g) <=> H_2(g) + C_2N_2(g)$ with Kp = $4.0\ X10^{-4}$ at 500°C. If some amount of HCN was placed in a 2.0L flask at 500°C and there were 1.5 X 10^{-3} moles of H_2 at equilibrium, what was the initial concentration of HCN in the container?

15. The solubility of $Ba(OH)_2$ in water is 0.837 grams per 100 mL of solution.

    A) Write the equilibrium reaction for the saturated solution.

    B) What is the molarity of each ion in the equilibrium solution?

    C) What is the value of the K_{sp} for the barium hydroxide solution?

    D) Another hydroxide, KOH, has a solubility of 107 g per 100 mL of solution. How much greater is the K_{sp} of KOH than that of $Ba(OH)_2$?

Chapter 15

The Chemistry of Solutes and Solutions

This chapter covers solutions, which are homogeneous mixtures of two or more substances. The discussion addresses the nature of and interactions of the solvent, the substance present in the larger amount, and solute(s), the substance(s) present in the smaller amount. It also includes quantitative descriptions of concentrations of solutions.

15.1 Describe how liquids, solids, and gases dissolve in a solvent.

15.2 Predict solubility based on properties of solute and solvent.

15.3 Interpret the dissolving of solutes in terms of enthalpy and entropy changes.

When a solution is formed, the solute-solute and solvent-solvent interactions are disrupted, and new associations are formed between solvent and solute. When the molecular interactions in the pure substances are similar in type and magnitude to the new interactions between solute and solvent, a solution is very likely to form. When two liquids do not dissolve in each other, they are said to be **immiscible**; those that mix in any proportions are **miscible**. Molecules that are hydrophobic, water-fearing, will not dissolve substances with molecules that are hydrophilic, water-loving. This is summarized in the rule "like dissolves like." Recall that breaking molecular associations costs energy and forming new associations releases energy. When the sum of these processes, the **enthalpy of solution** ΔH_{soln}, is negative, dissolving is favored. Entropy is another factor in the dissolving process. Since a mixture is generally more disordered than pure substances, ΔS usually favors formation of solutions. In some cases the importance of entropy overcomes the importance of enthalpy and a solution is formed even when ΔH is unfavorable.

15.4 Differentiate among unsaturated, saturated, and supersaturated solutions.

15.5 Describe how ionic compounds dissolve in water.

The **solubility** of a substance is the maximum amount that will dissolve in a given quantity of solvent at a given temperature. A solution containing this amount of solute is said to be a **saturated solution**; usually excess solid is visible and an equilibrium exists between the solid and hydrated solute. Thus a saturated solution can be described as one in which $Q = K_c$. An **unsaturated solution** is one in which more solute can be dissolved, or $Q < K_c$. In a **supersaturated solution**, an excess amount of solute has somehow been made to dissolve, so that $Q > K_c$. Supersaturated solutions are unstable. Energy considerations in the process of dissolving ionic compounds include the lattice energy, which must be overcome to break up the network of ions in the solid, and the enthalpy of hydration, which is the energy released when ions are hydrated by water molecules. The value of ΔH_{soln} depends on the sum of these values. Solubility is determined by the value of ΔH_{soln}, along with the value of ΔS, which usually favors dissolving of ions with charges of 1, but may not favor dissolving ions with charges of 2 or more.

15.6 Describe how temperature affects the solubility of ionic compounds.

15.7 Predict the effects of temperature and pressure on the solubility of gases in liquids.

Another factor influencing solubility is temperature. The solubility of gases decreases with temperature. The solubility of solids usually increases with temperature, but there are numerous exceptions. Henry's law is a quantitative description of the the effect on solubility of gases in liquids as pressure of the gas above the solution either increases or decreases.

- **Henry's law:** $S_g = k_H P_g$ where P_g and S_g refer to pressure of the gas over the solution and the solubility of the gas in the solution. The Henry's Law constant k_H has units of mol/L-mmHg and changes value with the solvent.

15.8 Describe the concentration of solutions in terms of weight percent, mass fraction, parts per million, parts per billion, parts per trillion, and molarity.

Terms such as saturated and unsaturated or concentrated and dilute are not always precise enough to communicate the concentration of a solution. Mass fraction, weight %, ppm, ppb, and ppt are quantitative descriptions of concentration based on the ratio of the mass of solute to the mass of solution. **Mass fraction** is a simple ratio, and the other units include factors of 100, 1000, 10^6, and 10^9, respectively. These units are independent of temperature, but if weighing solutions is awkward, the density must be known. Molarity, M, describes concentration in terms of moles solute per liter of solution. The volume of molar solutions is easy to determine, but the value of molarity is dependent upon temperature.

15.9 Interpret vapor pressure lowering in terms of Raoult's Law.

15.10 Use molality to calculate the colligative properties: freezing point lowering, boiling point elevation, and osmotic pressure.

15.11 Differentiate the colligative properties of nonelectrolytes and electrolytes.

15.12 Explain the phenomena of osmosis and reverse osmosis and calculate osmotic pressure.

Colligative properties of solutions are those properties which are determined by the number of solute particles, not their identity or other properties. **Raoult's law** describes the lowering of the vapor pressure of a solution compared to the pure solvent; the degree of change depends on the mole fraction of solvent in the solution. **Boiling point elevation** and **freezing point lowering** are calculated based on the molality of the solution and the appropriate constant for the solvent. **Molality**, m, is a concentration unit using moles of solute per kilogram of solution.

Calculations such as these must account for the number of particles formed when ionic compounds, which are electrolytes, dissolve. Thus the change in vapor pressure, boiling point, or freezing point for 1.0 M NaCl would be greater than for 1.0 M sucrose, because the NaCl provides two moles of solute particles per liter of solution.

In the process of **osmosis**, solvent molecules selectively flow through a **semipermeable membrane** from an area of low solute concentration to an area of high solute concentration. **Osmotic pressure** is the pressure required to resist this flow, and is another colligative property. **Reverse osmosis** uses a pressure greater than the osmotic pressure to force the flow of solvent molecules against the natural direction of flow. This method can be used to purify water.

Formulas for calculations of these colligative properties are:

- **Raoult's law:** $P_\ell = X_\ell P^o_\ell$ where P_ℓ and X_ℓ refer to vapor pressure and mole fraction of solvent in the solution and P^o_ℓ refers to the vapor pressure of the pure solvent.

- **b.p. elevation:** $\Delta T_b = i\,(K_b\,m_{solute})$, where m refers to molality and i is the number of particles per formula unit of solute.

- **f. p. lowering:** $\Delta T_f = i\,(K_f\,m_{solute})$, where m refers to molality and i is the number of particles per formula unit of solute.

- **osmotic pressure:** $\Pi = i\,(cRT)$, where Π is osmotic pressure, c is concentration in molarity, R and T are the same as in gas law calculations, and i is the number of particles per formula unit of solute.

15.13 Describe the various kinds of colloids and their properties.

15.14 Explain how surfactants work.

Colloids are not true solutions, but are mixtures composed of a **dispersed phase**, analogous to the solute, and a **continuous phase**, analogous to the solvent. The particles in colloids are much larger than in true solutions, and scatter a beam of light shined through the mixture. Except for particle size, colloids are more like solutions than suspensions. **Surfactants** are large molecules which have both hydrophobic and hydrophilic portions. Because of this property, surfactants enhance the solubility of hydrophobic substances in water. This occurs when the non-polar portion of the surfactant interacts with a non-polar material, breaking it up into smaller particles called **micelles**. Since the hydrophilic portion of the surfactant interacts with water, these micelles become dispersed in the water. Soaps, detergents, and emulsifiers are examples of surfactants.

15.15 Discuss the earth's water supply and the sources of fresh water.

15.16 Discuss how municipal drinking water is purified.

15.17 Describe what causes hard water and how it can be softened.

15.18 Explain how household wastes can contaminate groundwater.

Water on the earth's surface is found primarily in the oceans, but also as ice and snow, surface water, or ground water. For this water to be potable, or fit for human consumption, it must be purified. The general steps involved in purification of water are

- coarse filtration
- settling after addition of chemicals which allow coagulation of suspended particles
- fine filtration, usually with sand
- aeration
- disinfection, usually with chlorine or ozone
- storage and delivery to the consumer

Some water still contains impurities of Ca^{2+}, Mg^{2+}, Fe^{2+}, or Mn^{2+} after purification. These ions cause water to be "hard," which is undesirable because unwanted precipitates can form or unpleasant tastes may be present. Hard water can be "softened" by chemical reactions or by the use of ion exchange columns, in which the unwanted cations are replaced by sodium ions. Since much of our drinking water comes from groundwater, waste disposal practices which allow hazardous materials to leach into the ground, either by direct dumping or by indirect flow from landfills, are extremely dangerous.

Chapter Review - Key Terms

In order to make all the sentences below TRUE, insert the appropriate word or phrase from the list of key terms which best fits the context of the sentence.

NOTE: Any phrase or word from the list may used more than once.

LIST:		
boiling point elevation	hypertonic	parts per million
colligative properties	hypotonic	Raoult's law
colloid	immiscible	reverse osmosis
continuous phase	isotonic	saturated solution
detergents	lattice energy	semipermeable membrane
dispersed phase	mass fraction	solubility
emulsion	miscible	supersaturated solution
enthalpy of solution	molality	surfactants
freezing point lowering	osmosis	Tyndall effect
Henry's Law	osmotic pressure	unsaturated solution
hydration	parts per billion	weight percent

Characteristics of the Solution Process and Solubility:

A solution is a homogeneous mixture of solvent and one or more solutes that can only be formed under certain circumstances. To form a solution, the solute and solvent must have similar noncovalent forces to be (1) _____ with each other. If the solute-solvent forces are not as strong as either the solute-solute or solvent-solvent interactions, the solute will normally be (2) _____ with the solvent and will not dissolve to form a solution. When the solution is formed, energy must first be absorbed to create separate solute and solvent particles, but then energy is released when the solute-solvent interact to form the solution, resulting in a net energy change called the (3) _____, which can be positive or negative. When an ionic solute dissolves in water, the energy absorbed is equal to the (4) _____ of the solid and the energy released is the energy of (5) _____ of the ions, which when added together produce the (6) _____.

An equilibrium state is achieved between the dissolving solid into the solvent and the solid precipitating from the solvent, when a (7) _____ is formed. The maximum amount of solute that can dissolve in a given amount of solvent, to reach this state, is called the (8) _____ of the solute. If the concentration is below the (9) _____, then Q<K and the solution is called an (10) _____. If the reverse is true, that the concentration of dissolved solute is greater than the (11) _____ at a certain temperature, then Q>K and a (12) _____ has formed. The (13) _____, as with all equilibria, is temperature dependent and increases or decreases depending on the sign of the (14) _____. If the (15) _____ is endothermic, increasing temperature will increase the (16) _____, while an exothermic (17) _____ will cause the (18) _____ of the compound to decrease with increasing temperature.

When gases are the solutes in a liquid, increasing the temperature always decreases the (19) _____ of the gas. The (20) _____ is also directly proportional to the partial pressure of the gas over the solution. The mathematical equation that relates the partial pressure to the (21) _____ of the gas, expressed in molarity units, is called (22) _____.

Concentration Units Based on Mass Ratios:

Comparing the mass of the solute dissolved to the total mass of solution defines the (23) _____ of solute in the solution. A convenient way to describe the solute concentration in percentage units is to multiply the (24) _____ by 100 to define the (25) _____ of solute in the solution. A disadvantage of the (26) _____ is that the density of the solution must be known to convert this concentration to molarity, which is more convenient for chemical calculations. In cases of very dilute solutions, the concentration of solute can be expressed as (27) _____, the number of grams of solute in a million grams of solution or (28) _____, the number of grams of solute in a billion grams of solution which will result in simple whole number values for the concentration.

Concerning the Colligative Properties:

Physical properties of solution that depend only on the number of solute particles, not the chemical nature of the solute, are called (29) _____. One such property is the vapor pressure lowering that occurs when a nonvolatile solute is added to a solvent. A law called (30) _____ predicts the vapor pressure of the solution (P_1) will be equal to the mole fraction of the solvent in the solution, χ_1, times the vapor pressure of the pure solvent ($P_1°$). A related property that results when a nonvolatile solute is added to a solvent, is that the boiling point of the solution will always be higher than the normal boiling point of the solvent resulting in a (31) _____. Also the temperature at which the solvent molecules can solidify, or freeze out of the solution, will be less than the freezing point of the pure solvent, resulting in a (32) _____ for the solvent molecules. The magnitude of either change in temperature in degrees Celsius, ΔT, is equal to the concentration of solute, expressed as a (33) _____, the number of moles of solute in one kilogram of solvent, times a constant, which is specific to the solvent. If the nonvolatile solute is an ionic compound, however, the moles of solute particles added to the solution, must be multiplied by the van't Hoff factor, *i*, in all the equations to calculate the (34) _____, since the dissociation increases the effective (35) _____ of the solute particles in the solution.

The one-way transfer of solvent molecules across a (36) _____ from pure solvent into a solution of solute and the same solvent is called (37) _____. The applied pressure needed to stop the flow of solvent into the solution is called the (38) _____ for the solution, which has the symbol Π, units of atm or mmHg and always increases as the number of solute particles increases and is also one of the (39) _____. The value of (40) _____, in atmospheres, is equal to the molarity of the solute multiplied by RT, where R is the gas constant and T the temperature in degrees Kelvin. If an external pressure is applied to the solution side which is greater than the (41) _____ of the solution, then the solvent can be made to flow from the solution to the pure solvent side in a process called (42) _____. The (43) _____ process is used to produce purified water from natural waters that contain many dissolved solutes.

A biological membrane acts as a (44) _____ in a cell to control the flow of solutes and water into and out of the cell. If the cell is placed in a solution that has the same (45) _____ as the inside the cell, the solution is (46) _____ and no net flow of water across the membrane occurs. If the solution outside the cell has a higher solute concentration than inside the cell, called a (47) _____ solution, the cell will shrink in size as water flows from inside to outside the cell. The opposite type of flow occurs when the outside solution has a lower solute concentration, called a (48) _____ solution, which causes the cell to expand in size. Pure water in contact with a cell acts like a (49) _____ solution and can cause the cell walls to rupture due to the expansion. In each of these two situations the water is always flowing to the side with the higher (50) _____ of solute.

Concerning Colloidal Solutions:

Some solutions that appear homogeneous are actually heterogeneous when viewed at the macroscopic level. This type of solution is called a (51) _____ and can be formed in the solid, liquid or gas states. The solute-like component, consisting of relatively large particles, is called the (52) _____ and has a uniform distribution in the solvent-like component, called the (53) _____. Visible light passing through a transparent (54) _____ will be scattered and produce a characteristic effect known as the (55) _____. When both the (56) _____ and (57) _____ are liquids, a third type of molecule is required as a component in the solution, that has both polar (hydrophilic) and nonpolar (hydrophobic) regions within the molecule to form a specific type of (58) _____ called an (59) _____. One class of molecules that meet the structural requirements to form (60) _____ s, are called (61) _____, which are surface-active agents. One group within the (62) _____ are the (63) _____, soap-like molecules made from petroleum or coal sources, that are essential to form an (64) _____ between the nonpolar molecules such as grease, oil or fats and water in the cleaning process.

- **PRACTICE TESTS** -

After completing your study of the chapter and the homework problems, the following questions can be used to test yourself on how well you have achieved the chapter objectives.

1.   Tell whether the following statements would be always **True or False**:

A)   A true solution is a homogeneous mixture of at least two components.

B)   Solubility is temperature dependent and will always increase with temperature.

C)   Supersaturated solutions are stable, but unsaturated solutions are not.

D)   The hydrocarbon, toluene, would be a good solvent for magnesium chloride.

E)   Hydration is an important part of the dissolving process for ionic materials in water for both the energy and entropy changes.

F)   According to Henry's law, if the pressure of a gas over a solvent is increased the solubility of the gas in the solvent will be decreased.

G)   When radon gas is dissolved in water, heating the water will decrease the solubility and cause the radon gas to leave the water.

H)   For any aqueous solution, the molality, m, will either be equal to or less than the molarity of that solution.

I)   A 15% solution of ethanol in water would have the same freezing point as a 15% solution of propanol in water.

J)   A 100 ppm solution would have the same concentration as a 0.001% by mass solution of the same solute.

K)   Since the vapor pressure lowering is a colligative property, the mole fraction of the solute is multiplied by the vapor pressure of the pure solvent in Raoult's Law to give the new vapor pressure of the solvent over the solution.

L)   The molal freezing point constant, K_f, solvent ethanol is probably different than the K_f value for the solvent cyclohexane.

M)   The molar mass of large biological molecules is often determined by measuring the osmotic pressure.

N)   If you are being treated by injection of intravenous fluids, it is important that the solution be isotonic with respect to your body fluids.

O)   When two aqueous solutions are placed on either side of a semipermeable membrane that lets water flow through the membrane, the water always moves to the side with the lower osmotic pressure.

P)   If the enthalpy of solution for a particular solute is negative the solution will get colder as the solute dissolves in the solvent.

Q)   Colloidal particles are larger than particles in a suspension.

R)   Normal household waste materials can be hazardous wastes that can pollute groundwater when placed in landfills.

S)   Detergents and soaps have the same chemical structures and both function as surfactants.

2.   A)   How could you convert a saturated solution to one that is supersaturated without adding any more solute?

B)   When solid $CaCl_2$ is added to liquid water, the temperature rises, but when solid $CaCl_2$ is added to ice at 0°C, the ice melts and the temperature of the resulting solution decreases to less than 0°C. Explain why the two temperature changes are not inconsistent with each other.

C)   What is the function of an emulsifier and what structural characteristics must it have?

D)   A multivitamin tablet contains the recommended daily dose of each vitamin. Suppose a person decides to take three tablets instead of one per day, thinking this will be more beneficial to their health. Based on the discussion in this chapter, explain why this choice may not be beneficial to the person's health.

E) How is the process of reverse osmosis able to produce pure water? Can you also produce pure water by putting the water through a water softener?

3. Of the concentration units, *mass percent, molarity, or molality,* which:
   A) refers the solute to the certain volume of solution?
   B) divides moles of solute by a specific amount of solvent?
   C) requires that the density of solution to be known in order to convert this unit to the other concentration units?
   D) would be temperature-independent?
   E) is not used in the colligative property equations?

4. In order to prepare a 0.91M $Na_2S_2O_3$ solution, which has a density of 1.12 g/mL, from the solute $Na_2S_2O_3 \cdot 5H_2O$, you should add:
   A) 144 g of the solute to 1000 mL water   B) 120 g solute to 1000 mL water
   C) 226 g of the solute to 1000 mL water   D) 226 g solute to 894 mL water

5. Consider two solutions, one of fructose ($C_6H_{12}O_6$) in water and the other of Iron (II) nitrate, $Fe(NO_3)_2$ in water, both of which freeze at -1.5°C. Would the following statements be TRUE or FALSE. Both solutes have the same molar mass.

   A) The two solutions have the same osmotic pressure

   B) The two solutions have the same mass of solute per kilogram of water.

   C) The two solutions will have the same boiling point.

   D) The vapor pressure of water over the solution will be the same for the two solutions.

6, Acetonitrile, CH_3CN, can act as a solvent and dissolve solutes such as LiBr(s). The density of a 2.30M LiBr solution in CH_3CN is 0.932 g/mL.

   A) What is the molality of LiBr in the solution?

   B) What is the mass percent of LiBr in the solution?

   C) Explain what interactions allow LiBr to dissolve in CH_3CN, using the noncovalent attractions expected to be most important in LiBr(s), pure CH_3CN and the solution.

   D) Draw a nanoscale diagram showing a likely arrangement of the ions and CH_3CN the solution.

7. Cadmium, Cd, is a metal that is more toxic than lead. The allowed, safe level for Cd in drinking water is 5 ppb. Persons who smoke inhale Cd metal directly from the burning tobacco and it is estimated that a smoker will inhale as much as 3 micrograms Cd for each pack of cigarettes smoked.

   A) Calculate how many micrograms of Cd are ingested if a person drinks 2.0 L of water per day that contains the safe level?

   B) How many packs of cigarettes would a person have to smoke to inhale the inhale the same mass of Cd as in (A)?

8. A solution of propylene glycol, $CH_2OHCH_2CH_2OH$, in water is commonly used as an antifreeze for car engines.

   A) What percent mass of propylene glycol in water will be a solution that does not freeze until the temperature reaches -40°C?

   B) What structural characteristic allows the solubility of propylene glycol in water to be high, as in this solution?

9. Arrange the following solutions in order of smallest to largest freezing point depression:
   A) 0.20m $NaNO_3$ B) 0.30m CH_3OH C) 0.15m $HgCl_2$ D) 0.10m $Al_2(SO_4)_3$

10. If a bottle of ginger ale contains 33 g of sugar, $C_{12}H_{22}O_{11}$, in every 355 mL (12 ounces) of and the ginger ale also has a density of 1.05 g/mL. What is the:
   A)   molarity of the sugar in the soft drink?

   B)   osmotic pressure (in atm) of the ginger ale at 20°C, due to the sugar?

11. The solubility of $CO_2(g)$ in water is 0.161g CO_2 in 100 mL of water when the pressure of CO_2 over the solution equals 1.00 atm.
   A)   What is the Henry's Law constant for CO_2 in water?

   B)   If a ginger ale soft drink in Problem 10 is carbonated with a CO_2 partial pressure of 5.50 atm, what volume of CO_2 gas, measured at STP, has been dissolved in the ginger ale, assuming the k_H is the same as that in (A)?

12. Cyclohexane freezes at 6.5°C. A solution that was prepared by mixing 18.4g of methylbenzene ($C_6H_5CH_3$) with 500.0 g of cyclohexane (C_6H_{12}) freezes at - 2.04°C.
   A)   What is the molal freezing point constant of cyclohexane?

   B)   What would be the mass percent in a solution of benzene, C_6H_6, dissolved in cyclohexane that would have the same freezing point?

13. A solution is 18.0% $CaCl_2$ in water and has a density of 1.16 g/mL.
   A)   What is the molality of $CaCl_2$ the solution?

   B)   What is the temperature at which the solution freezes?

   C)   If the solubility of $CaCl_2$ in water at 0°C is 59.5g in 100 mL of water and the 18% solution was cooled to 0°C, would either the solvent or solute and leave the solution as a solid? Explain your choice.

14. A 1.90 g sample of conferin, a glycoside from trees, was dissolved in 48.68 g of water and the boiling point of the resulting solution was 100.06°C.

   A)   What is the molar mass of the conferin?

   B)   If the osmotic pressure of the same solution had been measured, what value of Π would have been measured at 20°C? (*Assume density solution = 1.00 g/mL*)

   C)   Tell which quantity, the boiling point elevation or Π, you think would be easier to measure accurately in the laboratory and why.

15. Consider the two solutions separated by a semipermeable membrane that will let only water flow through the membrane.

   Solution A: 20.0% $C_{12}H_{22}O_{11}$, density = 1.083 g/mL

   Solution B: 5.0 % NaCl, density = 1.036 g/mL

   A)   What is the osmotic pressure in each solution at 20°C?

   B)   In which direction will the net flow of water occur- from B to A or A to B?

Chapter 16

Acids and Bases

Acid-base chemistry in its simplest form is based on the Arrhenius definition of acids as substances that produce H_3O^+, the hydronium ion, in aqueous solution, and bases as substances that produce OH^-, the hydroxide ion, in aqueous solution. This chapter expands this definition, considers the quantitative aspects of acid-base chemistry in equilibrium terms, and describes applications of acid-base chemistry.

16.1 Describe water's role in aqueous chemistry.

16.1 Identify the conjugate base of an acid, the conjugate acid of a base, and the relationship between conjugate acid and base strengths.

16.2 Recognize how amines act as bases and how carboxylic acids ionize in aqueous solution.

According to the Bronsted-Lowry definition of acids and bases, an **Bronsted-Lowry acid** is a hydrogen ion donor and a **Bronsted-Lowry base** is a hydrogen ion acceptor. Note that this definition encompasses the Arrhenius definition. in order to accept hydrogen ions, bases must have an unshared pair of electrons. Water is amphiprotic because it can donate or accept hydrogen ions. In writing an acid-base reaction, the acid becomes a conjugate base, written with one less H^+ than on the reactant side. The base becomes a conjugate acid, written with an additional H^+. Species in a **conjugate acid-base pair** differ only by one H^+, with the acid having the additional H^+. This concept allows consideration of relative strengths of acids and bases.

A strong acid readily donates its H^+, and its conjugate base is weak. A weak acid does not readily donate H^+, and its conjugate base is strong. A strong base readily accepts hydrogen ions, while a weak base does not readily accept them. A list of acids and conjugate bases arranged in order of decreasing acid strength is useful to predict whether a particular acid-base combination will react. Organic compounds also show acid-base chemistry: the -COOH functional group readily loses H^+ to form -COO^- and amines readily accept H^+ using the unshared electron pair on the nitrogen atom.

16.3 Use the autoionization of water and show how this equilibrium takes place in aqueous solutions of acids and bases.

16,4 Classify a solution as acidic, neutral, or basic based on its concentration of H_3O^+ or OH^-.

16.4 Calculate pH (or pOH) given [H_3O^+], or [OH^-] given pH (or pOH).

Since water is amphiprotic, H^+ can be transferred from one molecule to another in an **autoionization** reaction, written as:

$$2\ H_2O\ (l)\ <=>\ H_3O^+(aq)\ +\ OH^-\ (aq)\quad \text{with } K_w = [H_3O^+][OH^-] = 1 \times 10^{-14} \text{ at } 25°C$$

where K_w is the **ionization constant for water**. This allows aqueous solutions to be classified as
- **acidic solution:** $[H_3O^+] > [OH^-]$ and $[H_3O^+] > 10^{-7}$ M
- **neutral solution:** $[H_3O^+] = [OH^-] = 10^{-7}$ M
- **basic solution:** $[H_3O^+] < [OH^-]$ and $[H_3O^+] < 10^{-7}$ M

The **pH** scale was invented to simplify calculations using hydrogen ion concentrations. $pH = -\log[H_3O^+]$. An acidic solution has a pH < 7.00; a neutral solution has a pH = 7.00, and a basic solution has a pH > 7.00. The pH is measured using an electronic pH meter or using colored dyes called indicators.

16.5 Estimate acid and base strengths from K_a and K_b values.

16.5 Write the ionization steps of polyprotic acids.

The ionization of any acid in water can be written as an equilibrium reaction, with an equilibrium constant, called the **acid ionization constant**, K_a:

$$HA \text{ (aq)} + H_2O \text{ (l)} \iff H_3O^+\text{(aq)} + A^- \text{ (aq)} \qquad \text{with } K_a = \frac{[H_3O^+][A^-]}{[HA]}$$

As the value of K_a or K_b increases, the strength of the acid or base increases. Although this reaction is usually not written for strong acids because $K_a \gg 1$, it is very useful for describing the reactivity and degree of ionization for weak acids. Values of K_a can be obtained from tables. An analogous reaction and the **base ionization constant**, K_b, expression can be written for bases.

$$B \text{ (aq)} + H_2O \text{ (l)} \iff BH^+\text{(aq)} + OH^- \text{ (aq)} \qquad \text{with } K_b = \frac{[BH^+][OH^-]}{[B]}$$

For **polyprotic acids**, a series of these reactions can be written for each ionizable H^+. Note that the successive values of K_a for a polyprotic acid become smaller.

16.6 Calculate pH from K_a or K_b values and solution concentration.

Calculations based on systems containing weak acids or bases use the same principles as previous equilibrium calculations. K_a or K_b values can be calculated from concentration and pH data, and pH or other concentrations can be calculated using values of K_a or K_b and some concentration data. Note that $K_a \times K_b = K_w = 10^{-14}$. When working with bases, remember to convert pH to pOH to determine $[OH^-]$.

16.7 Describe the relationships between acid strength and molecular structure.

16.7 Explain the nature of zwitterions.

Several factors affect the acidity of a particular species. In general, acid strength is **increased** by:
- increased polarity of the bond between H and the next atom;
- decreased bond energy between H and the next atom;
- in oxoacids, increased electronegativity of the central atom;
- in carboxylic acids, increased polarity of substituent groups as the R group in R-COOH.

Amino acids can do acid-base chemistry internally by transferring H^+ from the carboxylic acid group to the nearby amine group, forming a **zwitterion**. Depending on the chemical environment, amino acids may donate or accept additional H^+.

16.8 Describe the hydrolysis of salts in aqueous solution.

The acid-base properties of solutions of salts can be predicted by considering the hydrolysis reactions of the anion and cation formed by the acid and base that produced the salt. The cation of the salt formed is the conjugate of a base and can act as a weak acid, and the anion (a conjugate of an acid) can act as a weak base.

A **hydrolysis** reaction is one in which a weak acid (cation) or base (anion) from the salt react with water to produce OH^- or H_3O^+. The four possible combinations are:

(1)   Salt of a strong acid and a strong base: *neutral solution*

(2)   Salt of a strong acid & a weak base: *acidic solution*
- Need K_a, cation to calculate the pH of the solution.

(3)   Salt of a weak acid & a strong base: *basic solution*
- Need K_b, anion to calculate the pH of the solution.

(4)   Salt of a weak acid & a weak base: *cannot be predicted* without knowledge of the relative values of K_a, cation and K_b, anion.

16.9 Apply acid-base principles to the chemistry of antacids, kitchen chemistry, and household cleaners.

Applications of acid-base chemistry include the production and sale of antacids, designed to relieve the discomfort of excess stomach acid; the use of hydrogen carbonates in baked goods to cause them to rise; the use of basic solutions as cleaning products; and the use of strong bases to remove grease from drains and ovens by converting them to soluble soaps.

16.10 Recognize Lewis acids and bases and how they react.

The Lewis acid-base theory is more general, and less commonly used, than the Bronsted-Lowry theory. A **Lewis acid** is a substance that accepts an electron pair, and a **Lewis base** is a substance that donates an electron pair to form a new bond. Note that this definition includes amines, with their unshared pair of electrons, as bases.

The new bond, formed between the Lewis acid and base, is a **coordinate covalent bond**, formed with the pair of electrons from the base and an empty orbital from the acid. Metal cations are common Lewis acids because they often have empty orbitals and because their positive charge attracts electrons.

Chapter Review - Key Terms

In order to make all the sentences below TRUE, insert the appropriate word or phrase from the list of key terms which best fits the context of the sentence.

NOTE: Any phrase or word from the list may used more than once.

LIST:		
acid ionization constant	base ionization constant expression	Lewis acid
acid ionization constant expression	basic solution	Lewis base
acid-base reaction	Bronsted-Lowry acid	monoprotic acids
acidic solution	Bronsted-Lowry base	neutral solution
amines	conjugate acid-base pair	oxoacids
amphoteric	coordinate covalent bond	pH
autoionization	hydrolysis	polyprotic acids
base ionization constant	ionization constant for water	zwitterion

General Characteristics of Acid and Base Equilibria:

A chemical reaction in which a transfer of a proton, H^+, occurs between two reactants is an (1) _____. Using the most common definition, the reactant that donates the H^+ is called the (2) _____ and the reactant that accepts the H^+ is called the (3) _____. Key structural differences between the two are that the (4) _____ must have at least one H that is part of a strongly polar bond and the (5) _____ must have at least one unshared pair of electrons on an atom in the molecule. A group of compounds that typically function as (6) _____s are the (7) _____ which contain a N atom with three covalent bonds and one unshared pair of electrons. The reactants and products in an (8) _____ must always form two (9) _____s, which are two molecules related to each other by the loss or gain, of a single H^+. The (10) _____, represented as HA, and its product, A^-, are one (11) _____ and the (12) _____, which can be represented as B, and its product BH^+, are another.

Water molecules have the ability to function as either a (13) _____ or (14) _____ so that in pure water, a small number react with each other, in a reaction called (15) _____, to produce H_3O^+ and OH^- ions. The equilibrium constant for the reaction is called the (16) _____, with the symbol, K_w, and is defined as equal to the molarity of H_3O^+, $[H_3O^+]$, times the molarity of OH^-, $[OH^-]$. In any solution that contains water, as well as in pure water, the product of these two ion concentrations must always equal the value of the (17) _____, which is 1.0×10^{-14} at 25°C. An aqueous solution in which the $[H_3O^+]$ and $[OH^-]$ are equal is called a (18) _____. However, when a (19) _____ is added to water, it can donate a proton to water, producing H_3O^+ and A^- as products. In such a solution, the resulting $[H_3O^+]$ is greater than the $[OH^-]$, and a(n) (20) _____ is produced. A (21) _____ is produced when a (22) _____ reacts with water to produce BH^+ and OH^- and the resulting $[OH^-]$ is greater than the $[H_3O^+]$ in the solution. A very convenient measure of the $[H_3O^+]$ concentration in the solution is to use the (23) _____, which is the negative of the logarithm of the $[H_3O^+]$. A (24) _____ will have a (25) _____ = 7.0, while an (26) _____ has a (27) _____ below 7.0, and a (28) _____ has a (29) _____ value greater than 7.0.

The extent of reaction of a (30) _____ with water is given by the (31) _____, K_a, the equilibrium constant for the reaction. The mathematical expression for the equilibrium, the (32) _____, always has the same form in that the equilibrium concentrations for $[H_3O^+]$ and the conjugate base, $[A^-]$, appear in the numerator, while that for the un-ionized conjugate acid (HA) is in the denominator, and all the powers are one for the concentrations. For a strong acid, the value of the (33) _____ is greater than 1.0 and complete ionization of HA occurs. For a weak acid, the value of the (34) _____ is less than 1.0, and a significant portion of the acid will be un-ionized at equilibrium.

The extent of reaction of a (35) _____, B, with water is given by the (36) _____, K_b, where the products of the reaction are BH^+ and OH^-. The form of the (37) _____ is also always the same, with the equilibrium concentrations, $[BH^+]$ and $[OH^-]$ appearing in the numerator, un-ionized base [B] in the denominator and all powers for the concentrations equal to one. For a weak base, the value of the (38) _____ is less than 1.0, indicating a significant portion of the base molecules are un-ionized at equilibrium.

Concerning Molecular Structure and Acid-Base Behavior:

Acids that can donate a single H^+ are (39) _____, while acids that can donate more than one H+ are (40) _____ which have stepwise (41) _____s, one for each H^+ lost. Generally, the value of the (42) _____ gets smaller for each successive H^+ donated by the (43) _____, since it is more difficult to remove H^+ from a negative ion, instead of a neutral molecule. Acids in which the acidic H is bonded to an O in a H−O−Z− group, where Z is generally a nonmetal, are called (44) _____. The strength of the (45) _____ increase, indicated by higher values of the (46) _____, as the number of O's bonded to Z increases, so that H_2SO_4 is a stronger acid than H_2SO_3. The acid strength of (47) _____ also varies directly with the electronegativity of Z so that HClO is a stronger acid than HBrO, even though the number of bonded O's are the same. Amino acids have a structure that allows them to form a (48) _____, where a H^+ is internally transferred within the molecule to form charged - CO_2^- and $-NH_3^+$ groups, leaving the molecule with an overall zero charge.

Nearly all weak bases undergo an (49) _____ reaction with water, where the water molecule donates a H^+ to the base. Ionic compounds, the product of an exchange reaction between an acid and base, also undergo (50) _____ reactions with water. The anions, A^-, if conjugate bases of weak acids, can produce an (51) _____ by reacting with water. The value of K_b for the (52) _____ of the anion, A^-, is equal to K_w/K_a for the acid, HA, since $K_a \times K_b$ for a (53) _____ equals K_w.

The cations of the salt that have the form, BH^+, produce an (54) _____ in water. These BH^+ cations are conjugate acids of (55) _____, such as $CH_3NH_3^+$ from CH_3NH_2, and the K_a for the (56) _____ reaction of the cation, BH^+, is equal to K_w/K_b for the base, B.

Concerning Lewis Acid-Base Behavior:

A more general definition of acids and bases can be used to explain an (57) _____ in terms of bond formation when the Lewis structures of the reactants and products are considered. The acid, a (58) _____, accepts a pair of electrons into an empty atomic orbital and the base, a (59) _____, donates the pair of electrons to the bond. The bond formed between the (60) _____ and (61) _____ is called a (62) _____. This definition includes all reactions between a (63) _____ and (64) _____ where a H+ is transferred, but extends acid-base behavior to include other reactions, such as the reaction of metal cations with water that produce acidic solutions.

Some metal hydroxides are (65) _____ in that they can act as a (66) _____ in strong base solutions to form a (67) _____ with OH^- or as a base in acidic solutions, to form the metal cation. Some gaseous nonmetal oxides such as CO_2 and SO_2 can also act as (68) _____s in solution, by reacting with OH^- which acts as the (69) _____, to form HCO_3^- or HSO_3^- ions when dissolved in water.

- PRACTICE TESTS -

After completing your study of the chapter and the homework problems, the following questions, can be used to test yourself on how well you have achieved the chapter objectives.

1. Tell whether the following statements would be always **True or False**:

    A) Bronsted-Lowry acids and bases are defined in terms of transfer of a proton.

    B) The ion HSO_4^- can act as either a Bronsted-Lowry acid or base.

    C) The extent to which a Bronsted-Lowry acid acts as a acid in water is given by its K_a value.

    D) A strong base will have a weak conjugate acid.

    E) The conjugate base of $HClO_4$ is ClO_4^-.

    F) A strong acid will have an acid ionization constant equal to 1.0.

    G) Amines are typically weak acids.

    H) CH_3CO_2H and H_2SO_4 are both examples of polyprotic acids.

    I) The K_a value for first acid ionization constant of H_3PO_4 will be less than that of the second acid ionization constant.

    J) The pH scale is value limited to a range of 0 to 14.

    K) That K_w is equal to the $[H_3O^+] \times [OH^-]$ is true at any temperature.

    L) If a solution's pOH was 8, its pH value would be 6.

    M) A solution with $[H_3O^+] = 4.0 \times 10^{-3}$ would have a higher pH than a solution that has a $[H_3O^+]$ equal to 4.0×10^{-5}.

    N) If the pH is 5.0, a weak acid rather than a strong acid is present in solution.

    O) A basic solution with a pH of 10 contains more dissolved base than a basic solution with a pH 8.

    P) If an acid has a K_a of 1.4×10^{-4}, its conjugate base has a K_b of 7.1×10^{-9}.

    Q) HOCN and $CH_3CH_2CO_2H$ are examples of oxoacids.

    R) Fe^{+3} could be a Lewis acid, but $Fe(H_2O)_6^{+3}$ could not be.

    S) $HClO_3$ is a stronger acid in water than HClO.

T) If a strong acid and strong base react a salt formed.

U) All weak bases will undergo hydrolysis in water.

V) Compounds that are amines can form zwitterions

2. A) For the following reactions, fill in the missing reactant, tell whether you think the reaction is product or reactant-favored and briefly explain the reason(s) for your choice. (Consult Table 16.2 for K_a and K_b values as needed)

(a) _____ + H_2O <=> $HOCl$ + OH^-

(b) _____ + NH_3 <=> NH_4^+ + NO_3^-

B) For the following reaction, indicate which reactant is acting as the acid, which the base and whether the reaction will be product-favored or reactant favored and the reason(s) for your choice: H_2S + CO_3^{-2} <=> HS^- + HCO_3^-

C) Explain why adding solid $Zn(OH)_2$ to an aqueous solution of NaOH causes the solution to become more acidic, but adding solid $Zn(OH)_2$ to an HCl solution causes the solution to become less acidic.

3. Write the balanced equilibrium equations and K_a or K_b expressions for the following:
   A) Bronsted-Lowry acids   B) Bronsted-Lowry bases
   (a) $C_6H_5CO_2H$ (a) CH_3NH_2
   (b) $H_2AsO_4^-$ (b) IO_3^-

4. For the following chemical species, tell whether they would
   A) act as an strong acid, weak acid, strong base, weak base or neither in water
   B) If an acid or bases, give the <u>formula of the appropriate conjugate</u> that forms in water
   (a) $HClO_4$ (b) NH_4^+ (c) C_6H_5N (d) HCN   (e) H_2 (f) S^{-2}

5. Given the monoprotic acids below and their Ka's:
   A) Tell which is the strongest acid?
   B) Tell which would have the strongest conjugate base?
   C) Circle the acidic proton in each acid

   (a)   HCOOH $K_a = 1.8 \times 10^{-4}$ (b)   HBrO $K_a = 2.3 \times 10^{-9}$
   (c) $(CH_3)_2AsO_2H$ $K_a = 6.4 \times 10^{-7}$ (d) $BrCH_2CO_2H$ $K_a = 1.3 \times 10^{-3}$

6. The pOH of a solution that has a $[H_3O^+] = 3.4 \times 10^{-5}$ is:
   (a) 4.5   (b) 10.5   (c) 9.5   (d) 6.3   (e) None of these choices

7. If a 0.10M HF solution is 8.4% ionized, the pH of the solution must be:
   (a) 0.084   (b) 1.07   (c) 1.00   (d) 2.08   (e) 2.16

8. If the pH of a solution is increased from 3.0 to 6.0, which of the following statements is
   FALSE: A)   The initial $[H_3O^+]$ was 1.0×10^{-3} M
             B)   The $[H_3O^+]$ was decreased by a factor of 2
             C)   The $[H_3O^+]$ was increased by a factor of 1000
             D)   The pOH was lowered from 11.0 to 8.0
             E)   The molarity of $[H_3O^+]$ was 1.0×10^{-3} and then became 1.0×10^{-6} M
             F)   The solution became more acidic

9. For the following values in aqueous solutions:
   (a) pH = 2.30   (b) $[OH^-] = 2.2 \times 10^{-5}$ (c) pOH = 7.52
   A) What would be the $[H_3O^+]$ in the solution?
   B) Tell whether the solution is acidic, basic or neutral.

10. What would be the pH in a saturated solution of $Ca(OH)_2$ made by dissolving 0.463 g $Ca(OH)_2$ in enough water to make 250 mL of solution?

11. For 0.15M propionic acid, $CH_3CH_2CO_2H$, in water:

A) Write the chemical equation for the equilibrium for propionic acid.

B) Tell whether the equilibrium is a K_a or K_b and write out the form of the expression with its value from Table 16.2.

C) Calculate the pH of solution at equilibrium.

12. Trimethyl amine has a $K_b = 6.25 \times 10^{-5}$. For 0.25M trimethylamine, $(CH_3)_3N$, in water:

A) What is the chemical formula for the conjugate base of $(CH_3)_3N$?

B) Write the chemical equation for the equilibrium for trimethylamine.

C) Calculate the pH of solution at equilibrium.

13. Consider the acid, pyruvic acid, $HC_3H_5O_3$ with $K_a = 1.4 \times 10^{-4}$. What molarity of pyruvic acid would be needed to make 500 mL of solution with a pH = 2.50?

14. A 0.150M solution of $CH_3CH_2CH_2CO_2H$, butanoic acid, has a pH = 2.82, what is the K_a of butanoic acid?

15. Suppose you had 500 mL of each the following solutions in 5 beakers:

Beaker A: 0.202M HNO_3 Beaker B: 0.202M HNO_2 Beaker C: 2.02M KNO_3

Beaker D: 0.202M KOH   Beaker E: 2.02M KNO_2

(a) Tell whether each solution would be acidic, basic or neutral.

(b) Which is the solution with the highest pH?

(c) Which is the solution with the lowest pH?

(d) In which solution(s) would you need to use the appropriate K_a or K_b to calculate pH?

(e) Which would contain the largest number of moles of undissociated molecules?

(f) Which solution(s) contain(s) a salt that results from reacting a strong acid with a strong base?

(g) Which solution(s) contain(s) a salt that results from reacting a weak acid with a strong base?

16. It is desired to produce an aqueous solution with a pH of 8.75 by dissolving ONE of the following salts, either KClO, NH_4Cl, $KHSO_4$ or $NaNO_3$

a) Which one should you use ?

b) What would be the initial concentration you should use of the salt?

17. Good wine will turn to vinegar if left exposed to the air because the ethanol is converted to acetic acid by the reaction with oxygen gas: $C_2H_5OH(l) + O_2(g) => CH_3CO_2H(l) + H_2O(l)$ Suppose you have a 1.0 L bottle of wine that is 13% (by volume) ethanol and that <u>two-thirds</u> of the ethanol is converted to acetic acid. (density of ethanol = 0.785g/mL)

A) How many moles of acetic acid would be present in the 1.0L of wine after <u>two-thirds</u> of the ethanol is converted?

B) What then would be the pH in the wine from the acetic acid?

18. Draw out the Lewis structures for the molecules involved in the following acid base reaction: $BF_3(g) + F^-(g) <=> BF_4^-(g)$ and identify the Lewis acid and the Lewis base in the reaction.

Chapter 17

Additional Aqueous Equilibria

This chapter combines principles of equilibrium and of solution chemistry to consider several applications of aqueous chemistry, including buffer solutions, acid-base titrations, and precipitation of solids from solutions.

17.1 Explain how buffers maintain pH, how they are prepared, and the importance of buffer capacity.

17.2 Use the Henderson-Hasselbach equation to calculate the pH of a buffer and the pH change after acid or base has been added to the buffer.

A **buffer** system is a solution that resists change in pH upon addition of acid or base because it contains similar concentrations of a weak acid or base and its conjugate base or acid. The buffering action of an acidic buffer occurs because the weak acid neutralizes any base that is added and the conjugate base neutralizes any acid that is added. Basic buffers react in an analogous fashion. Buffers usually do not change more than one pH unit in either direction before their capacity is exceeded; a buffer's capacity is exceeded when the number of moles of added acid or base exceeds the number of moles of acid or base in the original buffer sample. The pH of a buffer can be calculated from the equilibrium expression given the appropriate K value and concentrations. Another method is to use the **Henderson-Hasselbach equation**:

$$pH = pK_a + \log \frac{[\text{conj. base}]}{[\text{conj. acid}]} = pK_a + \log \frac{[A^-]}{[HA]}$$

Note that variations in the form of this equation may result from the properties of logarithms. Note also that if the concentrations of acid and base are equal, the pH of the buffer is the same as the pK_a of the acid. To calculate the change in pH when a known amount of acid or base is added, the amounts of base and acid in the log term are adjusted appropriately and a new value for pH is calculated. For example, if x mol acid are added to a solution, the log term would become $\log \frac{[\text{ mol } A^- - x]}{[\text{ mol } HA + x]}$. Note that moles, instead of concentrations, are used here; the ratio between moles and concentrations is identical.

17.3 Interpret acid-base curves, and calculate the pH of the solution at various stages of the titration.

An acid-base titration is performed by determining the amount of a standard (of known concentration) solution, called the **titrant**, of acid or base is required to neutralize an unknown sample of base or acid. At the equivalence point, the number of moles of acid or base added is equal to the number of moles of base or acid present in the original sample. A colored indicator that changes color at or near the pH of the equivalence point is usually used for visual detection of the **endpoint**. A **titration curve** is a plot of pH versus volume of titrant added. Values of pH can be calculated initially, during the titration, at the equivalence point, and after the equivalence point. Careful analysis of the number of moles of acid or base reacted and of salt produced, along with a correction for the volume change as titrant is added, is necessary. Titration curves for strong acid--strong base (SA-SB) and weak acid--strong base (WA-SB) are similar, starting at low pH, with a sharp increase at the equivalence point, and a continued slow increase if additional titrant is added. However, the WA-SB curve starts at a higher pH than the SA-SB because a weak acid has a higher pH than a strong acid of the same concentration. A titration curve for strong acid--weak base (SA-WB) is the reverse of the WA-SB curve, showing a steady decrease in pH as acid is added, and a sharp drop at the equivalence point. The titration curve for a polyprotic acid will show an equivalence point for each ionizable hydrogen.

17.4 Explain how acid rain is formed and its effects on the environment.

Acid rain is formed when gaseous non-metal oxides in the atmosphere become dissolved in rainwater. Carbon dioxide has always caused precipitation to be slightly acidic, but since the Industrial Revolution, the amount of carbon dioxide, and of other gases including nitrogen dioxide and sulfur dioxide, has increased greatly. Environmental effects include acidification of lakes and streams and damage to vegetation and rock.

17.5 Relate a K_{sp} expression to its chemical equation.

17.6 Use the solubility of a slightly soluble salt to calculate its solubility product.

17.7 Describe the factors affecting the aqueous solubility of ionic compounds.

17.8 Apply Le Chatelier's principle to the common ion effect.

17.9 Use the solubility product to calculate the solubility of a sparingly soluble solute in pure water and in the presence of a common ion.

17.10 Describe the effect of complex ion formation on the solubility of a sparingly soluble salt.

Salts that are slightly soluble in water reach an equilibrium between the solid and the aqueous ions at very low concentrations of those ions. The equilibrium constant for these reactions, symbolized by K_{sp}, is written the same way as other K expressions. Recall that solids are not included in these expressions. The value of K_{sp} can be determined by measuring the concentrations of the ions in a saturated solution and substituting the values into the K_{sp} expression. A limitation of these calculated values is that other interactions between ions may affect their concentrations.

Besides temperature and ion-pair interactions, there are several other factors that affect the solubility of ionic compounds. pH is an important factor if one of the ions involved is a weak acid or base, or with carbonates, in which gaseous CO_2 is also a factor. The **common ion effect** describes the decrease in solubility of a compound if there is another source of one of its ions in the solution. According to Le Chatelier's principle, this common ion decreases the solubility of the salt. Calculations of solubility of a salt when a common ion is present require substitution of the larger concentration into the K_{sp} expression and solving for the concentration of the other ion. The formation of complex ions involving one of the ions of a sparingly soluble salt usually increases the solubility of the salt because it removes one of its ions from the solution.

17.11 Relate Q, the ion product, to K_{sp} to determine whether precipitation will occur.

17.12 Predict which of two ionic solutes will precipitate first.

The quantitative basis for the solubility rules used in Chapter 5 is the comparison of values of Q with K_{sp}. In this case Q is the ion product, an expression with the same form as K_{sp}, but using the original values of concentrations in the solution. As before, if $Q < K_{sp}$, the system has not reached equilibrium and no precipitate will form; if $Q = K_{sp}$, the system has reached equilibrium and the solution is saturated, with precipitation imminent; if $Q > K_{sp}$, the system is supersaturated and unstable, and a precipitate will form in order to decrease the concentrations. A mixture of two ions can be separated by selective precipitation if both ions form sparingly soluble salts with the same counter ion and if the difference in their values of K_{sp} is large enough. The salt with the smaller value of K_{sp} will precipitate first.

Chapter Review - Key Terms
In order to make all the sentences below TRUE, insert the appropriate word or phrase from the list of key terms which best fits the context of the sentence.
NOTE: Any phrase or word from the list may used more than once.

LIST:		
acid rain	complex ion	ion product
buffer	endpoint	solubility product constant, K_{sp}
buffer capacity	formation constant	titrant
buffer solution	Henderson-Hasselbalch equation	titration curve
common ion effect		

Concerning Buffers, Titrations and Acid Rain:

A (1) _____ is a solution that resists changes in pH when limited amounts of strong base or acid are added to the solution. In order to produce a (2) _____, both forms of a conjugate acid-base pair from a weak acid or base must be present in solution in approximately equal concentrations. Only if the ratio between the conjugate acid-base pair is in the range of 0.1 to 10 will a (3) _____ be produced and the allowed equilibrium concentration of H_3O^+, will be centered around the K_a value for the weak acid. Consequently, the pH range for a (4) _____ will be equal to the pK_a (the negative logarithm of the K_a) ± 1.0 pH unit for the weak acid form. The (5) _____, the equilibrium expression for a weak acid written in terms of pKa and pH, is a convenient equation to use to calculate the pH or ratio of the conjugate acid-base pair for a (6) _____.

The quantity of added strong acid or base that a (7) _____ can accommodate without significant change in pH, is the (8) _____ of the (9) _____. The ration of moles of conjugate base to conjugate acid in the (10) _____ sets the value of the (11) _____ for the solution.

Indicators are used in acid-base titrations to determine the (12) _____ of the titration and change color at a pH that is slightly higher than the pH of the equivalence point. A (13) _____, the graph of the pH versus volume of (14) _____ added, has a distinctive shape depending on whether a strong acid, weak acid or weak base is being titrated.

If rainwater has a pH less than 5.6 it is called (15) _____ and contains dis-solved gaseous oxides other than CO_2. (16) _____ is produced when oxides such as NO_2, SO_2 or SO_3 dissolve in water droplets in the atmosphere and react with the water to produce either the weak acids, HNO_2 or H_2SO_3 or the strong acids, HNO_3 or H_2SO_4.

Concerning Solubility Equilibria:

The solubility of ionic compounds is an equilibrium process and the (17) _____ describes the extent to which the solid solute dissolves to produce ions in solution at equilib-rium. The expression for the (18) _____ is also called the (19) _____, since only the ion concentrations appear in the expression, taken to the power equal to the subscript in the ionic compound. If the solution, to which the solid is added, already contains one of the ions in the (20) _____ expression, from another source, the solubility of the ionic compound in the solution will be significantly lower than what it would be in pure water due to the (21) _____.

The solubility can also be affected when a (22) _____ is formed by a metal ion reacting with a Lewis base to forma coordinate covalent bond. The equilibrium constant for this type of reaction is called the (23) _____, represented as K_f. If the Lewis base is the same as anion in the compound, such as $AgCl + Cl^- \iff AgCl_2^-$, the formation of the (24) _____ will increase the solubility in a solution containing the anion, opposite to the effect normally observed because of the (25) _____.

- PRACTICE TESTS -

After completing your study of the chapter and the homework problems, the following questions, can be used to test yourself on how well you have achieved the chapter objectives.

1. Tell whether the following statements would be always **True or False**:

   A) The pH of a buffer is always centered around the pK_a for an acid and pK_b for a base.

   B) Adding strong base to strong acid solution will produce a buffer solution.

   C) To prepare a buffer solution of NH_3, you need to add NH_4Cl to the solution in a nearly equal molar amount to that of NH_3.

   D) The pH of a buffer solution for a weak base will always be greater than 7.0 and a buffer solution of a weak acid has a pH less than 7.0.

   E) A solution made by combining 0.10 moles of HOCl with 0.05 mol of KOCl would have the same pH as a solution that made by combining 2.0 moles of HOCl and 1.0 mole of KOH.

   F) The buffer capacity of a solution that contained 0.20 M acetic acid and 0.20 M sodium acetate would be greater for the addition of strong base than one that contained 0.20 M acetic acid and 0.10 M sodium acetate.

   G) The pH of buffer solution that contained 0.20 M acetic acid and 0.20 M sodium acetate would be higher than for one that contained 0.20 M acetic acid and 0.10 M sodium acetate.

   H) According to the Henderson-Hasselbalch equation if the concentration of the salt of the conjugate base in a buffer solution is increased, the pH of the solution will decrease.

   I) Titration curves for weak acids and weak bases show a buffer region before the equivalence point..

   J) The titration curve for a weak base has an initial pH higher than 7.0 and the pH of the equivalence point will be at a pH less than 7.0.

   K) The pH at the equivalence point for a titration of a strong acid by a strong base is always at a pH equal to 7.0.

   L) The length of rise in the steep region of the titration curve depends on the Ka of the acid and increases with increasing values of Ka.

   M) The K_{sp} can be used to tell when a precipitate will form from a solution of ions.

   N) Due to the common ion effect, the solubility of $CaCl_2$ would be greater in 0.10 M HCl than in pure water.

   O) Formation of a complex ion requires that a Lewis acid react with a metal ion to form a new ion.

2. A) Why can't a solution prepared by dissolving a weak base in water be a buffer solution?

   B) What is the difference between end point and equivalence point in a titration?

   C) The pKa for two acids are 4.74 and 7.46. Which is the stronger acid and why?

   D) What are the two major contributors to acid rain? Since these are not acids themselves how do they make rain acidic? Why are lakes that have limestone deposits (CaCO3) not affected by the acid rain, as lakes surrounded by granite deposits (silicates) are?

3. Does the pH of the solution - *increase, decrease or stay the same-* when:

   A) 0.5 mol LiF is added to 1.0 L of 1.0M HF solution?

   B) 0.5 mol KCl is added to 1.0 L of 1.0M HCl solution

   C) 0.5 mol NH_4Cl is added to 1.0 L of 1.0M NH_3 solution

   D) 0.5 mol $KClO_4$ is added to 1.0 L of 1.0M KCl solution

4. Which of the following would be examples of a common ion effect on an equilibrium?

   A) Adding NaCl(s) to a HCl solution

   B) Adding KBr(s) to saturated AgBr solution

   C) Adding HOCl to a $Ca(OCl)_2$ solution

5. Which of the following could act as buffer solutions of formic acid (pK_a = 3.75)?

   A) Mixing 300.0 mL of 0.10M formic acid with 200 mL of 0.20M sodium formate

   B) Mixing 300.0 mL of 0.10M formic acid with 100 mL of 0.1M NaOH

   C) Adding 10.0 g of sodium formate to 500 mL of 1.0M formic acid

6. An environmental chemist needs a buffer of pH = 10.0, prepared from solid Na_2CO_3 and 0.20M $NaHCO_3$, to study the effects of the acidification of limestone-rich soils.

   A) What are the conjugate acid and base forms in the buffer?

   B) What equilibrium is used to calculate the pH of the buffer?

   C) The pK_a of the conjugate acid is needed for the preparation, what is its value?

7. Tell which of the following would be TRUE if 17.4 grams K_2SO_4 was added to 100.0 mL of 0.50M $BaCl_2$ (aq)

   A) $[Ba^{2+}]$ at equilibrium = 0.025 M   B) $[SO_4^{2-}]$ at equilibrium = 0.50 M

   C) KCl(s) will precipitate   D) Barium sulfate would precipitate

   E) $[K^+]$ would be 2.0 M   F) No reaction occurs

8. For the titration curve shown on the right:

   A)   Tell whether the curve represents a titration of :

      (a) a weak acid and strong base

      (b) a weak base and strong acid

      (c) a strong acid and strong base

   B) Indicate which points marked on the curve represent the pH of the solution

      (a) Before any titrant added

      (b) After titrant added, but before the equivalence point.

      (c) At the equivalence point

      (d) After the equivalence point

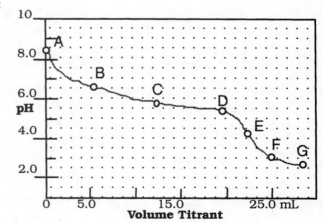

   C) Estimate the endpoint volume from the curve.

   D) Which indicator, bromothymol blue or methyl red, would be better as an indicator for this titration given the color ranges for each are:

         Bromothymol blue: blue ($pH \geq$ 8) $\rightarrow$ yellow (pH< 6.0)

         Methyl red: yellow (pH> 6.3) $\rightarrow$ orange (pH= 5.0) $\rightarrow$ red (pH< 4.0)

9. What is the pH of a buffer solution that contains 0.55 M formic acid, HCO_2H, and 0.63M sodium formate?

10. For the following conjugate acid-base pairs:
    (a) NH_4^+/NH_3 (b) HF/F^- (c) H_2S/HS^- (d) $B(OH)_3/B(OH)_4^-$
    A) Tell what the buffer range of pH's would be for each pair.

    B) How could you do to test that the solution in (c) was truly a buffer and not acting like pure water?

    C) Boric acid, $B(OH)_3$, does not lose a proton, yet is has a K_a. Show by writing the chemical reaction why it is the acid form for the conjugate acid-base pair.

11. For a solution that is 0.20M propionic acid ($pK_a = 4.87$) and 0.10M sodium propionate
    A) What is the pH of the solution?

    B) For the following changes:
            (a) Tell whether the *conjugate base- conjugate acid ratio* will increase, decrease or show no change when the change is made
                (b) If the ratio changes, give the value of <u>new</u> ratio and pH.
                    (1) 10.00 mL of 1.0 M HNO_3 added to 1.0 L of the buffer
                    (2) 500.00 mL of water added to the buffer
                    (3) 2.00 g of NaOH added to 1.0L of the buffer

12. Benzoic Acid, $C_6H_5CO_2H$, has a $K_a = 6.3 \times 10^{-5}$.
    A) How many grams of sodium benzoate must be added to 750 mL of 0.200M benzoic acid to make a buffer solution with a pH= 4.00?

    B) Suppose the chemistry stockroom doesn't have any sodium benzoate to make the solution in (A). Can either ammonium benzoate or lithium benzoate be used to prepare the buffer instead of sodium benzoate? Explain.

13. A) If you had a 0.10M solution of each of the following organic acids:
        (a) Acetic acid, $K_a = 1.8 \times 10^{-5}$ (c) Citric acid, $K_a = 1.4 \times 10^{-4}$
        (b) Lactic acid, $K_a = 8.7 \times 10^{-4}$ (d) Oxalic acid, $K_a = 5.9 \times 10^{-2}$
    A) Which solution of acid would produce a buffer of the lowest pH?
    B) Which solution of acid would produce a buffer of the highest pH??
    C) What would be the ratio of [conjugate base] to [conjugate acid] for solution (c), if the pH of the solution was 3.00? Would such a solution act as a buffer?

14. Suppose you had a solution that was 0.165M HOCl. Given that $K_a = 3.0 \times 10^{-8}$ for HOCl,
    A) what would be the pH of the solution ?

    B) what would the new value for the pH for the solution if 14.3 grams of $Ca(OCl)_2$ was added to 1.0 L of the solution?

    C) would the value of the $[H_3O^+]$ *increase, decrease, or stay the same* in the 0.165M HOCl solution if 0.74 grams of $Ca(OH)_2$ was added to 1.0 L of the solution?

15. Suppose you had 500 mL of each the following solutions in 5 beakers.
    Beaker A: 0.20 M HNO_3 Beaker B: 0.20 M HNO_2 Beaker C: 0.20 M KBr
            Beaker D: 0.20 M HOBr   Beaker E: 0.20 M KOH
    Which solutions, if any, could be combined as pairs to make:
        (a) a buffer solution with pH < 7.0 ?   (b) a buffer solution with pH > 7.0?

16. If the solubility of Cerium (IV) iodate, $Ce(IO_3)_4$, is 0.150 grams per 100 mL of solution.
    Assuming $d_{soln} = d_{water} = 1.00$ g/mL,
    A) what is the molarity of $Ce(IO_3)_4$ in the saturated solution?
    B) estimate the value of K_{sp} of Cerium (IV) iodate from this molarity.

17. If the K_{sp} of Barium hydroxide, $Ba(OH)_2$,in water is 3.0×10^{-4}
    A) What is the solubility of the barium hydroxide in moles per liter?
    B) What is the pH of the saturated barium hydroxide solution?
    C) Would adding 0.40 g NaOH to 100 mL of 0.05 M $BaCl_2$ solution cause the precipitation of Ba^{+2} as $Ba(OH)_2$ from the solution?

18. PbC_2O_4 and CaF_2 both have solubility products constants of 3.0×10^{-11}. Does this mean they have the same molar solubility in pure water and if not , which is higher?

19. Ascorbic acid (Vitamin C), $C_6H_8O_6$, acts as a monoprotic weak acid in water with a $K_a = 7.94 \times 10^{-5}$. It can be titrated with solutions of strong base, such as NaOH. If a tablet containing 500 mg of ascorbic acid is dissolved in 100 mL of water and titrated with 0.10 M NaOH:
    A) What will be the pH in the ascorbic acid solution before any NaOH is added?
    B) What volume of NaOH, in milliliters, should be needed to reach the endpoint of the titration?
    C) What is the pH of the ascorbic acid solution when one-half the endpoint volume of NaOH has been added to the solution?
    D) What will be the pH of the solution at the endpoint?
    E) Would phenolphthalein be a suitable indicator for this titration?

20. Gold can be reclaimed from various materials by reacting it with cyanide ion, CN^- to form the complex ion, $Au(CN)_2^-$.
    A) Write the formation equation and the form of the formation constant, K_f, for the complex ion, $Au(CN)_2^-$.
    B) One of the materials gold can be reclaimed from is to convert solid gold chloride, AuCl, by reacting it with CN^-.
        (a) Write the chemical reaction for the complex formation from AuCl(s)
        (b) Prove that this reaction is the sum of the solubility equilibrium reaction of AuCl(s) and the formation reaction of (A)
        (c) How would the K for this reaction be related to K_{sp} for AuCl(s) and K_f for $Au(CN)_2^-$?

Chapter 18

Thermodynamics: Directionality of Chemical Reactions

This chapter develops the full criteria for product-favored reactions by further defining the role of entropy and introducing the concept of Gibbs free energy, the net energy difference between entropy and enthalpy of a chemical reaction. The Gibbs free energy can then be used to predict values of the equilibrium constant for a reaction and whether the reaction will become more or less product-favored when the temperature changes. The difference between the thermodynamic stability and kinetic stability of the reactants is also discussed.

18.1 Understand and be able to use the terms product-favored and reactant-favored.

18.2 Explain why there is a higher probability that both matter and energy will be dispersed than that they will be concentrated in a small number of particles.

A product-favored reaction is one which, once started, proceeds such that the amount of products eventually is much greater than the amount of reactants. A reactant-favored reaction is one which only proceeds towards products when there is a continuous input of energy. Any reaction that is product-favored as written will be reactant-favored if written in the opposite direction. Energy tends to be dispersed over the largest possible number of particles because this allows many more possible distributions of energy. Likewise, matter tends to be arranged as randomly as possible because this allows many possible arrangements.

18.3 Calculate the entropy change for a process occurring at constant temperature.

18.4 Use qualitative rules to predict the sign of the entropy change for a process.

18.5 Calculate the entropy change for a chemical reaction, given a table of standard molar entropies for elements and compounds.

The disorder or randomness of matter at the nanoscale level is called entropy, S. For reversible processes occurring at constant temperature, entropy can be calculated using the formula $\Delta S = S_{final} - S_{initial} = \frac{q_{rev}}{T}$. For phase changes, the value of q_{rev} is the ΔH associated with that change. Entropy changes for a process can be predicted qualitatively using the following guidelines:

- In comparing the same substance in different phases, $S_{gas} >> S_{liq} > S_{solid}$.

- In comparing similar compounds, the entropy of a more complex molecule is greater than the entropy of a simpler molecule.

- In ionic solids, the entropy increases as the ionic attractions decrease.

- When a solid or liquid dissolves in a liquid, entropy usually increases.

- When a gas dissolves in a liquid, entropy decreases.

Absolute entropy values for elements and compounds can be found in tables. Note that the units of entropy are $J\ K^{-1}\ mol^{-1}$. These values can be used to calculate values of ΔS^o for any process according to the formula:

$$\Delta S^o = \Sigma\{mol\ products \times S^o\ products\} - \Sigma\{mol\ reactants \times S^o\ reactants\}$$

The guidelines for predicting entropy change can be used to determine whether a calculated value of ΔS^o is reasonable.

18.6 Use entropy and enthalpy changes to predict whether e reaction is product-favored.

18.7 Describe the connection between enthalpy and entropy changes for a reaction and the Gibbs free energy change; use this relation to estimate quantitatively how temperature affects whether a reaction is product-favored.

18.8 Calculate the Gibbs free energy change for a reaction from values given in a table of standard molar free energies of formation.

According to the **second law of thermodynamics**, which states that the total entropy of the universe is constantly increasing, a product-favored process is one in which $\Delta S_{universe}$ increases. $\Delta S_{universe}$ can be calculated as follows:

- Determine the dispersal of energy in the process using $\Delta S^{o}_{surroundings} = \dfrac{-\Delta H}{T}$.

- Determine the entropy change in the process, $\Delta S^{\circ}_{system}$, using values of standard molar entropies.

- Combine the two values using $\Delta S^{o}_{universe} = \Delta S^{o}_{surroundings} + \Delta S^{o}_{system} = \dfrac{-\Delta H}{T} + \Delta S^{o}_{system}$

This calculation can be expressed in simpler terms using **Gibbs free energy**, G. The commonly used form of this quantity is $\Delta G^{o}_{system} = \Delta H^{o}_{system} - T\Delta S^{o}_{system}$. Gibbs free energy is defined such that a negative value of ΔG° indicates a product-favored reaction. It can be seen from the formula for ΔG° that when entropy and enthalpy have opposite signs, temperature will determine whether the value of ΔG° is positive or negative. At high temperatures, a positive value of entropy will make the $T\Delta S^{\circ}$ term large enough to overcome the unfavorable ΔH° term.

Table 18.2 in the text can be expanded as follows:

Sign of ΔG^{o}	Sign of ΔH^{o}	Sign of ΔS^{o}	Product-favored?
Always negative	Negative (exo)	Positive	Always
Negative at low T	Negative (exo)	Negative	At low T
Negative at high T	Positive (endo)	Positive	At high T
Always positive	Positive (endo)	Negative	Never

Values of ΔG° can be calculated from ΔH° and ΔS° values determined by previous methods. In addition, ΔG° values can be calculated from standard molar Gibbs free energy data obtained from tables.

$\Delta G^{o}_{reaction} = \Sigma$ {mol products $\times \Delta G^{o}_{f}$ products} - Σ {mol reactants $\times \Delta G^{o}_{f}$ reactants}

Note that the value ΔG° of a pure element in its standard state is zero.

18.9 Relate Gibbs free energy and standard equilibrium constant for the same reaction and be able to calculate one from the other.

As a product-favored reaction proceeds, the value of ΔG continually decreases until it reaches the value of $\Delta G^{\circ}_{overall}$ for the reaction. In a system that reaches equilibrium as reactants mix and react, the graph of ΔG versus extent of reaction is a curve in which the lowest point is at Q = K. To the left of this point, Q < K, and the system continues moving toward equilibrium. To the right of this point, Q > K, showing that some products must decompose to return to equilibrium. This can be represented mathematically as $\Delta G^{\circ} = -RT \ln K^{o}$, where K^{o} is the **standard equilibrium constant** as defined in the previous chapter. To describe a system that is not at equilibrium, the relationship $\Delta G = \Delta G^{\circ} + RT \ln Q$ is used.

18.10 Describe how a reactant-favored system can be coupled to a product-favored system so that a desired reaction can be carried out.

18.11 Explain how biological systems make use of coupled reactions to maintain the high degree of order found in all living organisms; give examples of coupled reactions that are important in biochemistry.

The value of Gibbs free energy describes the maximum amount of useful work from a product-favored reaction that is available for other uses. Gibbs free energy also describes the minimum amount of work that must be done in order to force a reactant-favored reaction to occur. One way of using the Gibbs free energy from a product-favored reaction is to make that energy available to cause a desired reactant-favored reaction to occur. This joining of reactions is called "coupling." Note that the product-favored reaction includes an increase in entropy and the reactant-favored reaction involves a decrease in entropy.

In biological systems, coupled reactions are used to force reactions producing highly structured biochemical molecules to occur. A product-favored reaction is said to be **endergonic**, and a reactant-favored reaction, **exergonic**. The conversion of ATP to ADP is exergonic; the Gibbs free energy of this reaction is used to drive many necessary biochemical reactions. Free energy from other exergonic reactions is used to convert the low-energy ADP back into ATP to be used again. One example of coupled biochemical reactions is the breakdown of glucose (an exergonic reaction) and the formation of ATP from ADP (an endergonic reaction). Another example of a coupled reaction is the conversion of glucose to glucose-6-phosphate at the expense of ATP in the first step of glucose metabolism.

18.12 Explain the relationship between Gibbs free energy and energy conservation.

18.13 Distinguish between thermodynamic stability and kinetic stability and describe the effect of each on whether a reaction is useful in producing products.

The law of conservation of energy states that energy cannot be created or destroyed. However, energy conservation commonly refers to using forms of matter that contain a lot of Gibbs free energy in ways in which the maximum amount of useful work is obtained. No process of energy conversion is completely efficient, but conservation of energy means that as little free energy as possible is wasted as heat. Calculating values of free energy allows prediction of whether a reaction is product-favored. A value of $\Delta G^o < 0$ means a reaction is product-favored. This means that in terms of free energy, the products are more stable than the reactants. However, the study of kinetics shows that a reaction will not proceed unless additional energy, activation energy, is supplied. The situation just described is an example of thermodynamically unstable reactants that are kinetically stable. The favorable free energy of the products will not be released unless the activation energy is supplied. In summary, thermodynamic calculations predict whether a reaction is product-favored, and if so, the amount of free energy that can be produced, but kinetics predicts how long that reaction would take or how much energy must be supplied to get it started.

Chapter Review - Key Terms

In order to make all the sentences below TRUE, insert the appropriate word or phrase from the list of key terms which best fits the context of the sentence.

NOTE: Any phrase or word from the list may used more than once.

LIST:		
endergonic	Gibbs free energy	reversible process
energy conservation	metabolism	second law of thermodynamics
exergonic	nutrients	standard equilibrium constant
extent of reaction	photosynthesis	third law of thermodynamics

Characteristics of Entropy and Gibbs Free Energy :

Entropy changes correspond to a change in the dispersal of energy or matter, or both, in chemical systems. Phase transitions such as melting or boiling, which occur at a constant T and P are examples of a (1) _____, where a small change in conditions will alter the direction of the change. The entropy change for the (2) _____ is then calculated by dividing the enthalpy by the transition temperature. Two laws of thermodynamics further describe the entropy associated with physical or chemical changes. Entropy is directly related to molecular motion, so that lowering temperature of a substance will decrease its entropy. The zero level of entropy for any substance is defined by the (3) _____, that says the entropy for any perfect crystalline substance will be zero only at a temperature of 0 K, when no molecular motion is possible. Comparison of entropy changes for the surroundings (from dispersal of energy) and the system (from dispersal of matter), for physical changes or chemical reactions, leads to the (4) _____, which says that overall, the entropy of the universe increases in product-favored reactions which are then the natural or spontaneous changes.

For a chemical change, the change in (5) _____, represented by symbol ΔG, is a measure of the $\Delta S_{universe}$ caused by the change, but has the opposite sign, so that spontaneous or product-favored reactions always have a negative sign for the (6) _____ change. Positive signs for the (7) _____ change then indicate the reaction is reactant-favored.

The standard state value of the (8) _____, $\Delta G°$, is the difference between the $\Delta H°$ for the reaction and the standard molar entropy change, $\Delta S°$, times the temperature (in degrees K), both calculated from the standard tabled values. Reactions that have the same sign for $\Delta H°$ and $\Delta S°$ will have a (9) _____ that can change from product-favored to reactant-favored, or vice versa, at a certain temperature.

When a reaction occurs, the value of (10) _____ released or absorbed equals $\Delta G°$ only when all the reactants have been converted to products, under standard state conditions. When only partial conversion occurs, under standard state conditions, the amount of (11) _____ released or absorbed is determined by the (12) _____, the fraction of reactants converted to products at that point in the reaction, times $\Delta G°$. Under non-standard state conditions, the (13) _____ is the sum of the $\Delta G°$ and $RT\ln Q$, where Q is the reaction quotient. When the equilibrium state is achieved, the (14) _____ will be zero and Q equals the (15) _____, $K°$, which equals Kc for solution reactions for Kp for gaseous systems. The value of $\Delta G°$ defines the value of the (16) _____ for a reaction by the equation: $\Delta G° = - RT\ln K°$. Negative values for the or $\Delta G°$ mean that the (17) _____ will be greater than one, and likewise positive values for $\Delta G°$ produce a (18) _____ value less than one.

Free Energy and Biochemical Systems:

Two terms are commonly used to describe the free energy changes in biochemical reactions, to indicate the direction of flow, or sign, without giving the value. An (19) _____ reaction releases (20) _____, while an (21) _____ reaction uses up (22) _____.

(23) _____ refers to the coordinated series of reactions that convert food (24) _____, the chemical raw materials needed for survival, into (25) _____ and the chemical constituents of living cells. (26) _____ is similar to a combustion reaction or burning, in that compounds containing carbon and hydrogen are converted to CO_2 and H_2O, producing thermal energy in the process. However, it is different form burning, in that organisms couple the (27) _____ reaction(s), represented by the combustion, to (28) _____ reaction(s) that produce ATP, adenosine triphosphate, from ADP, adenosine diphosphate, with the help of enzymes. This results in better (29) _____, in that much more of the useful work gotten from the (30) _____ change, is employed in making other (31) _____ chemical changes occur, rather than being lost to the surroundings as heat. (32) _____, another process employing coupled biochemical reactions converts CO_2 and H_2O gases into glucose and $O_2(g)$, an (33) _____ reaction, but uses sunlight as the source of energy instead of an (34) _____ chemical reaction.

*After completing your study of the chapter and the homework problems, the following questions,
can be used to test yourself on how well you have achieved the chapter objectives.*

1. Tell whether the following statements would be always **True or False**:

    A) The entropy of a substance always increases when the temperature is increased.

    B) An exothermic reaction is always product-favored.

    C) Every substance is a solid and will have zero entropy at 0 K.

    D) The entropy of a substance increases as it changes from a liquid to the vapor state at any temperature.

    E) Ionic solids have larger entropies as the attraction between the ions become weaker.

    F) Reactions with positive ΔH and positive ΔS can never be made to produce a spontaneous reaction.

    G) Reactions with a $\Delta G°$ less than zero always have a K less than 1.0.

    H) When temperature is changed, $\Delta G°$ is unchanged and the ratio of products to reactants is unaffected.

    I) Reactions with negative ΔH and positive ΔS are always product-favored so K° does change with temperature.

    J) Recycling metals is a form of energy conservation.

    K) The Gibbs Free Energy is a measure of the energy available to do useful work for tasks that would not happen on their own.

    L) An endergonic reaction is the same as an endothermic reaction but occurs in a biological system.

    M) The ATP to ADP reaction is an exergonic reaction that is a major energy source for reactions occurring during metabolism.

    N) Photosynthesis is used by phototrophs and metabolism by chemotrophs as the major source of Gibbs free energy to drive endergonic reactions needed for survival or the organism.

2. A) What combination(s) of enthalpy and entropy will always ensure a reactant-favored reaction? Explain.

    B) Circle the type of energy: entropy, enthalpy or free energy- that would be the best fit for the situations below and briefly explain the reason(s) for your choice:

    (a) Getting to choose wherever you wanted to sit when you go to a class is an example of what atoms experience as *enthalpy*   *entropy*   *free energy*

    (b) Wanting to sit next to a person you have a strong attraction for is an example of what atoms experience as: *enthalpy*   *entropy*   *free energy*

    (c) The person you want to sit next to is sitting much closer to the front of the room than you would prefer and having to decide which is more important is an example of what atoms experience as: *enthalpy*   *entropy*   *free energy*

    C) Can a reaction be both exergonic and exothermic? Explain.

    D) The human body gives off heat at a rate of 100 joules per second, while resting or sitting. A change in entropy occurs in the surroundings (the room you are sitting in) because of this loss of heat for the 50 minutes while you listen to the chemistry lecture, when room temperature is at 20°C. Assuming its a reversible process, estimate the change in entropy that has occurred and explain what you used to make the estimation.

3. Which of the following reactions would have a POSITIVE value for ΔS?

   A) $2\ Ag_2O(s) \rightarrow 4\ Ag(s) + O_2(g)$ B) $I_2(s) \rightarrow 2\ I\ (g)$

   C) $2\ CO(g) + O_2(g) \rightarrow 2\ CO_2(g)$ D) $H_2O(g) \rightarrow H_2O(s)$

4. For the reaction: $2\ H_2S(g) + 3\ O_2(g) <=> 2\ H_2O(l) + 2\ SO_2(g)$

   A) What is the $\Delta H°$ for the reaction?

         (a) -562 kJ   (b) 832 kJ   (c) -1124 kJ   (d) -1420 kJ (e) none of these

   B) What is the sign of the $\Delta S°$ for this reaction?

5. For a certain chemical reaction, $K = 460$ at 25°C, but changes to $K = 0.021$ at T= 300°C. Which of the following is TRUE for the reaction.

   A) $\Delta H°$ is endothermic and $\Delta S°$ is positive

   B) $\Delta H°$ is exothermic and $\Delta S°$ is positive

   C) $\Delta H°$ is endothermic and $\Delta S°$ is negative

   D) $\Delta H°$ is exothermic and $\Delta S°$ is negative

   E) cannot be determined from information given

6. For the following changes in the particles (at constant T) depicted in the nanoscale diagrams below, tell whether the ΔS is positive, negative, zero, or cannot be determined from description only.

A)
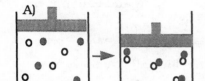
B)
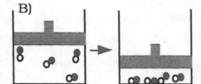
C)
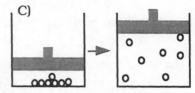

7. The free energy of formation, $\Delta G°_f$, of $SO_2(g)$ is - 300.2 kJ/mol and the $\Delta G°_f$ for $SO_3(g)$, - 371.1 kJ/mol. Which of the following is the value of $K°$ for the reaction:

$$2\ SO_2(g) + O_2(g) <=> 2\ SO_3(g)$$

A) 5.76   B) 7.18×10^{24} C) 1.24×10^{14}

D) 1.06   E) 3.29×10^2 F) none of these

8. Given that $\Delta S°$ for N_2, H_2 and NH_3 gases are 192, 130, and 192 J/ K, respectively, which of the following is TRUE for the reaction:

$$N_2(g) + 3\ H_2(g) <=> 2\ NH_3(g)\qquad \Delta H = - 92.2\ kJ$$

A) $K°$ would increase when T increases   B) $\Delta S°$ is - 390 J/K

C) The reaction is product-favored at 25°C   D) $\Delta S°$ is +390 J/K

E) The reaction is reactant-favored at 25°C   F) $\Delta S°$ is - 198 J/K

G) There will be a temperature at which the reaction changes direction

9. Using the diagram on the right, match the points (A) -(D) and the energy indicated by arrow E with the appropriate characteristic given below:

   (a) Represents the calculated value of $\Delta G°$ for the reaction.

   (b) Represents the equilibrium position for system, $Q = K°$.

   (c) Represents where $Q < K°$ and system shifting to right.

   (d) Represents where $Q > K°$ and system shifting to left

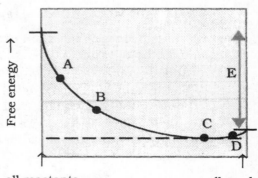

10. Mercury, water and bromine are all liquids at room temperature, but the observed molar entropies of each are $\Delta S°$ Hg(l) = 76.02, H_2O(l) = 69.91 and Br_2(l) = 159.82 J/mol-K .
    (a) What are the intermolecular forces in each of the liquids?
    (b) Using the intermolecular forces, explain the ordering of the $\Delta S°$ values.

11. In the stratosphere, ozone is produced from oxygen by the reaction: $3\ O_2(g) \Rightarrow 2\ O_3(g)$
    A) Using the tabled standard values calculate $\Delta H°$, $\Delta S°$ and $\Delta G°$ for the reaction, assuming standard conditions apply.
    B) Is the reaction product-favored or reactant-favored?
    C) What is the $K°$ for the reaction?
    D) Will $K°$ get smaller or larger at the much cooler temperature in the stratosphere, assuming $\Delta G°$ stays the same?

12. For the reaction; $4\ NH_3(g) +\ 5\ O_2(g) <=> 6\ H_2O(l) + 4\ NO(g)\ \ \Delta H° = -1169\ kJ$
    A) What is the value of $\Delta S°$ for the reaction?
    B) What is the value of $\Delta G°$ for the reaction?
    C) Considering your answer to (B) and the given information, is the reaction:
        (a) product-favored and exothermic   (b) product-favored and endothermic
        (c) reactant-favored and exothermic   (d) reactant-favored and endothermic
    D) Which of the following would be true for $K°$ of this reaction:
        (a) $K° = 1$ (b) $K°$ is very small   (c) $K°$ is very large   (d) $K° = 0$

13. Oxides of nitrogen can undergo a number of reactions. Two (unbalanced) reactions are:
    (a) $2\ NO(g)\ <=> ?\ N_2O(g) + \frac{1}{2}\ O_2\ (g)$ (b) $2\ NO(g)\ +\ O_2(g) <=> ?\ NO_2(g)$
    A)   Balance the reactions by determining the missing coefficient in each reaction
    B)   Using the tabled standard thermodynamic values, determine the $\Delta H°$, $\Delta S°$, and $\Delta G°$ for each reaction.
    C)   Which, if any, of the reactions will be product- favored at 298 K?
    D)   If the two reactions were coupled to produce an overall reaction, so that NO was an intermediate, which way would produce the greatest negative Gibbs free energy?
        (a) Add (a) to (b)   (b) Reverse (b), add to (a)   (c) Reverse (a), add to (b)
    E) Give the overall reaction for your choice in (D) and the value of $\Delta G°$ for the reaction.

14. One possible source of acid rain is the reaction between NO_2 from automobile exhaust and water droplets in the atmosphere $3\ NO_2(g) + H_2O(l) \Rightarrow 2\ HNO_3\ (aq) + NO(g)$
    A)   Using the tabled standard values what is the $\Delta G°$ for the reaction?
    B)   What is the $K°$ for the reaction?

15. At 700 °C, the Kp = 5.10 for the reaction: $CO(g) + H_2O(g) <=> CO_2(g) + H_2(g)$.
    A) What is the $\Delta G°$ at 700 °C for the reaction when Q = Kp?
    B) What is the value of free energy available when the extent of reaction is 0.33?

16. Wine will turn to vinegar if it is left exposed to the air because the ethanol is converted to acetic acid by the reaction: $C_2H_5OH(aq)\ +\ O_2(aq) \Rightarrow CH_3CO_2H(aq) + H_2O(l)$
    A)   Using the tabled values, what are the $\Delta H°$, $\Delta S°$ and $\Delta G°$ values for the reaction?
    B)   Which energy factor, $\Delta H°$ or $\Delta S°$, appears to dominate and determines whether the reaction is product-favored or reactant-favored at room temperature?
    C)   Would keeping the wine at a colder temperature than 25 °C decrease the amount of acetic acid formed at equilibrium, compared to the amount formed at 25°C? Explain the reason(s) for your choice.

17. The value of K_c equals 1.8×10^{-4} at 25°C for the decomposition of $NH_4HS(s)$:
$$NH_4HS(s) <=> NH_3(g) + H_2S(g)$$

   A) What is the value of $\Delta G°$ for the reaction at 25°C?

   B) Given that the $\Delta H°f$ for $NH_4HS(s)$ equals - 159.4 kJ/ mol, what will be the sign of the $\Delta H°$ for this reaction?

   C) Will there be a temperature at which this reaction will become product-favored? If so, should it be higher or lower than 25°C?

   E) If 10.0 g of $NH_4HS(s)$ is placed in a 1.0 L flask at 25°C, how much free energy will need to be absorbed to produce the equilibrium amount of products for the reaction?

18. For the following reactions, which, if any, will: (NOTE: Can be more than one or none)
   A) have a positive value for $\Delta G°$ at 25°C.

   B) have a negative value for $\Delta G°$ at 25°C.

   C) the magnitude of $T\Delta S°$ be larger than the magnitude of $\Delta H°$ at 25°C and determine the direction of the reaction.

   D) be product-favored at low temperatures, but not at high temperatures.

   E) be product-favored at high temperatures, but not at low temperatures.

   F) be product-favored at all temperatures.

Reaction	K_c at 25°C	$\Delta H°$(kJ)
(a) $CO(g) + Cl_2(g) <=> COCl_2(g)$	6.5×10^{11}	- 108.2
(b) $H_2(g) + Br_2(g) <=> 2\ HBr(g)$	2.0×10^9	- 72.5
(c) $CH_3OH(l) + CO(g) <=> CH_3CO_2H(l)$	0.315	- 37.8
(d) $2\ H_2(g) + CO(g) <=> CH_3OH(g)$	3.76	- 90.6
(e) $O_2(g) + N_2(g) <=> 2\ NO(g)$	1.7×10^{-3}	+ 180.2

Electrochemistry and Its Applications

Oxidation-reduction (redox) reactions are chemical reactions that involve a transfer of electrons. When the oxidation and reduction half-reactions occur at different locations, a flow of electrons occurs. **Electrochemistry** is the study of the relationship between electron flow and redox reactions.

19.1 Identify the oxidizing and reducing agents in a redox reaction.

19.2 Write equations for the oxidation and reduction half-reactions, and use them to balance the net equation.

A redox reaction can be recognized by **any one** of the following characteristics:
- the presence of chemicals known to be strong oxidizing agents (OA) or reducing agents (RA);
- recognition of a change in oxidation number;
- the presence of at least one pure element in the reaction.

Any one of the following observations is indicative of oxidation:
- <u>loss</u> of electrons;
- an <u>increase</u> in charge or oxidation number;
- electrons being <u>produced</u> (<u>right</u> side of reaction);
- causing another species to be <u>reduced</u>, also stated as "acting as a <u>reducing agent</u>."

If any one of these conditions is observed, all of the others are also present. In a reduction, observations that are the opposite of the underlined terms would be made. I n order to think about redox reactions in simpler terms, **half-reactions** are used. A net redox reaction can be balanced by considering each half-reaction separately, balancing it for atoms and charge, and then using a multiplier to make the number of electrons lost in oxidation equal to the number of electrons gained in reduction. The half-reactions are then combined and identical species appearing on both sides of the reaction eliminated. A list of reduction potentials and half-reactions may simplify balancing redox reactions.

19.3 Identify and describe the functions of the parts of an electrochemical cell; describe the direction of electron flow outside the cell and the direction of ion flow inside the cell.

Electrochemical or **voltaic cells** use product-favored reactions that produce electricity. In an electrochemical cell, the oxidation half-reaction and reduction half-reaction are arranged so that the reaction can only be completed by an external flow of electrons. A **salt bridge** is used to connect the two reactions chemically. Each **half-cell** consists of a pure element, which serves as the electrode, and a conducting solution. The salt bridge contains an unreactive ionic solution which allows flow of ions between the half-cells in order to maintain electrical neutrality. The external circuit is a wire that allows flow of electrons, and is connected to a meter or some other device that uses electricity. The oxidation half-reaction occurs at the electrode labeled as the **anode**. The electrons from this reaction flow through the external circuit to the **cathode**, which is the electrode where reduction occurs. Anions flow from the salt bridge toward the anode, and cations toward the cathode. Oxidation of the anode can be demonstrated by noting its loss in mass during an experiment; likewise the cathode gains mass. A shorthand notation for electrochemical cells uses the format: *anode| solution | | solution| cathode*, where the single vertical line indicates a phase boundary, and the double vertical line indicates a salt bridge.

19.4 Describe how standard reduction potentials are defined and use them to predict whether a reaction will be product-favored as written.

The force causing the flow of electrons in an electrochemical cell is called **electromo-**

tive force, or emf. It is equal to the number the potential energy difference in volts between the two electrodes. Another definition is the amount of work that must be done to move a particular charge. Recall that units of charge are coulombs, C, where C = amperes x seconds, and one joule = 1 V x 1 C. **Standard voltages**, E°, are described under standard conditions: 25ºC, 1 bar, and 1 M solutions. Cell voltages are calculated from $E^o_{cell} = E^o_{ox} + E^o_{red}$. A positive cell voltage defines a product-favored reaction. The reduction potential for half-reactions is determined by comparison with a standard reaction arbitrarily assigned a potential of 0 V. Values of reduction potentials are usually listed in tables with the most positive value, the reactant most easily reduced, at the top. An oxidation half-reaction is written as the reverse of reduction and $E^o_{ox} = -E^o_{red}$. Reduction tables can be used to predict whether a reaction will be product-favored as written. A combination of reactants from the upper left and the lower right will produce a product-favored reaction; the farther apart the two reactions, the greater the cell potential. Note that E°$_{cell}$ does not depend on the quantity of reactants, but only on the identity and concentrations of reactants used.

19.5 Calculate ΔGº from the value of E°$_{cell}$ for a redox reaction.

19.6 Explain how product-favored electrochemical reactions can be used to do useful work.

19.7 Explain how the Nernst equation relates concentrations of redox reactants to E°$_{cell}$.

19.8 Use the Nernst equation to calculate the potentials of cells that are not at standard conditions.

Standard cell potentials can be related to Gibbs free energy according to $\Delta G^o = -nFE^o_{cell}$, where n is the number of electrons transferred in the balanced redox reaction, and F = 96,485 C/mol, **Faraday's constant**, which relates charge to moles of electrons. A useful conversion factor is 1 C = 1 J/1 V. This relationship of ΔGº and E°$_{cell}$ also allows E°$_{cell}$ to be related to K, the equilibrium constant, using the formula $E^o_{cell} = \frac{RT}{nF} \ln K^o = \frac{0.0257\ V}{n} \ln K^o$. A variation of this formula at 25ºC is $E^o_{cell} = \frac{0.0592\ V}{n} \log K^o$. The connection between the values of K and E°$_{cell}$ allows calculation of E$_{cell}$, the cell potential when the concentrations of solutions in the cell are not 1 M. This formula, known as the **Nernst equation**, is $E_{cell} = E^o_{cell} - \frac{0.0592\ V}{n} \log Q$.

A **concentration cell** is one in which both the anode and cathode reaction are based on the same element, but the concentration of solution in each compartment is different. In a concentration cell, the value of the E°$_{cell}$ term is zero, so the voltage produced is dependent only on n and the ratio between the concentrations. A **pH meter** is a common application of a concentration cell.

19.9 Explain the source of the equilibrium potential across the membrane of a neuron cell.

19.10 Describe the chemistry of the dry cell, the mercury battery, and the lead-acid storage battery.

19.11 Describe how a fuel cell works, and indicate how it is different from a battery.

One example of a naturally occurring electrochemical cell is the function of nerve cells, which is based on the concentration cell. Under normal conditions, the concentration of various ions within nerve cells is different than in the extracellular fluid, and this difference is maintained by ion pump mechanisms fueled by ATP. When a nerve cell is stimulated, the membrane potential changes drastically from the normal potential. This generates an electrical current called the action potential, which then stimulates the next nerve cell to repeat the process.

Batteries are a portable source of energy using electrochemical reactions. **A primary battery** is used and then discarded; a **secondary battery** can be recharged and reused. The reaction for the common dry cell uses a zinc anode with zinc chloride, reacting with a graphite electrode in a mixture of ammonia, ammonium chloride, and manganese dioxide. The zinc is oxidized, and ammonium ions and manganese(IV) are reduced. Zinc is also the anode in a mercury battery, but the cathode is a mixture of graphite and mercury(II) oxide, with elemental mercury produced. Mercury batteries are useful because of their small size, but their disposal is hazardous. The lead-storage battery is a secondary battery based on the oxidation of lead and reduction of lead(IV) oxide. The product in both reactions is lead(II) sulfate, with the sulfate ions becoming available from sulfuric acid used as a conducting solution. The battery is recharged by oxidation and reduction of the lead(II) sulfate. **Fuel cells** also use chemical reactions to produce electricity, but the fuel must be continuously supplied from an external source. The most common example of a fuel cell uses gaseous hydrogen and gaseous oxygen.

19.12 Use standard reduction potentials to predict the products of electrolysis of an aqueous salt solution.

19.13 Calculate the quantity of product formed at an electrode during an electrolysis reaction, given the current passing through the cell and the time during which the current flows.

19.14 Explain how electroplating works.

Electrolysis is the use of electricity to force reactant-favored reactions to occur. A common application of electrolysis is the production of pure elements from their molten salts, such as NaCl. Another common application is the electrolysis of salts in aqueous solutions. However, these reactions can be complicated by the possibility of water reacting. In general, a metal can be reduced from aqueous solution if its reduction potential is greater than - 0.8 V, the reduction potential of water, and other species can be oxidized from aqueous solution if the oxidation potential is greater than -1.2 V, the oxidation potential of water. Overvoltage, the energy required to overcome the transfer rate of electrons at the electrode interface, means that the voltage required for such processes is always greater than predicted from the tables.

The amount of product, often metals, from an electrolysis can be predicted if the amperage, time, and identity of the reactant are given:
- Multiply the current (amps) by the time (seconds) given to determine the amount of charge (Coulombs) provided.
- Use Faradays constant (96,500 C/mol e⁻) to calculate the number of moles of electrons provided.
- Use the half-reaction to determine the number of moles of electrons required per mole of product.
- Convert moles of product to other units as needed.

Another application of electrolysis is electroplating, in which a surface is coated with a metal. This is often done for physical protection or for a more pleasing appearance. The object to be plated is used as the cathode in a solution of ions of the metal that is to be plated onto the surface.

19.15 Describe what corrosion is and how it can be prevented by cathodic protection.

Corrosion is the unwanted oxidation of a metal. Corrosion can result in structural damage if the metal is part of a bridge or the hull of a ship, or in a change in appearance if the metal is a protective coating. Structural damage poses a huge risk, as in collapse of bridges, and has a great economic impact. Corrosion can be prevented by protecting surfaces form contact with air or moisture by painting or by plating with a less reactive metal. The other common means of protection is the use of sacrificial metals, also called **cathodic protection**. In this method a more reactive metal is electrically connected to the surface to be protected.

This metal is preferentially (sacrificially) oxidized so that the protected metal acts as a cathode and remains intact.

Chapter Review - Key Terms

In order to make all the sentences below TRUE, insert the appropriate word or phrase from the list of key terms which best fits the context of the sentence.

NOTE: Any phrase or word from the list may used more than once.

LIST:

ampere	electrochemistry	pH meter
anode	electrode	primary battery
anodic inhibition	electrolysis	salt bridge
battery	electromotive force	secondary battery
cathode	emf	standard conditions
cathodic protection	Faraday constant	standard hydrogen electrode
cell voltage	fuel cell	standard reduction potential
concentration cell	half-cell	standard voltages(E°)
corrosion	half reaction	volt (V)
coulomb	Nernst equation	voltaic cell
electrochemical cell	neurons	

Components and Types of Electrochemical Cells:

(1) _____ is the study of chemical reactions that produce or consume electrical energy, which involve transfer of electrons between two chemical substances, or oxidation-reduction (redox) reactions. Each redox reaction can be thought of as the sum of two balanced (2) _____ s, one that describes the oxidation and the other, reduction. When the two (3) _____ s are separated from each other by external electrical circuitry, through which the electrons flow, an (4) _____ has been constructed and each (5) _____ is then referred to as a (6) _____. The oxidation reaction takes place in the (7) _____ called the (8) _____, and the reduction reaction occurs at the (9) _____. Each (10) _____ must contain an (11) _____, a solid conductor that provides for the flow of electrons to the external circuit. A (12) _____ or porous barrier must also be included, between the (13) _____ s, to allow ions to flow and prevent charge buildup in either cell.

When direct current electricity is produced, the cell is called a (14) _____ and the electron flow is always from the (15) _____ to the (16) _____. A (17) _____ is a portable version of a (18) _____. The force that "pushes" the electrons through the electrical circuitry in the cell is called the (19) _____, or (20) _____, and is the difference in electrical potential measured in units of (21) _____ s. The flow of electrons between the (22) _____ s can be measured as a current with units of (23) _____ s, which is the number of (24) _____ s of charge flowing per second through the electrical circuitry.

When the electrical current is made to flow in a direction that forces reactant-favored reactions to make large amounts of products, the cell is called an (25) _____ cell which is used to purify or plate out metals. In contrast to a (26) _____ cell, the (27) _____ consumes electricity rather than producing it. The current, in (28) _____ s, passed through a cell, multiplied by the time, in seconds, gives the total number of moles of electrons transferred. (29) _____, which states that there are 96,485 coulombs of charge for every one mole of electrons transferred, can then be used to calculate the number of moles of metal reduced or oxidized in an (30) _____ cell.

Measuring Voltages in Electrochemical Cells:

The (31) _____, E_{cell}, is the voltage difference measured between the two (32) _____s that make up the cell and has a magnitude which is affected by the conditions of measurement. For electrochemical measurements, (33) _____s are when all reactants are either pure liquid, pure solid, or gases at a pressure of 1 bar, and all solutes concentration are 1.0 M. For a full (34) _____, the (35) _____, is represented by $E°_{cell}$, where the superscript ° indicates (36) _____, and is equal to the sum of the oxidation, $E°_{ox}$ and reduction, $E°_{red}$, voltages from the (37) _____ s. The $E°_{cell}$ was then used to determine the specific voltage for each (38) _____, by measuring the voltage when it was combined with the (39) _____ assigned $E°_{ox}$ = 0.0 volts. The voltage measured for each (40) _____ under these conditions is tabled as the (41) _____. The (42) _____ occurring at the (43) _____ provides $E°_{ox}$ and the (44) _____ reaction from the table must be multiplied by -1 before adding it. The (45) _____ that is the (46) _____ is $E°_{red}$ and will have its (47) _____ added as it appears on the table. A (48) _____ always has a positive $E°_{cell}$ from $E°_{red}$ + $E°_{ox}$.

The standard free energy for the (49) _____ is equal to the maximum useful electrical work that can be obtained from a redox reaction and is related to the (50) _____ measured under (51) _____, by the equation, $\Delta G°$ = - $nFE°_{cell}$, where n equals the number of moles of electrons transferred in the redox reaction and F is the (52) _____. The (53) _____ describes how the (54) _____ changes when the cell is not under (55) _____ and the ratio of products to reactants is given by Q, the reaction quotient. The (56) _____ also shows that at equilibrium, when Q = K, the measured (57) _____ will be zero (so cell appears "dead") and $E°_{cell}$ and K are related by the equation: $E°_{cell} = \frac{0.0257 \text{ V}}{n} \ln K$.

When each (58) _____ contains the same ions, but at different concentrations, the cell is called a (59) _____ and the (60) _____ measured depends on the ratio of the two different concentrations of the ion. A (61) _____ is an instrument that uses a glass electrode, which is a (62) _____, since its voltage varies with the ratio of [H⁺] outside to [H⁺] inside of the electrode, as a reference electrode for pH. Changes in the voltages that result from concentration difference across a membrane are also what (63) _____, specialized cells that can generate electrical signals as a nerve impulse.

Concerning Applications of Electrochemical Cells:

All batteries are (64) _____ cells, but a (65) _____ uses a reaction that cannot be reversed, so that the battery cannot be recharged. A (66) _____ can be recharged, and its cell reaction is reversible, often because the products of the reaction stay on the electrode surface. A (67) _____, which produces electricity but is not a (68) _____, that converts the chemical energy of fuels directly into electricity, is called a (69) _____. Unlike a (70) _____, a (71) _____ does not need to be recharged since the reactants are continually introduced into the cell to produce electricity.

The oxidation of metal exposed to the environment is an electrochemical reaction known as (72) _____ in which the metal, as the (73) _____, is oxidized to an ion. Coating the metal with paint or anything that prevents the (74) _____ reaction from occurring can prevent the (75) _____ and is known as (76) _____. In a technique called (77) _____, two metals are brought into contact so that the first metal with the greater (78) _____, for which (79) _____ is to be prevented, becomes the (80) _____ and the second metal, with the lesser (81) _____, is oxidized (called the sacrificial anode) preventing the first metal from being oxidized.

- PRACTICE TESTS -

After completing your study of the chapter and the homework problems, the following questions, can be used to test yourself on how well you have achieved the chapter objectives.

1. Tell whether the following statements would be always **True or False**:

    A) Oxidation is more common than reduction.

    B) If silver metal acts as a reducing agent, silver ion is produced

    C) For a substance to act as an oxidizing agent it must undergo a gain of electrons.

    D) When an oxidation half reaction occurs, the element oxidized has a more positive oxidation number in the product form than in the reactant form.

    E) Fe^{2+} could be either an oxidizing and reducing agent.

    F) Reduction is the loss of electrons by an element to another element during a chemical reaction

    G) The chromium atom in $Cr_2O_7^{-2}$ has a higher oxidation number than Cr^{+3}.

    H) If one half-cell in a electrochemical cell has a atom X that loses two electrons per moles of X then the other half-cell must have an atom Y that will gain two electrons per mole of atoms Y.

    I) Cathode reactions are reduction half reactions.

    J) A porous membrane may function just like a salt bridge.

    K) Cell voltages are relative values set by the standard hydrogen electrode value.

    L) A cell with a negative $E°_{cell}$ value is a product favored reaction.

    M) Voltaic cells are systems that convert chemical energy into electrical energy.

    N) The Faraday constant is the charge on an electron.

    O) If $\Delta G°$ is negative, $E°$ will have a value greater than 1.0 V.

    P) The standard alkaline cell (battery) is an example of a secondary cell.

    Q) Both the cathode and anode in a lead storage battery have electrodes that are lead metal, which is why the cell reaction can be reversed for recharging.

    R) Corrosion of iron metal requires the reduction of H^+ ions in water, as the reduction half reaction, to occur.

    S) Electric water heaters often have a magnesium block ($E°= -2.37$ V) connected to the iron ($E°= -0.44$ V) heating element which provides cathodic protection to the heating element.

    T) To electroplate an object it must be made the cathode in a voltaic cell.

2. A) What is the difference between oxidation and oxidizing agent?

    B) For the reaction in hydrogen fuels cells: $2 H_2 + O_2 \rightarrow 2 H_2O$ tell:

        (1) which element is oxidized, which reduced,

        (2) which element is the oxidizing-agent, which the reducing agent and

        (3) why alkaline conditions are needed for the fuel cell to run.

    C) Why can't all redox reactions be used to construct batteries?

    D) What is the difference in the electrochemical reactions of a primary battery and a secondary battery?

    E) What type of substances typically function as an electrode, if a metal is not included in the half reaction and what is the basic criteria for the selection?

    F) Explain why a salt solution must be used instead of pure water in a salt bridge for an electrochemical cell.

    G) Would copper be a good metal to use for cathodic protection of iron? Explain.

3. For an electrochemical system at equilibrium, which of the following statements are FALSE:
    A) All reactants have been converted to products.
    B) The concentration of ions in the cell will appear constant.
    C) Current is flowing from anode to cathode
    D) Current is flowing from cathode to anode
    E) $E°_{cell}$ has become zero.

4 For the cell reaction: $2\ Br^-(aq) + F_2(g) <=> 2\ F^-(aq) + Br_2(g)$ If both gases, $F_2(g)$ and $Br_2(g)$, are kept at 1.0 atm, tell what will happen to the cell voltage, E_{cell} - will it *greater than, less than or stay the same as* $E°$ - when the concentration of:
    A) $[F^-]$ is double the $[Br^-]$ in solution?
    B) $[F^-]$ is one-half the $[Br^-]$ in solution??
    C) $[Br^-]$ and $[F^-]$ both equal 0.20 M?

5. Which of the choices below is TRUE for the cell reaction:
$$Zn(s) + 2\ HCl(aq) <=> ZnCl_2(aq) + H_2(g)$$
    A) K is less than 1.0 and $E°_{cell}$ greater than 1.0.
    B) K and $E°_{cell}$ are both less than 1.0
    C) K is greater than 1.0 and $E°_{cell}$ less than 1.0
    D) Zn^{2+} ions appear in the anode solution
    E) Cl is the oxidizing agent
    F) Hydrogen ion is reduced

6. What reaction products would you observe, if any, when:
    A) a clean strip of Fe(s) was immersed in an aqueous solution of $AgNO_3$?
    B) a stiff aqueous paste of mercury (I) chloride was rubbed on the surface of a clean strip of Aluminum metal?
    C) a clean strip of silver wire was placed in an aqueous solution of magnesium sulfate?

7. Which of the following statements about the electrochemical reaction below is FALSE?
$$Pb^{2+} + Zn(s) \rightarrow Pb(s) + Zn^{2+} \quad E°_{cell} = 0.637$$
    A) The Zn electrode is the cathode
    B) The reaction will go in the direction indicated
    C) The shorthand notation for the cell is : $Zn(s)|\ Zn^{2+}|\ Pb^{2+}|\ Pb(s)$
    D) If initially $[Pb^{2+}] = 1.0$ M and $[Zn^{2+}]$ is less than 1.0 M , E_{cell} is less than $E°$.
    E) The cell voltage would be 0.637 V only when all reactants have been converted to products.

8. Tell which of the following would be TRUE statements, given the following $E°_{red}$ for each:

Ion/Metal:	Mg^{2+}/Mg	Fe^{2+}/Fe	Sn^{2+}/Sn	Cu^{2+}/Cu	Zn^{2+}/Zn
$E°_{red}$	-2.37 V	- 0.44 V	- 0.14 V	+ 0.34 V	- 0.78 V

    A) Mg will act as a reducing agent for Zn^{2+}
    B) Cu will be oxidized in a solution containing Sn^{2+}
    C) Fe will be oxidized by Zn^{2+}
    D) Fe^{2+} will be reduced and H^+ oxidized in an acidic solution

9. What would the products would be formed in the electrolysis of:
    A) molten LiBr   B) RbBr(aq)

10. Given <u>acidic conditions</u> for the following unbalanced redox reaction, answer the questions
   given below: $NO(g) + MnO_2 (s) <=> Mn^{2+} + NO_3^-$
      A) For the <u>reduction</u>, (a) What element is being reduced?
                               (b) Initial and final oxidation numbers for the element?
                               (c) Write the balanced reduction half reaction
      B) For the <u>oxidation</u>, (a) What element is being oxidized?
                               (b) Initial and final oxidation numbers for the element?
                               (c) Write the balanced oxidation half reaction
      C) Write the balanced overall reaction and calculate the value of $E^°_{cell}$

11. For the following cell reaction: Cu^{2+} (3.0M) + Zn(s) <=> Cu(s) + Zn^{2+} (0.0001M)
      A) Diagram the cell, using the beakers given below and indicate:
         (a) which beaker is the cathode and which the anode
         (b) the chemical identity of both electrodes and any
            ions in the solutions
         (c) other part(s) needed for cell to function by draw-
            ing them in and labeling them.
      B) What is the correct magnitude for the cell voltage, E_{cell}?
      C) What is the value of K for this cell reaction? Would the concentrations above be near the
      start or near equilibrium for the reaction?

12. For the two half reactions: $AgI(s) + e^- <=> Ag(s) + I^-$ and $F_2(g) + 2 e^- <=> 2 F^-$
      A) Write the balanced overall reaction for a voltaic cell made
from the two half reactions.
      B) Calculate $E^°_{cell}$ and $\Delta G^°$ for the cell
      C) Using the figure on the right:
         (a) Diagram the cell by indicating and indicate the ALL chemi-
         cal species that must be present for the cell to function and
         their placement in the cell. Indicate the anode and cathode.
         (b) Draw in any missing chemical and electrical components
         needed for a functioning cell.

13. Given the information below:
      $Cr^{3+} + 3 e^- \rightarrow Cr(s)$ $E^°_{red}$ = - 0.74 V $Cu^{2+} + 2 e^- \rightarrow Cu(s)$ $E^°_{red}$ = 0.34 V
   $Cr_2O_7^{2-} + 14 H^+ + 6e^- \rightarrow 2 Cr^{3+} + 7 H_2O$ $E^°_{red}$ = 1.33 V $Cu^+ + e^- \rightarrow Cu(s)$ $E^°_{red}$ = 0.16 V
      A) Which form of Cr, Cr^{3+} or $Cr_2O_7^{2-}$, is the best oxidizing agent?
      B) Which form of Cu, Cu^+ or Cu^{2+}, is the best reducing agent?
      C) Using two of the half-reactions above, write the balanced overall reaction between a
      Cu and Cr half reaction that would produce the <u>largest</u> $E^°_{cell}$.

14. A cell filled with 1.0L of 2.00M KCl undergoes electrolysis. A current of 2.0 amperes is
   passed through the solution for 8 hours to oxidize Cl^- ions. If the reaction that occurs is:
         $2 H_2O + 2 Cl^- <=> H_2(g) + Cl_2(g) + 2 OH^-$
      A) What are the cathode and anode reactions that are occurring?
      B) What is the standard emf, $E^°_{cell}$ for the cell reaction?
      C) What would be the pH in the solution after the 8.0 hours?

15. A given amount of electricity is passed through two cells simultaneously. One cell contained Cu^{2+} ions and the other, Cr^{3+} ions. During the electrolysis the current applied was
   10 amperes to both cells and 2.542 g of Cu(s) was deposited.
      A) How many moles of electrons have passed through each cell?
      B) How many grams of Cr were also deposited in the second cell?
      C) How many minutes was the current run through the cells?

Chapter 20

Nuclear Chemistry

This chapter covers the topic of nuclear chemistry, which includes the results of changes in atomic nuclei, including spontaneous decay, transmutation, fission, and fusion; the particles and energy associated with such changes; the units used to measure radioactivity; and the applications of nuclear chemistry.

20.1 Characterize the three major types of radiation observed in radioactive decay: alpha (α), beta (β), and gamma (γ).

Radioactivity is the spontaneous emission of energy or particles from a nucleus. The **alpha particle** (α) is also symbolized by $^4_2$He. The superscript indicates that the mass number of the a particle is 4 and the subscript indicates that its atomic number is 2. Thus the alpha particle is the same as a helium nucleus. Note that the +2 charge is not usually indicated. Alpha particles are much more massive than the other particles, and therefore have relatively small penetrating power. Skin or clothing will prevent penetration by alpha particles. The **beta**

(β) particle is also symbolized by $^{0}_{-1}$e. Note that the beta particle has a mass number of 0; the subscript of negative one indicates that its charge is opposite that of a proton. A beta particle is an electron which is ejected from the nucleus at high speed. It is indistinguishable from other electrons. A beta particle has higher penetrating power than an alpha particle, but it can be stopped by a 5 mm layer of aluminum foil. **Gamma (γ) radiation** is a high energy form of radiation which has very high penetrating power.

20.2 Write a balanced equation for a nuclear reaction or transmutation.

20.3 Decide whether a particular radioactive isotope will decay by α, β, or positron emission or by electron capture.

20.4 Calculate the binding energy for a particular isotope and understand what this energy means in terms of nuclear stability.

Nuclear reactions can be written to illustrate the transformation of one isotope into another when nuclear decay occurs. In radioactive decay the reactant is called the parent nucleus and the product nucleus is called the daughter. The basic principle of balancing nuclear reactions is that the sum of the mass numbers (superscripts) on each side must be the same, and the sum of the atomic numbers (subscripts) on each side must be the same. The symbol for the daughter nucleus is determined from the Periodic Table after the atomic number has been calculated. Note that alpha emission results in a nucleus with two fewer protons, and a mass number four less than the original nucleus. In beta emission the atomic number of the daughter increases by one while the mass number remains the same.

Two other forms of radioactive decay are sometimes observed: positron emission and electron capture. In **positron emission**, a particle identical to the beta particle, except with a positive charge, is emitted from the nucleus. As a result the atomic number decreases by one while the mass number remains constant. As the name suggests, in **electron capture** an inner shell electron is captured by the nucleus. This results in a decrease of one in the atomic number with no change in the mass number.

In transmutation, a nuclear decay is caused by bombarding a nucleus with a small fast-moving particle, such as a or b particles, hydrogen nuclei, or neutrons. Transmutations usually produce one or more nuclei smaller than the original and one or more very small particles. The nuclear arithmetic of transmutations follows the same principles as natural decays.

Nuclear decay occurs because of excessive repulsive forces within the nucleus. Some observations that help to predict stability of isotopes are:

- Where Z = 20 or less, stable isotopes have equal numbers of protons and electrons;

- Where Z > 20, a ratio of neutrons to protons greater than one is needed for stability, and the required ratio increases as Z increases;

- All elements with Z > 83 are unstable and radioactive.

From these guidelines and a plot of number of neutrons versus number of protons for stable and unstable isotopes, predictions about the stability and probable type of decay can be made. Another way of predicting stability is by calculating the value of the binding energy of a nucleus. **Binding energy** is the energy that would be released if all the nuclear particles in a nucleus were separated. Its magnitude is related to the mass defect, the discrepancy between the actual mass of a nucleus and the mass predicted on the basis of the number of protons and neutrons. **Binding energy per nucleon** is the value of the binding energy divided by the number of protons plus neutrons. The larger the value of the binding energy per nucleon, the more stable the nucleus will be.

20.5 Use the equation $\ln \dfrac{N}{N_o} = -kt$, which relates (through the decay constant k) the time period over which a sample is observed (t) to the number of radioactive atoms present at the beginning (N_o) and end (N) of the time period.

20.6 Calculate the half-life of a radioactive isotope ($t_{1/2}$) from the activity of a sample, or use the half-life to find the time required for an isotope to decay to a particular activity.

Radioactive decay follows first-order kinetics, and the formulas developed in Chapter 13 can be used. Half-life is the amount of time required for 1/2 of a given sample to undergo radioactive decay and is often used to express the relative stability of an isotope. The formula relating half-life and decay constant is $t_{1/2} = \dfrac{0.693}{k}$. Half-life and decay constant are easily interconverted using this formula. Note that an isotope with a long half-life has a relatively small value of k and is relatively stable. The half-life formula is derived from the integrated rate law for first order reactions, $\ln \dfrac{N}{N_o} = -kt$. This more general formula can be used to predict amount of radioactivity after a certain time period or to calculate the time required to reach a certain level of activity. The usual unit of activity is the **becquerel, Bq**, which is one disintegration per second, or the **curie, Ci**, which is 3.7×10^{10} disintegrations per second. In the rate law equation, concentration, mass or activity can be used as units of N.

20.7 Describe nuclear chain reactions, nuclear fission, and nuclear fusion.

20.8 Describe the basic functioning of a nuclear power reactor.

Nuclear fission is the splitting of large nuclei into smaller ones by bombardment with neutrons, producing two or more additional neutrons. Since fission produces at least two neutrons, the process can be propagated, resulting in a chain reaction. Thus fission reactions are useful for producing large amounts of energy with a small initial investment of energy. When this process is controlled, it is called a **nuclear reactor**; uncontrolled it is called an atomic bomb. A nuclear reactor uses uranium-235 as a fuel. When the fuel is bombarded with neutrons, three more neutrons are produced, along with daughter nuclei and a huge amount of energy. The neutrons can cause additional fissions, and the energy is absorbed by a primary coolant. The primary coolant then transfers its heat to a secondary coolant, which is used to heat water, generating steam to drive a turbine to produce electricity. The reaction can be moderated as necessary by inserting control rods, which are mad of materials that readily absorb neutrons, into the fuel rod arrangement.

Nuclear fusion is the joining of very small nuclei (usually hydrogen or helium) to form larger nuclei. This is the reaction that produces the energy of the sun and other stars. Practical considerations in the development of fusion as an energy source include the large energy needed to initiate the reaction and the extremely high temperatures associated with the process.

20.9 Describe some sources of background radiation and the units used to measure radiation.

Humans are constantly exposed to naturally occurring radiation, commonly called background radiation. Sources include cosmic radiation and decay of naturally occurring isotopes such as potassium, thorium, uranium, and in some localities, radon. Artificial sources include medical procedures, consumer products, and nuclear wastes. Units used to measure human exposure to radioactivity are given in Table 1 below.

Table 1: Units for Measuring Human Exposure to Radioactivity

Unit	Symbol	Definition	Comments
roentgen	R	93.3×10^{-7} J/g tissue	Usually used to measure X-rays and γ rays
radiation absorbed dose	rad	1.00×10^{-2} J/kg	Usually used to measure X-rays and γ rays
gray	Gy	1 J/kg; 1 Gy = 100 rad	Usually used to measure X-rays and γ rays
roentgen equivalent applied to mammals	rem	quality factor $\times$ dose in rads	Allows for differing effects of different particles
sievert	Sv	quality factor $\times$ dose in grays; 1 Sv = 100 rem	Allows for differing effects of different particles

Rems are most commonly used in the nuclear industry to measure exposure.

20.10 Give examples of some uses of radioisotopes.

In addition to nuclear fuels, other uses of radioisotopes include:

• *Carbon-14 dating* This is based on the fact that living things have the same C-12 to C-14 ration in their tissue as the atmosphere, but when they die the ratio changes because new carbon is no longer being incorporated into tissues. Measurement of this ratio allows calculation of the amount of time since death.

• *Food irradiation* This uses gamma rays to kill bacteria, molds, and yeast, which contribute to food spoilage.

• *Radioactive tracers* to study chemical or biological processes. Since radioactive elements have the same chemical behavior as non-radioactive elements, chemicals can be "tagged" with an isotope and followed through various steps of a process by using appropriate detection equipment.

• *Medical imaging.* This uses radioactive isotopes that are selectively absorbed by certain organs to allow visualization of that organ in order to study its function.

• *Medical treatment.* Since radiation disrupts cell function, it can be used to kill cancer cells by disrupting their function.

Chapter Review - Key Terms

In order to make all the sentences below TRUE, insert the appropriate word or phrase from the list of key terms which best fits the context of the sentence.

NOTE: Any phrase or word from the list may used more than once.

LIST:

activity	curie (Ci)	nucleons
alpha (α) particles	electron capture	plasma
alpha radiation	gamma radiation	positron
background radiation	gray(Gy)	rad
becquerel(Bq)	nuclear fission	radioactive series
beta (β) particles	nuclear fusion	rem
beta radiation	nuclear medicine	roentgen (R)
binding energy	nuclear reactions	sievert
binding energy per nucleon	nuclear reactor	tracers
critical mass		

Concerning Natural Decay of Radioactive Atoms:

Unstable atoms, also called radioactive atoms, can spontaneously change their chemical identity through (1) _____, where the nucleus of the atom emits radiation and/or a particle and converts to an atom with a different set of chemical properties. This phenomena was first observed when minerals containing uranium and thorium were shown to emit at least two types of invisible radiation, (2) _____ and (3) _____ which contained high energy particles that responded differently to an applied electric field. The (4) _____ was attracted to the negative part of the field and contained particles called (5) _____, later identified as positively charged helium nuclei. (6) _____ had a negative charge and consisted of (7) _____ which proved to have a charge and mass identical to an electron, but came from the nucleus of the atom. A third type of radiation was also identified that was higher in energy and unaffected by an electrical field called (8) _____.

(9) _____ is different from (10) _____ and (11) _____ in that is has zero mass and far greater energy and penetrating ability than either of the first two types. Less frequent types of nuclear transformations are those of (12) _____ emission, where a positively-charged electron is emitted from the nucleus, and (13) _____, where an inner shell electron is captured by the nucleus of the atom.

Both protons and neutrons represent the (14) _____, the particles that make up the nucleus of any atom. When all the products and reactant (15) _____ are counted in nuclear transformations, it is apparent that the total number of nucleons does not change, so that (16) _____ are conserved in (17) _____, like atoms are conserved in chemical reactions. Most unstable atoms, like Uranium-238, undergo a sequence of successive transformations, called a (18) _____, before a stable nucleus is produced.

A extremely small change in mass accompanies all (19) _____, even though the number of (20) _____ does not change. It is presumed the change in mass is converted to energy and is related to the binding force holding the (21) _____ together in the nucleus which is an extremely strong, but very short-range force. The (22) _____, ΔE, a measure of the binding force can be calculated from the change in mass, Δm, and the equation, $\Delta E = \Delta mc^2$, where c is the speed of light. To calculate the (23) _____ for formation of a particular atom, the change in mass for the formation of nucleus can be calculated by taking the difference between the measured mass of the isotope and the total mass of the separate protons and neutrons that make up the nucleus. The (24) _____, can then be calculated by dividing the formation (25) _____ by the total number of

(26) _____ in the chemical element, a value which shows why some elements are more stable than others.

The actual rate of radioactive decay of atoms follows first order kinetics and can be measured as the rate of disintegrations per unit time called the (27) _____ for the atom. The unit (28) _____ is used to represent one atom disintegrating per second (1 dps). A more commonly used unit, the (29) _____ represents the rate observed for 1.0 g of radium-226, which is 3.7 X 10^{10} disintegrations per second. In addition, the half-life of the (30) _____ for decay can be used for dating ancient objects or for estimating the time needed for the amount of a radioactive substance to decay to a certain level.

Concerning Other Nuclear Reactions and Units for Exposure:

Besides decay, atoms can undergo either (31) _____, where the nucleus splits into two unequal parts and some neutrons, and (32) _____, where two small nuclei combine to form a larger nuclei. Either reaction that releases energy in much larger quantities than normal radioactive decay which originated in the (33) _____ of nucleons . Electricity can be generated by a (34) _____ which uses the releaseof (35) _____ from a (36) _____ type of reaction to provide thermal energy, which is then converted to electricity. The amount of fissionable material in the fuel rod is important in that a (37) _____ of the radioactive isotope must be present in the (38) _____ to produce a self-sustaining reaction. Consequently, some fuel rods must be periodically removed from the (39) _____ and replaced to obtain the (40) _____in the reactor.

(41) _____ reactions occur naturally on the surface of stars, like the Sun, but are very difficult to make happen and control, compared to (42) _____ reactions. (43) _____ reactions require extremely high temperatures, in millions of degrees, so that the (44) _____ can form a (45) _____ state, where the separated nuclear particles then can combine to form new, larger atoms, releasing the (46) _____.

Radiation from (47) _____ on the Sun, natural decay of atoms and some man made nuclear activities result in everything on the Earth's surface being exposed to radiation, called (48) _____. To measure the exposure, several units have been developed that gives the amount of energy of the radiation absorbed or administered per unit mass of the recipient object To measure the radiation administered for something such as an X-ray, the unit called (49) _____ is used where 1 R = 93.3 X 10^{-7} joule per second. Since not all the radiation administered will be absorbed, a second unit the (50) _____, short for radiation absorbed dose, is used which equals the absorption of 0.01 joule per kilogram of material. The SI unit for absorbed dose is the (51) _____ equal to 1 joule per kilogram, which is the same as absorbing 100 (52) _____s.

Since the biological effect has been found to vary with the type of radiation, the dose in (53) _____s needs to be multiplied by a quality factor to get the adjusted, effective dose measured in (54) _____s, which is short for radiation equivalent applied to mammals. The quality factor is about one for exposure to (55) _____ or (56) _____, but increases to 10-20 for exposure to (57) _____. The (58) _____ is the most useful unit for measuring exposure, but is too large a unit for normal exposure, so that the milli (59) _____ is more often used. The typical level for (60) _____, the natural unavoidable level of exposure, results in each person being exposed to less than 0.5 (61) _____ per year, or less than 500 millirems, of radiation. The (62) _____ is the SI unit for the effective dose and is equal to 100 (63) _____.

Radioactive isotopes are useful diagnostic or therapeutic tools in (64) _____.
Because the chemical properties of the radioactive isotope are identical to that of the nonradioactive isotopes, these atoms can often be used as (65) _____ to determine the location
of tumors, follow blood circulation, or other processes. The small quantity of radiation emitted
by the short-lived (66) _____ makes it possible to detect their exact position while in
the body or to follow the progress of their elimination from the body.

- PRACTICE TESTS -

*After completing your study of the chapter and the homework problems, the following questions,
can be used to test yourself on how well you have achieved the chapter objectives.*

1. Tell whether the following statements would be always **True or False**:

    A) The radioactive characteristics of an atom undergo a major change when the
        element is reacted to form a compound.
    B) Alpha radiation is the least penetrating of the common types of radiation.
    C) Only gamma radiation has the ability to cause extensive ionization, but not
        alpha and beta radiation.
    D) Loss of a beta particle results in a new isotope that has a lower mass.
    E) The final product of the radioactive series for Uranium-238 is Radon-222.
    F) When carbon-14 decays by beta emission the new isotope is of the element
        nitrogen.
    G) When an atomic mass is given in parentheses in the periodic table all isotopes of
        that element are radioactive.
    H) The oxygen-16 atom is less stable than an aluminum-27 atom in terms of
        binding energy per nucleon.
    I) The proton to neutron ratio must be greater than 1.0 for an atom to be stable.
    J) The greater the change in mass when an atom is formed from nucleons, the
        more stable the nucleus will be.
    K) The source of nuclear energy is the binding energy of the nucleons.
    L) Both nuclear fission and fusion result in the production of atom(s) with more
        stability than the original atom.
    M) Transuranium elements are artificial elements and are lanthanides.
    N) Uranium-238 is fissionable and is used as fuel in a nuclear reactor.
    O) Nearly pure fissionable fuel is required for the nuclear reactor.
    P) The fusion process requires a plasma state to occur to a significant level and
        nuclear fission requires a critical mass.
    Q) A stated half-life values refer only to a specific isotope and not to all isotopes of
        an element.
    R) Radioactive decay processes have half lives that do not change with concentra-
        tion or number of atoms.
    S) When considering the activity of a radioactive sample, all you have to consider is
        the half-life to determine the activity of the sample.
    T) A plasma occurs at extremely high temperatures and can occur in a nuclear
        reactor.
    U) The radiation unit "rad" is a measure of radiation absorbed.
    V) Background radiation is measured in rems which is the dose in rads multiplied
        by quality factors.
    W) Radon gas can contribute to background radiation.

X) Radioactive tracers are useful in nuclear medicine because they have the same chemical properties as their nonradioactive isotopes.

Y) An alpha particle is an electron and a proton from the nucleus

2. A) Compare alpha, beta and gamma radiation in terms of:
    (a) penetration ability.
    (b) which is the most dangerous if emitted when an atom is inside the body? Give the reason(s) for your choice.

    B) Explain how chemical and nuclear reactions differ in the:
    (a) magnitude of energy change?
    (b) effect on rate of increasing temperature by $100°C$?
    (c) the source of the energy released?

    C) Describe the main factor(s) that determine whether an isotope will be stable or radioactive?

    D) Describe transmutation and how it is different from nuclear fission or decay.

3. Which of the following statements would be TRUE for Uranium-238?
    A) Its chemical properties will be the same as those of Uranium-235.
    B) Its mass will be slightly less than that of Uranium-235.
    C) It contains a different number of protons than Uranium-235.
    D) It has a greater relative abundance than U-235.

4. To complete a cherished dream of the alchemist's, an atom of Mercury-198 was converted to Gold-197 using neutron bombardment (that is, reacting the mercury atom with a neutron). What was the other product from the reaction?

5. Fill in the blanks for the following nuclear reactions and define the type of decay that the reaction represents.

   A) $_{88}^{226}\text{Ra} \rightarrow \,_{2}^{4}\text{He} +$ ___ _____   B) $_{17}^{34}\text{Cl} \rightarrow \,_{-1}^{}\text{e} +$ ___ _____

6. For the following reaction, which of the statements below would be TRUE:

   Reaction: $_{7}^{14}\text{N} + \,_{2}^{4}\text{He} \rightarrow \,_{8}^{17}\text{O} + \,_{1}^{1}\text{H}$

   Mass: 13.992   4.0015   16.9986   1.0073 g/mole

    A) The energy change for the reaction is about 1.12×10^{12} kJ per mole of particles.
    B) The energy change for the reaction is about 1.12×10^{9} kJ per mole of particles.
    C) The energy change for the reaction is about 1.12×10^{15} kJ per mole of particles.
    D) This is an example of a transmutation reaction.
    E) This is an example of alpha decay reaction.

7. Write the balanced nuclear reaction for the following nuclear reactions. Show all components of the reactions in proper symbol notation.
    A) Decay of ^{157}Eu by beta particle emission
    B) Bromine-75 undergoes positron emission
    C) Cadmium-114 undergoes electron capture
    D) Decay of Plutonium- 239 by alpha particle emission.

8. When a ^{27}Al nucleus absorbs an alpha particle, it emits a neutron. What is the atomic number, atomic mass and chemical name for the nucleus left?

9. Does the neutron/proton ration increase, decrease or stay the same when an atom:
    A) emits a beta particle?   B) emits a positron?   C) emits a alpha particle?

10. Argon-41 undergoes beta decay. A sample of ^{41}Ar has an initial activity of 5.12×10^5 Bq and is the activity changes to 5.28×10^4 Bq after 6.0 hours have elapsed. What is the half life of ^{41}Ar in hours?

11. Ytterbium-164 has a half-life of 75.0 minutes. After 6 hours, what weight of Ytterbium-164 will be left of a 80.0 gram sample?

12. Sodium-24 has a half-life of 15 hours and is used to study blood circulation. If a patient is injected with a ^{24}NaCl solution with an activity of 2.5×10^9 Bq, what will be the activity after 4 days?

13. The maximum safe level of radon-222 is set at 4.0 picocuries per liter. At this level, how many atoms of radon-222 would be disintegrating per second in a room that was 8 ft X 10 ft X 8 ft? (*1 ft^3 = 28.3 L*)

14. A patient weighing 65 kg was given an intravenous dose of 15.0 mL of a 2.5×10^{-9} M Technetium-99 (^{99}Tc) a radioactive isotope that decays by gamma emission. The energy of the gamma ray from the decay is $1.35 \times 10^{+7}$ kJ/mol and the half-life of ^{99}Tc is 6.0 hours.

    A) How many atoms of Technetium-99 have been introduced into the patient?

    B) How many gamma rays would be emitted over a period of 6.0 hours?

    C) If it was assumed that only 1% of the energy released by the decay is absorbed by the patient's body, how many rads did the patient absorb in 6.0 hours?

15. Polonium-210 has a half life equal to 138.4 days and is an alpha emitter.

    A) Suppose 10.0 g of ^{210}Po is placed in a container and kept for two years. What percentage of the original mass of ^{210}Po will be left in the container?

    B) Will the amount of solid appear about the same or significantly less in the container after the 2.0 years? Briefly explain the reason(s) for your choice.

16. Strontium-90 (Sr-90) is a beta-emitter and considered one of the more hazardous by-products of nuclear fission since it can mimic Calcium ion and be stored in bones of humans and animals. The half-life of Strontium-90 is 28.1 years,

    A) Write the rate law for the Strontium-90 decay, and define each term in the law.

    B) What is the value of the decay constant in sec^{-1}? (Use 1 year= 365.25 days)

    C) What will be the initial activity of Strontium-90 in becquerels from 1.00 gram of Sr-90? What would it be in curies?

    D) By what percent would be the activity of the 1.00 gram Sr-90 sample be decreased after 100 years have elapsed?

Chapter 21

Chemistry of Selected Main Group Elements

This chapter covers formation, purification and properties of selected main group elements, which are the elements in the "A" columns of the Periodic Table.

21.1 Give a general explanation of how elements form in stars.

Nuclear burning (not the same as combustion) is the general term for the process in which elements up to iron are formed in stars. These elements are formed by fusion reactions which are given specific names according to the reactant atoms. For example, **helium burning** is the fusion of two $_2^4 He$ nuclei to produce $_4^8 Be$. Nuclei with odd values for Z or A are formed when a $_1^1 H$ is emitted after the burning reaction. Elements heavier than iron are formed by neutron capture in one of two ways:

- *s process*, in which neutrons are captured slowly, or
- *r process*, in which a succession of neutrons are captured rapidly.

21.2 Know the principal elements in the earth's crust.

21.3 Describe the general structure of silicates.

21.4 Identify the general methods by which elements are extracted from the earth's crust.

The ten most common elements in the earth's crust, in order of abundance by mass, are O, Si, Al, Fe, Ca, Na, K, Mg, H, and Ti. These elements are never found as pure elements, but as **minerals**, naturally occurring inorganic compounds with definite crystal structures. More than 75% by mass of the earth's crust is composed of oxygen and silicon, usually in the form of silica, SiO_2, or combined with other elements as silicates. Silicates contain the SiO_4^{2-} ion, in which the Si is surrounded tetrahedrally by the four oxygen atoms, which can be shared in order to form network solids composed of chains, sheets, or rings. **Ores** are mineral deposits from which it is economically feasible to extract pure elements. Physical methods, electrolytic redox reactions, and chemical redox reactions are common methods of obtaining pure elements from their mixtures or compounds.

21.5 Identify the major components of the atmosphere.

21.6 Explain how elements are obtained by the liquefaction of air.

The major elements found in the atmosphere are, in order of abundance, nitrogen, oxygen, and argon. These can be separated by fractional distillation of liquid air using the Linde process. First the air is liquefied by lowering the temperature and raising the pressure until condensation occurs. Then the nitrogen is allowed to boil off, followed by the oxygen. Oxygen produced by this process is used in steel-making, as an oxidizing agent for other processes, and in rocket propulsion. Pure nitrogen is used primarily in the manufacture of fertilizers by the Haber-Bosch process, and in providing an inert atmosphere. Both liquid oxygen and liquid nitrogen are used as **cryogens** in applications where extremely low temperatures are required.

21.7 Describe the Frasch process for obtaining sulfur.

21.8 Explain how sulfuric acid is produced.

Sulfur is found in underground deposits as a pure element, and is usually mined using the **Frasch process**. In this process superheated steam and compressed air are forced down a shaft into the deposit, which melts and then is forced to the surface by the compressed air. Sulfur can also be recovered from combustion of high-sulfur fossil fuels. Sulfur has two common allotropes, rhombic and monoclinic. It is important in protein chemistry as a component of the amino acids cysteine and methionine. Sulfuric acid is produced by the contact process, in which sulfur is oxidized stepwise to SO_2 and SO_3, and then reacted with water in a two-step process. Waste SO_2 from combustion can be used directly to manufacture sulfuric acid.

21.9 Describe how electrolysis is used to obtain sodium, chlorine, magnesium, and aluminum.

Halogens and reactive metals can be extracted from their compounds using electrolysis. In the Downs cell, molten NaCl is electrolyzed to produce pure Na and Cl_2. The **chlor-alkali** process is also used to obtain chlorine by electrolysis of aqueous NaCl. NaOH, an important industrial chemical, is also produced. Magnesium is one of the few elements that can be economically extracted from seawater. The Dow process first uses the insolubility of magnesium hydroxide to form a precipitate which is readily converted to magnesium chloride, which can be electrolyzed. Aluminum is purified from bauxite using the Hall-Heroult process, in which the Al_2O_3 is dissolved in molten cryolite, Na_3AlF_6. This mixture is electrolyzed using graphite anodes, producing pure aluminum and carbon dioxide. Since this process is so energy-intensive, recycling of aluminum makes especially good sense.

21.10 Explain how chemical redox reactions are used to extract bromine, iodine, and phosphorus from compounds.

Phosphorus is extracted from phosphate ores by heating in the presence of coke, which acts as a reducing agent. Thus the phosphate is reduced to P while the C is oxidized to CO. Phosphorus is used mainly in the production of phosphoric acid for use in fertilizers. It is important in biochemistry as a component of DNA, RNA, and ATP, and as a component of bones and teeth. As a pure element phosphorus has three allotropes. Bromine and iodine are produced by displacement from solutions of seawater or brine Chlorine gas is used to oxidize the anions to the pure elements.

Chapter Review - Key Terms

In order to make all the sentences below TRUE, insert the appropriate word or phrase from the list of key terms which best fits the context of the sentence.

NOTE: Any phrase or word from the list may used more than once.

LIST:		
chlor-alkali process	helium burning	nitrogen fixation
cryogen	hydrogen burning	nuclear burning
Frasch process	mineral	ores

Concerning Formation of Elements and Processes to Obtain or Convert Some Elements:

Chemical elements were formed, and are still being formed, in stars in a process called (1) _____, which releases large amounts of heat and light energy, but involves fusing small nuclei together to form larger nuclei, not oxidation of a fuel. The first step of the formation process occurs within a star and is called (2) _____, where four hydrogen-1 nuclei fuse to form a helium-4 nucleus, beta particles, gamma radiation and energy. After billions of years, (3) _____ takes place, as a second phase, where helium-4 nuclei fuse with each nuclei to form larger atoms.

Many elements are found combined with other elements as solid inorganic compounds that have a characteristic crystalline structure and fixed composition in the Earth's crust called (4) _____s. (5) _____ are (6) _____s that contain a high enough concentration of element to make extraction of the element profitable.

Elements such as O_2 and N_2 make up most of the atmosphere and can be condensed from air into pure liquids, which can be kept at normal pressures but have temperatures well below -70°C. Such liquid elements are called (7) _____s, because of their extremely low temperatures and are commonly used in medicine or industry for a variety of purposes.

Nitrogen as N_2 in the atmosphere is a very unreactive element but is the source of the N atoms essential for living organisms. Bacteria in the soil and water, which have a primary role in the process of (8) _____, chemically convert the N_2 to NH_3 and to various oxides and then NO_3^- as the final product, which are the forms needed for plants to incorporate N. (9) _____ is one-half of the nitrogen cycle, in that other bacteria chemically convert dead organic matter into N_2 ,then release it to the atmosphere, so that all forms, N_2, NH_3 and NO_3^-, are constantly replenished.

Sulfur is an essential element for living substances that is also found largely pure in the natural world. It must be converted to forms such as H_2S and SO_4^{-2} ions through a different chemical cycle for use by organisms and plants. Extraction of sulfur from deposits in the Earth's crust uses superheated steam and compressed air to drive the molten sulfur to the surface in a process called the (10) _____. Extraction of sulfur from natural gas and petroleum, which aids in reducing air pollution, has replaced extraction with the (11) _____ as the main source for sulfur.

Chlorine gas, Cl_2, also has many uses in the chemical industry and for disinfecting water supplies, but is not found in a pure form in the natural world. It can be produced from the electrolysis of Cl^- in a brine (concentrated NaCl) solution in a process called the (12) _____ . The name, (13) _____, gives both products of the electrolysis since Cl_2 is produced at the anode of the electrolysis cell and OH^- at the cathode, from the reduction of H_2O into H_2.

- PRACTICE TESTS -

After completing your study of the chapter and the homework problems, the following questions, can be used to test yourself on how well you have achieved the chapter objectives.

1.   Tell whether the following statements would be always **True or False**:
   A)   Silicon is created from the fusion of two carbon nuclei in a star.
   B)   Two hydrogen and one oxygen nuclei can fuse to form water in nuclear burning.
   C)   Some elements such as the radioactive actinides are made during an explosive stage of the star
   D)   The most abundant metal in the Earth's crust is iron.
   E)   Silicates are minerals largely composed of silicon dioxide units in chains.
   F)   Asbestos is a silicate, as is quartz.
   G)   Clays result from the weathering of rocks and are mainly aluminum oxides.
   H)   Sulfur is one of the "aloof" elements.
   I)   LOX is short for liquefied oxygen and is a cryogen.
   J)   NO(g) can be produced naturally from lightning in the atmosphere and is a source of nitrates in rain.
   K)   Both the Haber-Bosch process and the NH_3 to HNO_3 process developed by Ostwald are example of how to use LeChatelier's Principle to affect yield.
   L)   Airbags inflate because of the N_2 produced from the KNO_3 reaction.
   M)   Sulfur is oxidized to form $SO_2(g)$, but then is reduced by water to form H_2SO_4.
   N)   The main source for magnesium, bromine and iodine is seawater.
   O)   Aluminum was a precious metal, like silver, in the 1800's and is also therefore an "aloof" metal.
   P)   White phosphorus is extremely toxic, but red phosphorus is not and the difference is due to the amount of oxygen in the forms.

2.　A)　Describe the feature(s) that helium burning and alpha emission by a nucleus share and what is(are) the differences between them.

B)　Explain why recycling aluminum makes good economic sense.

C)　What main group elements are extracted by electrolysis?

D)　Explain the role that each of the three components sodium azide, potassium nitrate and silicon dioxide play in the deployment of an airbag.

E)　Explain what the text meant when it said that NH_3 is the most reduced form of N and NO_3^- is the most oxidized form.

F)　Nitrogen is primarily found in the atmosphere while phosphorus is only found in the soil and water and not in the atmosphere. Using the common forms of the two elements explain why this is the case.

3. H_3PO_4 can form three sodium salts, but H_3PO_3 forms only two sodium salts. Explain why this is the case using the Lewis structures of the two compounds.

4. Pure liquid HNO_3 decomposes easily to produce N_2O_4, O_2 and H_2O
(a) What's the coefficient of N_2O_4 in the balanced chemical reaction describing the decomposition?
(b) This reaction is an example of an internal redox reaction. What's being reduced and what oxidized in the decomposition?

5. The following UNBALANCED redox reaction is an important part of the nitrogen cycle to replenish $N_2(g)$　　　__ $NH_3(g)$ + 6 NO_2 (g) → __ $N_2(g)$ + 12 $H_2O(g)$
A) How many N_2 molecules are produced in the balanced reaction?
B) Referring the tabled $\Delta G°$ values, would this reaction be product-favored at 25°C?

6. Sulfuric acid is a very good dehydrating agent because it easily forms hydrates such as $H_2SO_4 \cdot H_2O$. A common chemistry demonstration is to "dehydrate" a monosaccharide which has the formula, $C_x(H_2O)_y$ [Chapter 3] by combining it with concentrated H_2SO_4.
A) Write the balanced reaction for glucose, $C_6H_{12}O_6$ and H_2SO_4, showing the formation of the $H_2SO_4 \cdot H_2O$ and a second product.
B) When H_2SO_4 spills on paper or cotton cloth, a hole appears within a short while where the acid was in contact. Other acids do not produce the same reaction. Do you think that this is the same type of reaction as with the glucose. Explain your reasoning.

7. The dimerization: of $NO_2(g)$ <=> $N_2O_4(g)$ becomes product-favored at temperatures below 0°C
A)　Complete the Lewis structures for each molecule given the molecular skeletons in the box on the right.
B)　Show two resonance forms for each molecule.
C)　Calculate the $\Delta H°$ and $\Delta S°$ for the dimerization. Which factor, enthalpy or entropy, do you think is most important in making the reaction reactant-favored at room temperature even though NO_2 is a free radical and N_2O_4 is not.

8. Calculate the %P in superphosphate $Ca(H_2PO_4)_2 \cdot H_2O$ + 2 $CaSO_4 \cdot H_2O$ and compare it against the % P in Calcium phosphate, $Ca_3(PO_4)_2$. Also compare the solubilities of $Ca(H_2PO_4)_2$ and $Ca_3(PO_4)_2$ estimated from the K_{sp} (Appendix H). Why do you think it's called superphosphate because of the %P or because of the solubility? Explain.

9. All the phosphorus trihalides are unstable in water and undergo reactions such as:
　　PCl_3 (g) + 3 $H_2O(l)$ <=> $H_3PO_3(aq)$ + 3 HCl(aq)
A)　Estimate the $\Delta G°$ for the reaction, given the $\Delta G°_f$ $H_3PO_3(aq)$ = - 760 kJ/mol.
B) What would be the pH in 500 mL of solution if 15.0 g of PCl_3 reacts with the H_2O?

10. Ammonium nitrate, NH_4NO_3, is an excellent fertilizer for plants. However, if heated the compound decomposes explosively and has been used for making homemade bombs. The decomposition can take place as one of the following reactions:

$$NH_4NO_3 (s) \rightarrow N_2O(g) + 2\ H_2O(g) \qquad \Delta H° = -\ 36.0\ kJ, \quad \Delta S° = +\ 446\ J/K$$
$$NH_4NO_3 (s) \rightarrow 2\ N_2(g) + 2\ H_2O(g) + O_2(g) \qquad \Delta H° = +\ 124.0\ kJ, \quad \Delta S° = +\ 626\ J/K$$

A) Is the first reaction, product- or reactant-favored at room temperature? Which factor, the $\Delta H°$ and $\Delta S°$ is keeping the second reaction less likely to occur at low temperatures than the first reaction?

B) At what temperature will the second reaction start to become product-favored?

C) What factor(s) do you think allow bags of the fertilizer be kept at room temperature without explosion.

11. To simulate seawater for the laboratory, marine biologists have to mix 26.5 g NaCl, 2.40 g $MgCl_2$, 3.35 g $MgSO_4$, 1.20 g $CaCl_2$, 1.05 g KCl, 0.315 g $NaHCO_3$ and 0.098 g NaBr in enough water to make 1.0 kg of solution.

A) How many moles of Cl^- are present in one liter of the mixture?

B) If the mixture were electrolyzed in a Downs cell at 7.0 V and 4.0×10^4 amperes and assuming the same reaction occurred (Cl^- oxidized, H_2 reduced), how long would the cell have to be run to convert all the Cl^- in 10.0 L of the mixture to $Cl_2(g)$? (Assume 100 % efficiency)

Chapter 22

Chemistry of Selected Transition Elements and Coordination Compounds

This chapter gives an overview of the electron configurations and the chemical and physical properties of the transition metals. Transition metals of economic significance and coordination compounds are also described.

22.1 Recognize the general properties of transition metals.

22.2 Write electron configurations and orbital box diagrams for transition metals and their ions.

22.3 Explain how most transition metals have multiple oxidation states.

22.4 Explain trends in sizes of transition metal atomic radii.

The transition metals are located in columns 3B to 2B in the Periodic Table. They are also called the d-block elements because much of their chemistry involves electrons from the d orbitals. Their physical properties, with a few exceptions, include the usual metallic properties of high electrical and thermal conductivity, ductility, malleability, and silvery color. Properties that differ from main-group metals include the tendency to form brightly colored compounds; paramagnetism or ferromagnetism; and the tendency to have multiple oxidation states. Many of the properties of transition metals occur because they do their chemistry with the electrons from the ns and (n-1)d orbitals. The ns electrons are lost first in forming ions; this explains the existence of +2 ions for almost all transition metals. Any or all of the d electrons may be lost, accounting for the existence of multiple oxidation states for many transition metals. Because half-filled orbitals are especially stable, there are some anomalies in the pattern of filling orbitals. The decrease in atomic radius across a row of transition metals is less pronounced than in main group elements because d-electrons do not shield the nucleus as effectively as p electrons.

22.5 Describe how iron ore is processed into iron and then into steel.

Iron is the most abundant transition metal in the earth's crust. It is purified from oxide ores using **pyrometallurgy**, or chemical reactions carried out at extremely high temperatures. In a blast furnace the ore, coke (a form of carbon that acts as a reducing agent) and limestone are heated and then subjected to a blast of hot air. The carbon is oxidized; the Fe^{n+} is reduced; and impurities react with the limestone to produce slag. The slag floats on the molten iron making the separation fairly simple. Iron is further processed into steel using the basic oxygen process, in which a blast of oxygen is forced into molten iron to oxidize the excess carbon impurity. Small amounts of other transition metals are then added to produce alloys with the desired properties.

22.6 Discuss how copper is extracted from its ores and purified.

Although pure copper is found in nature, demand for larger amounts requires extraction from various ores. Depending on the ore, it may be crushed and roasted by heating in air. Then it is melted and subjected to a blast of hot oxygen, which produces blister copper, a crude form of the element, and sulfur dioxide. In an electrorefining process, blister copper is used as the anode in a bath of $CuSO_4$, with very pure copper used as the cathode. Copper is oxidized from the anode and plates out on the cathode, leaving impurities in solution or forming anode sludge, which can be further treated to extract other metals. Bronze, a mixture of Cu and Sn, and brass, made from Cu and Zn, are two important alloys of copper.

22.7 Discuss the chemistry of gold and silver.

Gold and silver both occur as pure elements in nature. These two elements are especially valued for coins and jewelry because they are unreactive. They can be extracted from their ores by formation of complexes with the cyanide ion. The complexes are then reduced using zinc. Cyanide waste is extremely toxic, so special care must be taken to dispose of mine wastes properly.

22.8 Explain the coordinate covalent bonding of ligands in coordination compounds and complexes.

22.9 Interpret the names and formulas of coordination complex ions and compounds.

22.10 Discuss isomerism in coordination compounds and complex ions.

Transition metals readily form **coordinate covalent bonds** by using vacant d-orbitals to allow sharing of a pair of electrons from a Lewis base, usually called a ligand. **Ligands** can be molecules or ions, but must have at least one unshared pair of electrons. The resulting complexes are called complex ions, or, if neutral, **coordination compounds**. Complex ions can form compounds with other counter ions. Complex ions are named using the following guidelines:

- Ligands are given identifying names with endings slightly different than their usual name;
- The number of each ligand involved is identified using Greek prefixes;
- The names of ligands are listed in alphabetical order, ignoring prefixes;
- The charge on the metal is calculated using the charges on the ligands and the overall charge on the complex;
- The metal is named last, with a Roman numeral in parentheses to indicate its charge. Additional details and exceptions are noted in your text.

The **coordination number** of a metal in a complex is the number of coordinate covalent bonds it forms with ligands. Coordination numbers are usually even. Ligands can be **monodentate**, forming one bond with the metal; **bidentate**, forming two bonds; or **hexadentate**, forming six bonds. Note that the possibility of **polydentate** ligands means that the coordination number is not necessarily the same as the number of ligands. The coordination number helps to determine the geometry of the complex, which can assist in prediction of magnetic properties and of color.

Isomerism occurs when there is more than one possible arrangement of ligands around metals. Constitutional isomers occur with a ligand such as SCN⁻, which can bond either through the N or the S. Geometric isomers occur when ligands can be arranged in two distinct ways that cannot be interconverted without breaking bonds. A common form of geometric isomerism is *cis-trans*; in a *cis isomer*, two identical ligands are adjacent to each other, and in the *trans isomer*, these two ligands are opposite each other.

22.11 Give examples of coordination compounds and their uses.

One example of coordination compounds already discussed is the formation of cyanide complexes in the extraction of gold and silver from their ores. Formation of complexes allows the metals to dissolve and to be separated from other materials. Another example is cisplatin, a complex of platinum which is used in cancer treatment; its trans form is not biologically active. Many biomolecules, including the component of hemoglobin which contains the iron ions are coordination compounds. Complexes of cobalt, zinc, and copper are also important in human metabolism.

Chapter Review - Key Terms

In order to make_all the sentences below TRUE, insert the appropriate word or phrase from the list of key terms which best fits the context of the sentence.

NOTE: Any phrase or word from the list may used more than once.

LIST:		
bidentate	coordination number	monodentate
chelating ligands	hexadentate	polydentate
coordinate covalent bond	lanthanide contraction	pyrometallurgy
coordination compound	ligands	steel

Concerning Properties of Transitions Metals:

Transition metals have properties and trends that differ from main group elements. One difference is in the trend for atomic radii of neutral atoms. Because of the (1) _____, most of the atoms in the third row of the transition metals (period 6) are the same size as the transition metals in the row before (period 5). Pure forms of the transition metals are extracted from ores using (2) _____, since the extracting chemical reactions require high temperatures to proceed to a reasonable extent. Iron, needed for many structural materials, is one of the metals extracted from ores using (3) _____ in a blast furnace. The form of iron needed for many applications is (4) _____, which is an alloy of nearly pure iron with a small percentage of other elements such as carbon, that has more desirable characteristics than pure iron.

Another difference between main group elements and transition metals is the ease of formation of complex ions by transition metals. The transition metals, or their ions, have empty atomic orbitals that can accept an unshared pair of electrons from other ions or molecules, called (5) _____ , to form (6) _____s between the metal and the (7) _____. The neutral compound formed between the complex ion and oppositely charged ion(s) is called a (8) _____, in which the complex ion can be the anion or the cation. The number of (9) _____s formed to a single metal in a complex ion is the (10) _____ for that metal and is usually 2,4 or 6.

Some molecules are capable of donating only one unshared pair of electrons to the metal and such (11) _____ are called (12) _____. To act as a (13) _____, where one molecule forms two (14) _____s to the same metal atom, the molecule must have two atoms with unshared electron pairs separated by several intervening atoms. A single molecule that can form more than two (15) _____s to one metal ion is called a (16) _____, which can then act as (17) _____, in that they appear to grab the metal and form cage around it in solution. One of the commonly used (18) _____, which donate six pairs of electrons to a single metal, which means it is a (19) _____, is ethylenediaminetetraacetate ion, or EDTA^{-4}.

- PRACTICE TESTS -

After completing your study of the chapter and the homework problems, the following questions, can be used to test yourself on how well you have achieved the chapter objectives.

1. Tell whether the following statements would be always **True or False**:

   A) One of the general properties of transition metals is that they have higher melting points than main group elements and are less dense.

   B) Transition metals are effective catalysts because with partially filled p orbitals, they can either accept or donate electrons easily.

C) Nickel and nickel (II) ion both have eight electrons in the 3d subshell.

D) The process of producing steels from pig iron involves the oxidation of Fe.

E) Blister copper, a slightly impure form of copper, acts as the cathode in the electrolysis to refine copper.

F) Bronzes are alloys of copper and tin, while brasses are alloys of copper and zinc.

G) Gold, silver, and platinum are also recovered from the electrorefining of copper.

H) The metal is always named last in a complex ion.

I) Coordination number gives the number of ligands in the complex ion.

J) A common geometry for the ML_6^{2+} complex would be square planar.

K) It is possible to have both cis and trans isomers of octahedral complex ions.

L) Hemoglobin contains four Fe^{2+} ions in separate complex ion units that can each form a coordinate covalent bond with an oxygen molecule.

M) Linkage isomerism results from a bidentate ligand forming either a *cis* or *trans* isomer.

2.   A)   Explain what characteristics a coordinate covalent bond and a covalent bond between main groups elements would share and what would be different about them.

      B)   Can a molecule such as hydrazine, H_2NNH_2, act as a bidentate ligand? Explain your reasoning.

      C)   Describe how hemoglobin is affected by exposure to CO(g) and why CO poisoning is difficult to reverse.

      D)   Explain why AgCl(s) is only very slightly soluble in water, but dissolves readily in HCl, but not HNO_3 of the same concentration.

3. Anhydrous $NiCl_2$ is a yellow solid. When dissolved in water and NH_3(aq) is added, a pale green precipitate forms. If more NH_3(aq) is added, the precipitate dissolves and a deep blue colored solution appears. Explain these changes in terms of complex ions that could form and indicate what ligands the complexes should contain.

4. For the following transition metals, and their ions, give the electron configuration of the valence electron, using the orbital box diagrams below.

A) Mn   [Ar] 4s ☐ 3d ☐☐☐☐☐ 4p ☐☐☐

B) Mn^{+3} [Ar] 4s ☐ 3d ☐☐☐☐☐ 4p ☐☐☐

C) Mn^{+4} [Ar] 4s ☐ 3d ☐☐☐☐☐ 4p ☐☐☐

D) Zn   [Ar] 4s ☐ 3d ☐☐☐☐☐ 4p ☐☐☐

E) Zn^{+2} [Ar] 4s ☐ 3d ☐☐☐☐☐ 4p ☐☐☐

F) Ag^+ [Ar] 4s ☐ 3d ☐☐☐☐☐ 4p ☐☐☐

5. In the following complex ions:

A) Give the name of the metal in the complex ion.

B) Give the charge on the metal in the complex ion.

C) Give the coordination number for the metal.

D) Identify each type of ligand and give its molecular formula and charge, if any.

E) Tell whether the ligand is a monodentate, bidentate or polydentate ligand.

(a) $[Co(Cl)_6]^{3-}$ (b) $[Fe(C_2O_4)_3]^{3-}$ (c) $[Fe(CO)_4]^{2+}$ (d) $[Ru(CN)_5(Cl)]^{2-}$

6. Write the formula for the coordination compound that corresponds to the names below:
    A) Sodium tetrachlorodiamminecobaltate(III)
    B) Dibromotetramminecobalt(III) chloride
    C) Monochloropentammineplatinum(IV) chloride
    D) Isothiocyanatoiron(II) chloride
    E) Potassium tetracyanodiaquaferrate(III)

7. For the following descriptions of complex ions:
    A) Give the formula and net charge on the complex ion.
    B) Indicate whether K^+ or Cl^- ion is needed to balance the net charge on the complex ion and how many of these ions would be needed.
    C) Write the formula for the coordination compound.
    D) Give the proper name for the coordination compound.

    (a) A central cobalt +3 ion bonded to two ethylenediamine molecules and two fluoride ions.

    (b) A central manganese +2 ion with four water molecules and two hydroxide ions

    (c) A central copper +2 ion is bonded to three chloride ions and one ammonia molecule

8. Name the following compounds of complex ions:
    A) $[Mn(NH_3)_3 CN] Cl$ B) $K_2[Ni(Cl)_4]$ C) $[PdCl_2(NH_3)_2]$ D) $K_2[CoF_6]$

9. Given the name and geometry of the complex ions below:
    A) Make a sketch of the complex ion, showing the placement of the ligands.
    B) For those that could have geometric isomers, sketch the structure of each isomer and label each one as the cis or trans isomer.
        (a) dichlorogold(I) ion, linear
        (b) tetrachlorocuprate(II) ion, square planar
        (c) tetrachlorodiammineplatinum(IV), octahedral
        (d) diaquadibromocopper(II), square planar
        (e) monobromotriaquazinc(II) ion, square planar

10. Portraits can be toned a sepia color by replacing the Ag(s) in the photograph with Au(s) in the reaction: $Ag(s) + [AuCl_2]^- \rightarrow AgCl(s) + Cl^- + Au(s)$
    A) What is the name of the complex ion used as the reactant?
    B) If 2.1 mg of Ag(s) is oxidized on a particular photograph, how many milligrams of Au(s) have precipitated on the same photograph?

11. If the total surface area of a piece of metal that is to be chrome-plated is 1750 in^2 and the depth of the Cr(s) must be 0.0003 inches. Using a Cr^{+4} solution and a time period of 1.0 hour, what would be the minimum current needed, in amperes, to produce the amount of Cr(s) needed, assuming the process was 70% efficient?

ANSWERS - PRACTICE EXERCISES/TESTS -
Chapter 1

Answers to Key Terms Exercise:

The Way Science is Done:

1) hypothesis   2) quantitative   3) qualitative   4) law   5) law
6) hypothesis   7) law   8) theory   9) theory   10) model   11) hypothesis
12) law   13) theory   14) qualitative   15) quantitative

Chemistry and Matter- Properties and Transformations:

16) Chemistry   17) matter   18) physical   19) chemical
20) matter   21) physical properties   22) physical property   23) density
24) physical change   25) temperature   26) chemical change   27) physical change   28) physical property(ies)   29) melting point   30) boiling point
31) Celsius   32) density   33) proportionality factor   34) conversion factor
35) density   36) proportionality (or conversion) factor   37) dimensional analysis

Working with Chemical Transformations:

38) chemical change   39) chemical reaction   40) chemical properties
41) reactant   42) product   43) reactant   44) chemical reaction
45) products   46) Law of Conservation of Mass   47) Chemical reaction
48) energy   49) energy   50) chemical reaction   51) chemistry

Classification of Types of Matter:

52) mixture   53) Heterogeneous mixture   54) homogeneous mixture   55) solution
56) physical properties   57) homogeneous mixture   58) chemical property(ies)
59) physical property (ies)   60) chemical reaction   61) element
62) compound   63) element   64) compound   65) compound
66) mixture   67) solid   68) liquid   69) gas   70) solid   71) liquid
72) gas   73) Gas   74) homogeneous mixture

Dealing with Matter on the Nanoscale Level:

75) macroscale   76) microscale   77) nanoscale   78) nanoscale
79) macroscale   80) microscale   81) kinetic molecular theory   82) nanoscale
83) atom   84) macroscale   85) atom   86) chemical properties   87) element
88) compounds   89) element   90) Law of Conservation Mass   91) chemical reaction
92) atom   93) chemical change   94) chemical properties   95) chemical reaction
96) nanoscale   97) atom   98) compound   99) Law of Multiple Proportions
100) Law of Constant Composition   101) element   102) element   103) metal
104) solid   105) Nonmetal   106) physical properties   107) metal
108) nonmetal   109) metalloid

Describing Chemical Substances:

110) nonmetal   111) molecule   112) element   113) molecule   114) atom
115) diatomic molecule   116) chemical formula   117) atom   118) molecule
119) compound   120) element   121) chemical formula   122) Allotropes   123) element

Answers to True/False

1.   A) **False**   Testing a hypothesis can lead to laws <u>or</u> theories. It is not limited to one or the other.

     B) **True**   The physical state refers to whether a substance is a liquid, solid, or gas and these do depend on temperature.

     C) **False**   Solutions are homogeneous but can be either made in the solid, liquid or gaseous states.

     D) **False**.   The situation describes a chemical compound, but elements can also be composed of molecules, such as the gases O_2 and N_2.

     E) **True**.   Energy changes do accompany chemical changes.

F) **False**. Atom rearrangement occurs in chemical, not in physical changes.

G) **False**. Physical and chemical properties both change when the substance changes since the set is unique to the substance.

2. a) **Theory** It a broad statement that can explain laws and hypothesis. Atoms cannot be seen, so this cannot be proven to the point of being a law.

b) **Law** There are no exceptions to this statement.

c) **Theory** It cannot be a law, since it cannot be proven true or false quantitatively. People may believe this to be true and that it explains many different phenomena which fits the characteristic of a theory. It is also too broad a statement to be a hypothesis.

d) **Hypothesis** It cannot be a law, since it is not always going to be true, unless you can specify the place being talked about. It is very specific, unlike a theory, so it would be an hypothesis.

e) **Law** This is a broad statement that can be proven and measured quantitatively. If true there would be no exceptions. Therefore it fits the characteristic of a **law**, rather than a theory.

f) **Theory** This is a broad sweeping statement that cannot be proven, but its acceptance could be used to explain a lot of the observed behavior in the universe.

3. It would mean that the scientific method has been applied to whatever is being stated. The statement has been tested, proved true and reproduced by many different scientists in many laboratories.

4. a) A **physical property**   • describes a physical change, the melting point
   b) A **physical property**   • describes a change in shape only
   c) A **chemical property**   • describes how gold interacts with water.
   d) A **chemical property**   • indicates gold does not react with other natural elements and remains pure.

5. A) **(c)**   Need to use: (T(°F) - 32)/1.8) = T(°C).
               (70- 32)/1.8 = 38/1.8 = 21°C
   B) **(b)**   Need to use: (T(°C) × 1.8) + 32 = T(°F)
               (15° × 1.8) = (- 27°) + 32 ≈ + 5°C , so (d) closest

6. A) A **chemical change** since bromine and hydrogen will be are bonded together and no longer by themselves when a compound is formed.

B) A **physical change** since this describes a condensation process. The water molecule stays intact through the changes.

C) A **physical change** takes place as the sugar dissolves and a mixture is formed. The mixture is still tea, but now the sweet taste of sugar added to that of tea. Each component retains its own chemical properties.

7. A) **Qualitative** data
   B) **Quantitative** data   • since measured numeric data is given.
   C) **Qualitative** data   • the physical state and odor listed are not numeric
   D) **Quantitative** data   • measured numeric data given.
   E) **Qualitative** data   • solubility described, but value not given.
   F) **Qualitative** data,   • a property, no numeric data given.

8. **Not the same compound**   *If same compound, ratio of Phosphorus to Oxygen equal*

Ratio (#1): $\dfrac{2.58\ g\ P}{3.322\ g\ O} = 0.777$ Ratio (#2): $\dfrac{3.718\ g\ P}{2.881\ g\ O} = 1.29$ so different compounds.

9.   A) **a, b and d** represent physical changes where the chemical properties of substance A would stay intact.

   B) **Compound**   If A, as in (e), can be decomposed it is not an element.

10.   Substances changing: Type of change:
   A) charcoal   burning = chemical change
        marshmallows   chemical change, toasting is like the caramelization
   B) margarine   melting = physical change
        eggs   cooking the eggs changes chemical properties, since
                                            product has different characteriestics than beginning
                                            substance, so primarily a chemical change
   C) mulberry leaves   converted to silk, so chemical change
   D) wool   wool retains properties so physical change

11.   A) If the substance cannot be purified further, it cannot be a mixture but could be either a compound or an element. Further chemical reactions to determine if the substance could be decomposed would have to be conducted.

   B) The kinetic molecular theory. Changes in state, described as changes in the motion, are best explained by this theory.

12. A)

Silver	Gold	Copper	Tin	Lead	Mercury	Iron	Sulfur	Carbon

Symbol: Ag   Au   Cu   Sn   Pb   Hg   Fe   S   C
Type: metal   metal   metal   metal   metal   metal   metal nonmetal nonmetal

   B) Arsenic (As) is a metalloid, since it touches the staircase, but Thallium (Tl) is a metal

13.   Sequence should be **(c) → (d) → (b) → (a)**

14.   A) Use density of each to convert volumes to mass:

$$5.0 \text{ mL Bi} \times \frac{9.8 \text{ g Bi}}{1.0 \text{ mL Bi}} = \underline{49.0 \text{ g Bi}} \text{ and } 3.5 \text{ mL Hf} \times \frac{13.4 \text{ g Hf}}{1.0 \text{ mL Hf}} = \underline{46.9 \text{ g Hf}}$$

   The 5.0 mL of Bi has the greatest mass.

   B) **Looking for:** (mass of 5.0 mL Bi) + (mass of 5.0 mL Hf) = ? g

   Know: 5.0 mL Bi = 49.0 g from (A) and $5.0 \text{ mL Hf} \times \frac{13.4 \text{ g Hf}}{1.0 \text{ mL Hf}} = 67.0 \text{ g Hf}$

   So **Mass of mixture = (49.0 g) + (67.0 g) = 116 g total**

   C) **If a compound were formed the mass would be the same.** The modern atomic theory explains this best since it says that the atoms would all still be present, just in a different arrangement.

15.   A) **The mixture should be heterogeneous** since the phase boundaries between the sand and butter should be visible to the naked eye or under a microscope.

   B) **The density would be less than that of sand**. The volumes of sand and butter mixed would be 0.44 mL and 1.6 mL, so 2.00 g/2.0 ml would give a density of about 1.0g/ml for the mixture.

   C) **The volumes would add together.**   Both of the substances are solids and the particles are already very close together. Consequently, the sand particles cannot squeeze into empty spaces, but must instead push aside the butter molecules.

Chapter 2

Answers to Key Terms Exercise:

Internal Structure of the Atoms:
1) radioactivity   2) atomic structure   3) nucleus   4) electrons
5) Scanning tunneling microscope   6) proton   7) neutron   8) proton   9) electron
10) neutron   11) neutron   12) proton   13) proton   14) electron   15) electron
16) ion   17) mass spectrophotometer   18) isotope   19) mass spectrum
20) atomic number   21) proton   22) mass number   23) protons
24) neutron   25) nucleus   26) atomic mass units

Symbols and Organization of the Elements:
27) mass number   28) atomic weight   29) isotope   30) isotope   31) neutron
32) proton   33) isotope   34) percent abundance   35) periodic table   36) period
37) group   38) Law of Chemical Periodicity   39) group   40) lanthanides
41) actinides   42) transition elements   43) main group elements
44) alkali metals   45) alkaline earths   46) noble gases   47) halogens

Units for Chemistry
48) metric system   49) conversions factor   50) mass   51) mole   52) mass
53) mole   54) molar mass   55) Avogadro's Number   56) mole
57) atomic mass units   58) molar mass

Answers to True/False:

1.   A) **True**   Radioactivity results from atom either losing particles or energy from the nucleus, which is why it results from nuclear, not chemical reactions.

    B) **False**   Lanthanides and actinides are not considered main group elements, but neither are transition elements, which are part of the metals.

    C) **True**   Only the atomic number is needed to identify an element.

    D) **True**   A mass spectrophotometer detects ions based on the mass to charge ratio. Since the masses of the isotopes will be different, but not the charges, the mass spectrophotometer can trace them.

    E) **False**   The nucleus accounts for nearly all the mass of the atom, but the electrons set the size.

    F) **True**   Because the number of protons is the same, the chemical properties are identical. Physical properties related to mass could be different.

    G) **False**   The position is shown using this method, but the interaction is with the electrons, not the nucleus.

    H) **False**   The Law says these properties will vary with the atomic number, not the mass of the atom.

2. Matching Exercise:

A) l, Pu   B) g. Sr   C) d, Ar   D) b, O   E) e, Br   F) k, Po
G) a, N   H) h, Rh   I) f, Rb   J) i, In   K) c, Si   L) j, Bi

3. A) To compare, need same volume units, so convert 5.0 quarts to liters, 5.0 qt → ? L
    • *Conversion intersystem, know 1.0 L = 1.057 quarts*
    • *Use conversion factor with Liter units on top, so quarts cancel.*

$$5.0 \; quart \times \frac{1.0 \; L}{1.057 \; quart} = \underline{\underline{4.73 \; L}}$$

The 5.0 L beaker is _bigger_ than the 5.0 quart beaker.

3. B) To compare lengths must be in same units, so convert 10 km → ? miles
    • *Conversion: 1.0 km = 0.6214 mile.*

$$10 \; km \times \frac{0.6214 \; mile}{1.0 \; km} = \underline{\underline{6.21 \; miles}}$$

The 10 km is longer than 5.0 miles in length

3. C) **(a) Plan:** Convert $150 \text{ lb} \rightarrow ?$ stones   **(b) Plan:** Convert $150 \text{ lb} \rightarrow ? \text{ g} \rightarrow ? \text{ kg}$
   • *Conversion 1.0 stone = 14.0 lb*   • *Conversions: $1.0 \text{ lb} = 454 \text{ g}$, $1.0 \text{ kg} = 1000 \text{ g}$*

$$150 \text{ lbs} \times \frac{1.0 \text{ stone}}{14.0 \text{ lbs}} = \underline{\underline{10.7 \text{ stones}}} \qquad 150 \text{ lbs} \times \frac{454 \text{ g}}{1.0 \text{ lb}} \times \frac{1.0 \text{ kg}}{1000 \text{ g}} = \underline{\underline{68.1 \text{ kg}}}$$

4. A) Conversion needed: $3010 \text{ cm}^3 \rightarrow ? \text{ m}^3$

   • *Conversion type is volume to volume, within metric system, but no direct conversion given on Table 2.2.*

   Know: Table 2.2 gives conversion for L to m^3 as $1000 \text{ L} = 1 \text{ m}^3$ and since $1 \text{ mL} = 1$ cm^3 then $1.0 \text{ L} = 1000 \text{ cm}^3$ **Plan:** $3010 \text{ cm3} \rightarrow ? \text{ L} \rightarrow ? \text{ m}^3$

$$3010 \text{ cm}^3 \times \frac{1.0 \text{ L}}{1000 \text{ cm}^3} \times \frac{1.0 \text{ m}^3}{1000 \text{ L}} = \underline{\underline{3.01 \times 10^{-3} \text{ m}^3}}$$

   As an alternative, can derive a direct conversion. Can convert length to volume conversion by cubing both sides. Know $100 \text{ cm} = 1 \text{ m}$ then $(100 \text{ cm})^3 = (1\text{m})^3$ gives $1.0 \times 10^6 \text{ cm}^3 = 1\text{m}^3$. Using this as the single conversion factor results in the same answer.

4. B) Conversion needed: $50 \text{ yd} \rightarrow ? \text{ km}$

   • *Type is length to length, but intersystem. Stated conversion for intersystem length on Table 2.2 inches to cm, so need to connect yards to in, cm to km.*

   Know: $1 \text{ yd} = 36 \text{ in}$, $1 \text{ in} = 2.54 \text{ cm}$, $1 \text{ m} = 100 \text{ cm}$ and $1\text{km} = 1000 \text{ m}$.

   Plan: $50 \text{ yards} \rightarrow ? \text{ in} \rightarrow ? \text{ cm} \rightarrow ? \text{ m} \rightarrow ? \text{ km}$

$$50 \text{ yd} \times \frac{36.0 \text{ in}}{1.0 \text{ yd}} \times \frac{2.54 \text{ cm}}{1.0 \text{ in}} \times \frac{1 \text{ m}}{100 \text{ cm}} \times \frac{1 \text{ km}}{1000 \text{ m}} = \underline{\underline{4.6 \times 10^{-2} \text{ km}}}$$

4. C) Conversion needed: $10 \text{ ft}^3 \rightarrow ? \text{ m}^3$

   • *Type is volume to volume, intersystem, but no direct conversion given on Table 2.2.*
   • *Stated intersystem conversion on Table 2.2 is for length, but can convert <u>any</u> length conversion to volume conversion by cubing both sides.*

Know:	*Rewrite as volume conversion:*
$1 \text{ ft} = 12 \text{ in}$	$1 \text{ ft}^3 = (12)^3 \text{ in}^3 = 1728 \text{ in}^3$
$1 \text{ in} = 2.54 \text{ cm}$	$1 \text{ in}^3 = (2.54)^3 \text{ cm}^3 = 16.39 \text{ cm}^3$
$1 \text{ m} = 100 \text{ cm}$	$1 \text{ m}^3 = (100)^3 \text{ cm}^3 = 1.0 \times 10^6 \text{ cm}^3$

   Plan: $10 \text{ ft}^3 \rightarrow ? \text{ in3} \rightarrow ? \text{ cm}^3 \rightarrow ? \text{ m}^3$

$$10 \text{ ft}^3 \left[\frac{1728 \text{ in}^3}{1.0 \text{ ft}^3} \right]\left[\frac{16.39 \text{ cm}^3}{1.0 \text{ in}^3} \right]\left[\frac{1.0 \text{ m}^3}{1.0 \times 10^6 \text{ cm}^3} \right] = \underline{\underline{0.283 \text{ m}^3}}$$

5. A) **Looking for:** Number of gallons water that will fill water bed

   Know: Dimensions of bed, length = 8.0 ft, width = 7.0 ft, height = 8.0 inches
   Length × width × height = volume of bed = volume of water needed.
   • *If all dimensions in feet, volume, $\text{ft}^3 \rightarrow ? \text{ gal}$*
   • *Table 2.2 indicates gallons and liters related, through quarts, $\text{L} \rightarrow ? \text{ qt} \rightarrow ? \text{ gal}$*
   • *Can use technique used in 4C above to convert ft^3 to L, $\text{ft}^3 \rightarrow ? \text{ in3} \rightarrow ? \text{ cm}^3 \rightarrow ? \text{L}$*

   Plan: volume, $\text{ft}^3 \rightarrow ? \text{ in}^3 \rightarrow ? \text{ cm}^3 \rightarrow \text{L} \rightarrow ? \text{ qt} \rightarrow ? \text{ gal}$

$$\text{volume, ft}^3 = 8.0 \text{ ft} \times 7.0 \text{ ft} \times 8.0 \text{ in} \left[\frac{1.0 \text{ ft}}{12.0 \text{ in}} \right] = 42 \text{ ft}^3$$

$$42.0 \text{ ft}^3 \left[\frac{1728 \text{ in}^3}{1.0 \text{ ft}^3} \right]\left[\frac{16.39 \text{ cm}^3}{1.0 \text{ in}^3} \right]\left[\frac{1.0 \text{ L}}{1000 \text{ cm}^3} \right]\left[\frac{1.057 \text{ qt}}{1.0 \text{ L}} \right]\left[\frac{1.0 \text{ gal}}{4.0 \text{ qt}} \right] = \underline{\underline{\textbf{310 gal}}}$$

5. B) Conversion needed: 310 gal → ? lbs water
   - *Type volume to mass conversion, so will need density of water.*
   - *As given, density converts from mL (or cm^3) to grams, 1.0 mL H$_2$O = 1.0 g H$_2$O*
   - *So must add: gal →? mL, <u>before</u> density conversion and g → ? lb conversion <u>after</u>.*

   Plan: 314 gal → ? qt   →? L   →? mL   →? g   →? lb

$$310 \ gal \left| \frac{4.0 \ qt}{1.0 \ gal} \right| \left| \frac{1.0 \ L}{1.057 \ qt} \right| \left| \frac{1000 \ mL}{1.0 \ L} \right| \left| \frac{1.0 \ g}{1.0 \ mL} \right| \left| \frac{1.0 \ lb}{454.0 \ g} \right| = \underline{\underline{\mathbf{2600 \ lbs \ H_2O}}}$$

6. A)   (a) ^{31}**Si and** ^{28}**Si** are isotopes   (b) ^{31}**Si and** ^{33}**S** both have 17 neutrons

   B)   (a) ^{40}**Ca,**   *60/3 = 20 = maximum number for protons, neutrons and electrons.*
       Since Z = 20, element Ca and atomic mass = 20+ 20 = 40.

   (b) ^{234}**Th**
   Know: 234 = mass of protons + mass of neutrons, 5 protons for every 8 neutrons
       - *Combination of 5 protons + 8 neutrons has mass = 13 amu,*
       - *Divide 234 amu by 13 amu, 234/13 = 18, means must be 18 units of the 13 particles (5 protons, 8 neutrons) in the nucleus.*
       - *To get the atomic number (total number of protons) take 5 × 18 = 90 protons total in nucleus, which identifies element as Thorium, Th.*

7. A)   Atomic numbers equals the number of protons which must always be a whole number, since fractions of protons not possible.   The atomic masses of an element can vary since changing neutrons does not change the chemical nature of element, only the atomic mass.   Taking an average sample of an element means atoms of different masses included and the mass is average is a decimal value.

   B)   If the average is 63.546, then it is most likely that none of the atoms of Cu would have this value. All should be either above or below this value. If the all the atoms had same mass, the value would be a whole number, not a decimal.

   C)   The first and third values convey the same information since giving both the atomic number and the symbol for the element is not necessary. The second symbol only identifies the element, but does not tell its mass number.

8.   **Looking for:** Average atomic mass and identity of element

$$Average \ atomic \ mass = \left[\frac{\% \ isotope \ \#1}{100} \times (\ mass \ isotope \ \#1) \right] + \left[\frac{\% \ isotope \ \#2}{100} \times (\ mass \ isotope \ \#2) \right]$$

   Know: 80.92 amu isotope, 49.31 %, so 78.92 isotope= 100 - 49.31 = 50.69 %

$$Average \ atomic \ mass = \left[\frac{49.31}{100} \times (80.92 \ amu) \right] + \left[\frac{50.69}{100} \times (78.92 \ amu) \right] = \underline{\underline{79.90 \ amu}}$$

   The element that has the average atomic mass equal to 79.90 is Bromine, Br.

9. A)   Two possibilities exist. Since the average value = 32.064, then the largest amounts could be of the isotopes 31.972 and 32.971, which would average out to about 32.0 for the mass. Only small amounts of the higher mass isotopes would then exist in this case. Alternatively, there could be a larger amount of the lowest weight isotope and smaller amounts of the higher mass isotopes to get an average of 32.0 amu.

10. A)   **(a)< (c)< (b)**   • *Need to use same mass scale to compare, such as amu.*
       (a) 2 K atoms = 78.2 amu, (b) 2 Br atoms = 160 amu   (c) 1 Cs atom = 133 amu

   B)   **231 and 233 g/mol**   • *Mass 1.0 mole= mass of Avogadro's number of atoms:*
       3.836542×10^{-22} g/atom $\times 6.02 \times 10^{23}$ atoms/mol = 231.25 = **231 g/mol**
       3.869792×10^{-22} g/atom $\times 6.02 \times 10^{23}$ atoms/mol = 233.05 = **233 g/mol**

11. A) *Need to calculate number of protons, neutrons and electrons for each isotope:*

Isotope:	^{24}Mg	^{52}Cr	^{124}Sn	^{59}Co	^{35}Cl
A (atomic mass)	24	52	124	59	35
Z (atomic number)	12	24	50	27	17
A- Z = no. neutrons	12	28	74	32	18
No. electrons (= Z)	12	24	50	27	17

Answer: (a) ^{24}Mg   **(b)** ^{124}Sn

B) *Need to calculate the moles of each element using atomic mass of isotope:*

Isotope:	^{24}Mg	^{52}Cr	^{124}Sn	^{59}Co	^{35}Cl
Moles of isotope:	0.500	0.400	0.343	0.423	0.343

Answer: (a) 42.5g of ^{124}Sn and 12.0 g of ^{35}Cl   (b) 12.0 g of ^{24}Mg

12. A) **Looking for:** moles Cu in sample of chalcopyrite

Know: 125 g chalcopyrite, chalcopyrite is 34.67% Cu,

• *Write percentage as mass ratio, use as a proportionality factor, to convert mass of chalcopyrite to mass of Cu. Then need molar mass Cu to convert to moles Cu.*

Plan: 125 g chalcopyrite $\xrightarrow{\text{% Cu}}$? g Cu $\xrightarrow{\text{Atomic mass Cu}}$? moles Cu

$$125 \text{ g chalcopyrite} \times \frac{34.67 \text{ g Cu}}{100 \text{ g chalcopyrite}} \times \frac{1 \text{ mole Cu}}{63.55 \text{ g Cu}} = \textbf{0.682 moles Cu}$$

12. B) **Looking for:** Number of grams chalcopyrite containing certain amount of S

Know: 3.0 moles S in sample, chalcopyrite is 34.94% S

• *Write percentage as mass ratio, use as a proportionality factor, to convert mass of S to mass of chalcopyrite. To use will need to convert moles S to g S with atomic mass.*

Plan: 3.0 moles S $\xrightarrow{\text{Atomic mass S}}$? g S $\xrightarrow{\text{% S}}$? g chalcopyrite

$$3.0 \text{ mole S} \times \frac{32.1 \text{ g S}}{1.0 \text{ mole S}} \times \frac{100 \text{ g chalcopyrite}}{34.94 \text{ g S}} = \textbf{280 g chalcopyrite}$$

13. A) **Looking for:** grams Al in box of foil

Know: Dimensions of foil, length = 1.0 m, width = 304 mm, height = 0.60 mm
Length × width × height = volume of foil, density Al = 2.70g/cm^3

• *To use density requires volume in cm^3. So first step to convert all dimensions to centimeters, and calculate the volume in cm^3, then use density to convert volume to mass.*

$$\text{Wt. Al} = 1.0 \text{ m}\left[\frac{100 \text{ cm}}{1 \text{ m}}\right] \times 304 \text{ mm}\left[\frac{1 \text{ cm}}{10 \text{ mm}}\right] \times 0.60 \text{ mm}\left[\frac{1 \text{ cm}}{10 \text{ mm}}\right] = 182 \text{ cm}^3 \times \frac{2.70 \text{ g Al}}{1.0 \text{ cm}^3} = \textbf{490 g}$$

B) **Looking for:** moles Al in foil
Conversion needed: g Al → ? mole Al

$$490 \text{ g Al} \times \frac{1 \text{ mole Al}}{27.0 \text{ g Al}} = \textbf{18.0 mol Al}$$

C) **Looking for:** mass Au in same volume foil
Know: Volume foil = 180 cm^3 = volume Au
Density Au: 19.31 g/cm^3
Conversion needed: vol Au → ? g Au

$$182 \text{ cm}^3 \times \frac{19.31 \text{ g Au}}{1.0 \text{ cm}^3 \text{Au}} = \textbf{3500 g Au}$$

D) **Looking for:** No. atoms Au/ 1 atom Al if two types of foil form mixture

• *The mole ratio between any two substances is always the same as the atom ratio:*

$$\frac{\text{No. atoms Au}}{1.0 \text{ atoms Al}} = \frac{? \text{ mole Au in foil}}{18.0 \text{ mole Al in foil}} = \frac{3500 \text{ g Au} \times \frac{1.0 \text{ mole Au}}{197.0 \text{ g Au}}}{18.0 \text{ mole Al in foil}} = \frac{0.99}{1.0} \approx \frac{\textbf{1 atom Au}}{\textbf{1 atom Al}}$$

Chapter 3

Answers to Key Terms Exercise:

Formulas and Types of Compounds

1. chemical bond
2. molecular formula
3. chemical bonds
4. structural formula
5. condensed formula
6. molecular weight
7. molecular formula
8. inorganic compound
9. organic compound
10. ionic compound
11. molecular compound
12. molecular formula
13. molecular compound
14. ionic compound
15. molecular compound
16. binary molecular compound
17. Binary molecular compound
18. hydrocarbon
19. alkane
20. molecular formula
21. molecular formula
22. isomer
23. alkane
24. constitutional isomer
25. alkane
26. alkyl group
27. condensed formula
28. structural formula
29. alkyl group
30. alkane
31. alcohol
32. functional group
33. functional group
34. condensed formula
35. structural formula

Comparing Ionic to Molecular Compounds

36. ionic compound
37. cation
38. anion
39. crystal lattice
40. ionic compound
41. molecular compound
42. monatomic ion
43. monatomic ion
44. halide
45. polyatomic ion
46. polyatomic ion
47. oxoanion
48. ionic compound
49. electrolyte
50. strong electrolyte
51. molecular compound
52. nonelectrolyte
53. ionic compound
54. cation
55. anion
56. formula unit
57. formula unit
58. molecular compound
59. formula weight
60. molecular weight
61. molecular compound
62. crystal lattice
63. ionic compound
64. ionic hydrate
65. water of hydration
66. molecular formula
67. formula unit
68. water of hydration
69. percent composition by mass
70. percent composition by mass
71. empirical formula
72. formula unit
73. ionic compound
74. molecular compound
75. molecular formula
76. empirical formula
77. percent composition by mass
78. empirical formula

Biologically Important Compounds

79. dietary minerals
80. dietary minerals
81. major minerals
82. trace elements
83. carbohydrates
84. monosaccharide
85. alcohol
86. disaccharide
87. monosaccharides
88. polysaccharide
89. carbohydrates
90. hydrocarbon
91. carboxylic acid
92. functional group
93. alcohol
94. alkane
95. carbohydrates

Answers to True-False Exercises:

1. A) **False** The difference in crystal lattice means the compounds have different ratios between the ions and are not the same compounds. The arrangement in a crystal lattice of ionic compounds doesn't allow for isomers.

    B) **False** True of Cl⁻ but not other halides (Br⁻, F⁻ or I⁻) which are trace minerals.

    C) **True** A functional group has this characteristic.

    D) **True** The formula unit (the simplest whole number ratio) is always used.

    E) **False** The compounds would have the same empirical formula, but could have very different structural and molecular formulas.

    F) **True** The group numbers help only with the monatomic ions.

    G) **False** For ions of equal charge the force of attraction decreases with increasing distance between the two nuclei.

    H) **False** Metalloids are not as predictable as metals. They may or may not form an ionic compound with a nonmetal.

    I) **True** Nonelectrolytes come from the class of molecular compounds.

    J) **False** Alcohols and alkanes have similar names, but contain O as well as C and H so cannot be classified as a hydrocarbon.

K) **False**   The molar mass for an ionic compound is from the empirical formula, but a molecular compound which requires the actual formula.

G) **False**   Carbohydrates are molecular and do not have molecules of water in a crystal lattice as in ionic compounds. The formula just indicates the ratio between H and O in the carbohydrates is always 2 to 1.

2. A) (a) **No.**   This information will reliably produce the empirical formula, but not the molecular formula.

(b) **Yes.**   From this data can calculate the empirical formula and if the actual number for one of the atoms is given, can multiply all the subscripts as needed to get the molecular formula.

(c) **Yes.**   Can calculate the empirical formula from the %'s and comparing to the molecular weight determines the multiplier for the subscripts.

(d) **Yes.**   The structural formula will always tell you the molecular formula.

B) (a) **Yes.**   Once the ions are identified, the simplest combination that will produce zero charge can be determined

(a) **No.**   This is enough for a binary ionic compound, but if a polyatomic ion was in the compound there would be several possibilities for the same elements.

3.   A) **All pure elements must be neutral in charge**. When Na becomes a positive, it must have other ions of negative charge around it. Sodium ions with -1 charge are not stable so pure sodium composed of neutral atoms only.

B) **The molecular weight** of the molecule is one of the most important factors for these molecules to determine the physical state.

C) **The solution contains fewer Cu ions.** The molar mass of $CuCl_2$ is 134.5, and that of $CuCl_2 \cdot 2H_2O$ is 170.5 g/mole due to the water molecules. To have the same number of moles of copper ions in solution, the moles of the compound dissolved must be the same. The mass 14.5/170.5 produces fewer moles dissolved than the 14.5/134.5 of $CuCl_2$.

D)   **No, it has its usual charge.** Within $Ca_5(PO_4)_3F$, there are 5 Ca^{+2} ions and 3 PO_4^{-3} ions in the compound. The total positive charge is $5 \times 2 = +10$ and the three PO_4^{-3} ions produce $3 \times -3 = -9$ charge, so F ion must have a -1 charge.

4. A)   Structural

H-C-C-C-C-C-H (with H's and methyl branch)

Condensed

$CH_3CH_2CHCH_2CH_3$
        CH_3

• *Longest chain, pentane= 5 Carbons.*
• *methyl CH_3- group added to third C.*
• *Fill in H's needed to make 4 bonds to each C.*

4. B)   Structural

Condensed

$CH_3CHCH_2CH(CH_2)_4CH_3$
        CH_3 CH_3

• *The same structure will result if counting starts from right, instead of left side.*
• *-CH_2- groups in chain placed in parentheses in condensed formula.*

5. **(e) 2,5 dimethyloctane**
   • *Longest continuous chain = 8 carbons*
   • *The numbering shown on right produces the lowest number combination*
   • *Need "2,5- dimethyl" rather than "2-methyl-5-methyl" to use proper name.*

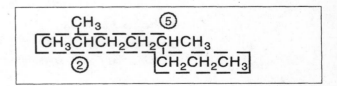

6. A) (a) **2-methyl pentane**   Longest chain = 5 carbons, number from left side.

   (b) **2,2 dimethylbutane**   Longest chain = 4 carbons, number from right side

   (c) **2-methylhexane**   Longest chain = 6 carbons, number from right side

   (d) **2,3 - dimethylpentane**   Longest chain = 5 carbons. number from right side

   B) constitutional isomers: **(a) and (d)** with the molecular formula C_6H_{14}

   and **(c) and (d)** with the molecular formula, C_7H_{16}.

7. A) **Only (c) applies**   The compounds have the same molecular formula and name. The structures are two ways to draw the same compound and not isomers.

   B) **Only (a) applies**   The compounds have the same molecular formula but would be named differently.

8. A) **(d), (e) and (f)**   B) **(d) and (c)**   C) **(a) and (b)**

   D) **Only (a) and (g)** since no more than 2 different elements make up the compound

   E) **(a) and (f)**, producing $Mg(H_2PO_4)_2$

   F) Two possible: **(b) with (c)** to make NH_4HS, **(b) with (g)** forming $(NH_4)_2S$

9. A) The **cation = K^+**, potassium ion and the **anion = PO_4^{-3}**, phosphate ion

   B) **Plan:** Convert 56.0 g K_3PO_4 → moles K_3PO_4 → formula units K_3PO_4 → ions K^+

   Conversions needed: molar mass   Avogadro's Number   3 ions K^+/1 K_3PO_4

   Molar mass K_3PO_4= [3×(39.1) + 1×(31.0) + 4×(16.0)] = 212.3 g/mol

   $$56.0\ g\ K_3PO_4 \left[\frac{1\ mole\ K_3PO_4}{212.3\ g\ K_3PO_4}\right]\left[\frac{6.02 \times 10^{23}\ units}{1\ mole\ K_3PO_4}\right]\left[\frac{3\ ions\ K^+}{1\ unit\ K_3PO_4}\right] = \underline{\underline{4.44 \times 10^{23} K^+ ions}}$$

   C) The picture drawn should look something like that shown on the right. There should be three potassium ions for each phosphate and the positive ions should be in between the negative ions.

10. A) **69 atoms** = 1 Ge + 4 S + 20 C + 44 H

    B) Molar mass = [1×(72.6) + 4×(32.0) + 20×(12.0) + 11×(1.0)] = **484.6 g/mol**

    C) % S = $\dfrac{128\ g\ S}{484.6\ g} \times 100 = \underline{\underline{26.4\ \%\ S}}$ D) $56.0\ g\ compd \left[\dfrac{1\ mole\ compd}{484.6\ g\ compd}\right] = \underline{\underline{0.00413\ mole}}$

11. | A) *Compound Formula* | B) *Type:* | C) *Ions:* |
    |---|---|---|
    | (a) $(Fe^{+3})_2(SO_4^{-2})_3 \rightarrow Fe_2(SO_4)_3$ | **ionic** | Fe^{+3} , SO_4^{-2} |
    | (b) NCl_3 | **molecular** | **No ions** |
    | (c) $(Li^+)_3(P^{-3})_1 \rightarrow Li_3P$ | **ionic** | Li^+ , P^{-3} |

    • *Li (group 1A) has +1 charge and P (group 5A) will have a charge of 5- 8= - 3*

(d) **Heptane**	**molecular**	**No ions**

    • *Formula indicates a hydrocarbon and alkane, since number H = 2(7)+ 2 = 14.*
    • *Since more than 4 carbons the Greek prefix of "hepta" used*

(e) **Chromium(II) nitrate**	**ionic**	Cr^{+2} , NO_3^-
(f) **Disulfur trioxide**	**molecular**	**No ions**

    • *Molecular binary compound, greek prefixes used.*

12. A) *Conversion needed:* mole compd $\rightarrow$? mole O $\rightarrow$? g O

Sample calculation: $3 \text{ mole } C_2H_5OH \left[\dfrac{1 \text{ mole O}}{1 \text{ mole } C_2H_5OH} \right] = 3 \text{ mole O} \left[\dfrac{16.0 \text{ g O}}{1 \text{ mole O}} \right] = 48.0 \text{ g O}$

	(a) C_2H_5OH	(b) CO_2	(c) KIO_3	(d) Na_2O
Moles in sample:	3 mole	2 mole	2 mole	3 mole
Mole O in sample:	3 mole	4 mole	6 mole	3 mole
mass O in sample:	48.0 g	64.0 g	96.0 g	48.0 g

Answer = 2 mole KIO_3 (c) contains the greatest mass of Oxygen.

 • *Could also compare just moles of O, since the greatest moles of O = greatest mass O.*

B)   Need % O: *Calculate from molecular formula or data in (A) with molar mass*

	(a) C_2H_5OH	(b) CO_2	(c) KIO_3	(d) Na_2O
molar mass:	46.0	44.0	214.1	62.0 g/mol
% O	34.8%	72.7%	22.8%	25.8%

Answer = (b) CO_2 contains the greatest %O in the compound.

 • *% 0 based on formula or sample size always gives same result.*

Based on molecular formula: *Based on data from (A) and molar mass:*

$\% \text{ O in } CO_2 = \dfrac{2(16.0 \text{ g O})}{44.0 \text{ g } CO_2} \times 100 = 72.7 \% \text{ O}$ $\% \text{ O in } CO_2 = \dfrac{64.0 \text{ g O}}{2(44.0 \text{ g}) CO_2} \times 100 = 72.7 \% \text{ O}$

C) **Oxygen monatomic ion in (d) Na_2O only.**

 • *(c) and (d) are both ionic, but O part of polyatomic ion, not monatomic, in (c)*

13. A) Conversion needed: 0.585 g lithium bromide $\rightarrow$? moles lithium bromide

 • *Need the molar mass for this conversion, so need to determine formula for compound.*
 • *Compound is ionic and based on group numbers charges are (Li^+) Br^-*
 • *Molar mass LiBr = 6.94 + 79.9 = 86.8 g/mole*

$0.585 \text{ g } LiBr \left[\dfrac{1 \text{ mole } LiBr}{86.8 \text{ g } LiBr} \right] = \underline{\underline{0.00674 \text{ mole } LiBr}}$

B) Conversion needed: 58.5 g potassium nitrate $\rightarrow$? number of potassium ions

 • *Conversion 3-step, need the molar mass , Avogadro's number and ratio K^+/KNO_3.*
 • *Formula: potassium nitrate= ionic compound, ions are K^+ (NO_3^-), so KNO_3 = formula*
 • *Molar mass KNO_3 = (39.1) + (14.0) + 3(16.0) = 101.1 g/mole*

$58.5 \text{ g } KNO_3 \left[\dfrac{1 \text{ mole } KNO_3}{101.1 \text{ g } KNO_3} \right]\left[\dfrac{6.02 \times 10^{23} \text{ units}}{1 \text{ mole } KNO_3} \right]\left[\dfrac{1 \text{ ion } K^+}{1 \text{ unit } KNO_3} \right] = \underline{\underline{3.48 \times 10^{23} K^+ ions}}$

C) Conversion needed: 5.85 moles $CH_2OHCH_2OH \rightarrow$? grams CH_2OHCH_2OH

 • *Molar mass $CH_2OHCH_2OH= C_2H_6O_2 = 2(12.0) + 6(1.00) + 2(16.0) = 62.0$*

$5.85 \text{ mole } C_2H_6O_2 \left[\dfrac{62.0 \text{ g } C_2H_6O_2}{1 \text{ mole } C_2H_6O_2} \right] = \underline{\underline{363 \text{ g } C_2H_6O_2}}$

D) Conversion needed: 5.85 mol of chloroform, $CHCl_3 \rightarrow$? volume of $CHCl_3$.

 • *Conversion two-step, need the molar mass, mol $\rightarrow$? g and density, g$\rightarrow$? mL.*
 • *Molar mass $CHCl_3$ = 1(12.0) + 1(1.00) + 3(35.5) = 119.5 g/mole*

$58.5 \text{ mol } CHCl_3 \left[\dfrac{119.5 \text{ g } CHCl_3}{1 \text{ mole } CHCl_3} \right]\left[\dfrac{1.0 \text{ mL } CHCl_3}{1.48 \text{ g } CHCl_3} \right] = \underline{\underline{472 \text{ mL } CHCl_3}}$

14. • Can compare the % by mass of Ca in each compound to answer.

• % Ca = $\dfrac{(\text{subscript Ca in compound}) \times (40.0\ g\ Ca)}{\text{molar mass compound}} \times 100$

• Since the starting weights same (1.0g), the highest % Ca^{2+} = highest mass Ca^{2+}

Compound:	$CaCO_3$	$CaSO_4$	$Ca_3(PO_4)_2$	
molar mass:	100.0	136.0	312.0	g/mol
% O	40.0 %	29.4%	38.4%	
g Ca/1.0 g	0.400 g	0.294 g	0.384 g	

   Answer = 1.0 g CaCO₃

15. **Looking for:** Identity of M by finding atomic mass, M
   Know: % M = 65.6%, (then also true % O = 34.5%)
      Molar mass MO_2 = (atomic mass M) + 2 (atomic mass O) = **(?)**
   Plan: Can find molar mass of MO_2 if the % and identity of one element in MO_2 known:

   $$\dfrac{\% X}{100} = \dfrac{(\text{subscript X in compound}) \times (\text{atomic mass X})}{\text{molar mass compound}}$$

   • Can use %O, since know % _and_ atomic mass to rearrange and solve for molar mass

   $$\dfrac{34.5}{100} = \dfrac{(2) \times (16.0)}{\text{molar mass } MO_2} \rightarrow \text{molar mass } MO_2 = \dfrac{(2) \times (16.0) \times 100}{34.5} = 95.5\ g/mole$$

   Then: **Atomic mass M** = 95.5 - 2(16.0) = **63.5 , and M = Cu** as closest match

16. A) **Looking for:** Empirical formula   **Know:** % composition
   Plan: Convert % to mass of element → ? moles each element → ? empirical formula
   • Assume 100 g compound

   $35.0\ g\ N \times \dfrac{1\ mole\ N}{14.0\ g\ N} = 2.50\ mole\ N$ / 2.5 = 1.0 N × 2 = 2 N

   $5.00\ g\ H \times \dfrac{1\ mole\ H}{1.00\ g\ H} = 5.00\ mole\ H$ / 2.5 = 2.0 H × 2 = 4 H

   $60.0\ g\ O \times \dfrac{1\ mole\ O}{14.0\ g\ O} = 3.75\ mole\ O$ / 2.5 = 1.5 O × 2 = 3 O

   > empirical formula
   > = **N₂H₄O₃**

   • needed to multiply by two to get whole numbers for all subscripts

   B)   Since compound is **an electrolyte it must an ionic compound.**
   C)   Given Molar mass = 80.0 g/mol and molar mass $N_2H_4O_3$ = 80.0, so **$N_2H_4O_3$ is** the molecular formula. Choosing common ions that involve only N,H and O produces **NH₄NO₃** as the combination of ions that would produce this molecular formula.

17. A) **Looking for:** Empirical formula   **Know:** % composition
   Plan: Same as in (16), but add step to compare empirical formula mass to molecular formula mass, to get multiplier for the subscripts.
   • Assume 100 g compound

   $85.69\ g\ C \times \dfrac{1\ mole\ C}{12.0\ g\ C} = 7.14\ mole\ C$ / 7.14 = 1.0 C   So empirical formula = **CH₂**

   $14.31\ g\ H \times \dfrac{1\ mole\ H}{1.00\ g\ H} = 14.31\ mole\ H$ / 7.14 = 2.0 H

   > Molar mass/ Empirical mass=
   > 55.0 amu/14.0 amu = 4
   > so molecular formula= **C₄H₈**

   B)   The compound is **a hydrocarbon, but not an alkane.**   To be an alkane, the molecular formula has to be C_nH_{2n+2}. With 4 carbons, the alkane formula would be C_4H_{10}.

Chapter 4

Answers to Key Terms Exercise:

1. decomposition reaction   2. combination reaction   3. combination reaction
4. decomposition reaction   5. oxide   6. combination reaction   7. oxide
8. combination reaction   9. combustion reaction   10. combustion analysis
11. displacement reaction   12. exchange reaction   13. balanced chemical reaction
14. stoichiometry   15. stoichiometric coefficient   16. mole ratio
17. stoichiometric factor   18. theoretical yield   19. actual yield
20. theoretical yield   21. actual yield   22. theoretical yield
23. percent yield   24. limiting reactant   25. limiting reactant
26. mole ratio   27. limiting reactant

Answers to True/False:

1. A) **False**   It is possible to have stoichiometric amounts of reactants (just enough of one to completely react the other) so this is not always true.

   B) **True**   The empirical and molecular formulas would always produce the same ratio of grams of CO_2 to 1.0 g of the compound.

$$C_3H_6O_3 \quad \frac{3 \text{ mole } CO_2}{1 \text{ mole } C_3H_6O_3} = \frac{3(44.0 \text{ g})}{(90.0 \text{ g})} = 1.47 \text{ g } CO_2/\text{g compound}$$

$$C_6H_{12}O_6 \quad \frac{6 \text{ mole } CO_2}{1 \text{ mole } C_6H_{12}O_6} = \frac{6(44.0 \text{ g})}{(180.0 \text{ g})} = 1.47 \text{ g } CO_2/\text{g compound}$$

   C) **True**   Moles of atoms is just another way to count atoms and the number of atoms must be the same on both sides of the reaction arrow

   D) **False**   The actual yield must be compared to theoretical yield of product, which is calculated from the limiting reactant, not equal to it.

   E) **True**   Always true that two or more formed.

   F) **False**   There are always two reactants, but if there are more than 2 elements in the compound burned, you may get more than two oxides as products.

2. A) **False**   The ratio of 3 mol S/1 mol SO_2 produces a mass ratio of 96g S/48 g SO_2 which is not 3/1

   B) **False**   The ratio is 3 mol S/2 mol H_2S, not 3 mol S/1 mol H_2S

   C) **False**   In the balanced reaction, 5 moles of product produced for every 4 moles of reactant, so they will never be equal. Although the total mass of product equals total mass of reactant, the number of moles can differ.

   D) **True**   The ratio is 2 mol H_2O/2 mol H_2S, which reduces to a 1:1 mole ratio.

   E) **True**   Two of the atoms of S come from H_2S in the balanced reaction.

3. **Balanced reaction:**   **Type:**

   A) $2 H_2O_2 \rightarrow O_2 + 2 H_2O$ decomposition

   B) $3 HgCl_2 + 2 Al \rightarrow 2 AlCl_3 + 3 Hg$ displacement

   C) $UO_2 + 4 HF \rightarrow UF_4 + 2 H_2O$ exchange

   D) $PI_3 + 3 H_2O \rightarrow H_3PO_3 + 3 HI$ exchange

   E) $Cl_2 + 3 F_2 \rightarrow 2 ClF_3$ combination

   F) $Br_2 + H_2O \rightarrow HBr + HBrO$ exchange

   G) $Cl_2O_7 + H_2O \rightarrow 2 HClO_4$ combination

   H) $2 HAuCl_4 \rightarrow 3 Cl_2 + 2 Au + 2 HCl$ decomposition

   I) $Zn(NO_3)_2 + Na_2S \rightarrow ZnS + 2 NaNO_3$ exchange

A- 13

4.   A) Refering to the filled table, the ratio (2mole/1 mole) true of CO_2 and $(CH_3)_2N_2H_2$, H_2O and N_2O_4

B) From the table, ratio of 1.2g to 1.0 g only true of the ratio of H_2O to $(CH_3)_2N_2H_2$.

$(CH_3)_2N_2H_2$	N_2O_4	CO_2	N_2	H_2O
1 molecule	2 molecules	2 molecules	3 molecules	4 molecules
1 mole	2 mole	2 mole	3 moles	4 mole
60.0 g	184 g	88.0 g	84.0 g	72.0g

5. A) **reactant to reactant** (1): $\dfrac{1\ mole\ CO_2}{2\ mole\ KO_2}$ **product to product** (1): $\dfrac{3\ mole\ O_2}{2\ mole\ K_2CO_3}$

reactant to product (4): $\dfrac{3\ mole\ O_2}{4\ mole\ KO_2}\quad \dfrac{2\ mole\ CO_2}{3\ mole\ O_2}\quad \dfrac{1\ mole\ CO_2}{1\ mole\ K_2CO_3}\quad \dfrac{1\ mole\ K_2CO_3}{2\ mole\ KO_2}$

B) $1.38\ g\ CO_2\left[\dfrac{1\ mole\ CO_2}{44.0\ g\ CO_2}\right]\left[\dfrac{3\ mole\ O_2}{2\ mole\ CO_2}\right]\left[\dfrac{32.0\ g\ O_2}{1\ mole\ O_2}\right]=1.51\ g\ O_2$

Yes, reaction does meet criteria since 1.38 g sample of CO_2 produces 1.51 g of O_2

6. A) $BaCO_3 \rightarrow CO_2 + BaO$ • Metal carbonates decompose to produce metal oxide, CO_2

B) $2\ ZnS + O_2 \rightarrow 2\ ZnO + 2\ S$ • Product is an ionic compound where Zn stays as Zn^{2+} ion, an O becomes O^{2-} ion.

C) $3\ Mg + N_2 \rightarrow Mg_3N_2$ • Ionic compound formed. Metal, $Mg \rightarrow Mg^{2+}$ and nonmetal, $N \rightarrow N^{3-}$ so compound formed $= (Mg^{2+})_3(N^{3-})_2$

D) $WO_3 + 3\ H_2 \rightarrow W + 3\ H_2O$ • Tungsten(VI)oxide $= WO_3$ and displacement of metal occurs

E) $ZnO + H_2SO_4 \rightarrow ZnSO_4 + H_2O$ • The anions, O^{2-} and SO_4^{2-} exchange cations.

7. A) $2\ Ag_2CO_3\ (s) \rightarrow 4\ Ag(s) + 2\ CO_2(g) + O_2(g)$

B) $56.2\ g\ Ag\left|\dfrac{1\ mole\ Ag}{108\ g\ Ag}\right|\left|\dfrac{2\ mole\ Ag_2CO_3}{4\ mole\ Ag}\right|\left|\dfrac{276\ g\ Ag_2CO_3}{1\ mole\ Ag_2CO_3}\right| = \underline{\textbf{71.8 g Ag}_2\textbf{CO}_3}$

8.   *Convert mole ratios for O_2 to reactant into mass ratios for each reaction to compare:*
A) $2\ NH_4NO_3(s) \rightarrow 2\ N_2(g) + 4\ H_2O(l) + O_2(g)$

$\dfrac{1\ mole\ O_2}{2\ mole\ NH_4NO_3} \Rightarrow \dfrac{1\ mole\ O_2 \times (32.0\ g/mol)}{2\ mole\ NH_4NO_3 \times (80.0\ g/mol)} = \dfrac{0.200\ g\ O_2}{1.00\ g\ NH_4NO_3}$

B) $2\ N_2O(s) \rightarrow 2\ N_2(g) + O_2(g)$

$\dfrac{1\ mole\ O_2}{2\ mole\ N_2O} \Rightarrow \dfrac{1\ mole\ O_2 \times (32.0\ g/mol)}{2\ mole\ N_2O \times (44.0\ g/mol)} = \boxed{\dfrac{\textbf{0.363 g } O_2}{\textbf{1.00 g } N_2O}}$

Reaction B produces the most O_2 per 1.00 g of reactant decomposed.

C) $2\ Ag_2O(s) \rightarrow 4\ Ag\ (s) + O_2(g)$

$\dfrac{1\ mole\ O_2}{2\ mole\ Ag_2O} \Rightarrow \dfrac{1\ mole\ O_2 \times (32.0\ g/mol)}{2\ mole\ Ag_2O \times (232.0\ g/mol)} = \dfrac{0.0689\ g\ O_2}{1.00\ g\ Ag_2O}$

9. A) $2\ H_2S + 3\ O_2 \rightarrow 2\ SO_2 + 2\ H_2O$

B) $0.985\ g\ H_2S\left|\dfrac{1\ mole\ H_2S}{34.0\ g\ H_2S}\right|\left|\dfrac{3\ mole\ O_2}{2\ mole\ H_2S}\right|\left|\dfrac{32.0\ g\ O_2}{1\ mole\ O_2}\right| = \underline{\textbf{1.39 g } O_2}$

10. A) $3\ Ca + N_2 \rightarrow Ca_3N_2$ *Reacting a metal, Ca (group 2A) $\rightarrow Ca^{2+}$ with nonmetal, N (group 5A) $\rightarrow N^{3-}$ so compound formed $= (Ca^{2+})_3(N^{3-})_2$*

10. B) *This is a limiting reactant situation- starting amounts given for each reactant.*

Plan: $3\ Ca$ + N_2 → Ca_3N_2

Know: 54.9 g   43.2 g   ? g

Need molar mass Ca ↓   *Need molar mass N_2* ↓   ↑ *Need molar mass Ca_3N_2*

? mol Ca   ? mol N_2 — $\dfrac{mole\ ratio}{Ca_3N_2/N_2}$ →   ? mol Ca_3N_2

$\dfrac{mole\ ratio}{Ca_3N_2/Ca}$ →   ? mol Ca_3N_2

| The smaller of the two is used to calculate the mass of product formed |

Mathematical solution:

$$54.9\ g\ Ca \left|\frac{1\ mole\ Ca}{40.1\ g\ Ca}\right|\left|\frac{1\ mole\ Ca_3N_2}{3\ mole\ Ca}\right| = 0.456\ mol\ Ca_3N_2 \left|\frac{148.3\ g\ Ca_3N_2}{1\ mole\ Ca_3N_2}\right| = \mathbf{67.7\ g\ Ca_3N_2}$$

$\boxed{Ca\ is\ limiting}$

$$43.2\ g\ N_2\left|\frac{1\ mole\ N_2}{28.0\ g\ N_2}\right|\left|\frac{1\ mole\ Ca_3N_2}{1\ mole\ N_2}\right| = 1.54\ mol\ Ca_3N_2$$

11. A) **Balanced reaction is:** $\mathbf{3\ A_2 + B_2 \to 2\ A_3B}$

 B) **B_2 limiting.** The " have" ratio of reactants: $7\ A_2/2\ B_2 = 3.5\ A_2/1.0\ B_2$ is gretaer than the "need" ratio: $3\ A_2/1.0\ B_2$

12. A) $2\ C_4H_{10} + 9\ O_2 \to 8\ CO_2 + H_2O$ which is combustion reaction
   • *Only the O_2 in air reacts with the C_4H_{10} for a burning reaction.*

 B) $1.20\ mole\ C_4H_{10}\left|\dfrac{8\ mole\ CO_2}{2\ mole\ C_4H_{10}}\right|\left|\dfrac{44.0\ g\ CO_2}{1\ mole\ CO_2}\right| = \mathbf{211\ g\ CO_2}$

13. A) **Second product is Br_2** $SrBr_2 + Cl_2 \to SrCl_2 + Br_2$ B) **displacement**
   C) Wt .$SrCl_2$ produced **always less than** the wt. of $SrBr_2$ reacted.

$$\frac{1\ mole\ SrCl_2}{1\ mol\ SrBr_2} = \frac{158.6\ g\ SrCl_2}{247.4\ g\ SrBr_2} = \frac{0.641\ g\ SrCl_2}{1.00\ g\ SrBr_2}$$

 D) $14.5\ g\ SrBr_2\left|\dfrac{0.641\ g\ SrCl_2}{1.00\ g\ SrBr_2}\right| = \mathbf{9.30\ g\ SrCl_2}$

14. A)   A **combination reaction**:   B) $Na_2SO_4 + 10\ H_2O \to Na_2SO_4\bullet 10\ H_2O$

 C)   **Plan:**   *Need:* $\dfrac{10\ mole\ H_2O}{1\ mole\ Na_2SO_4}$, *Have:* $\dfrac{?\ mole\ H_2O}{?\ mole\ Na_2SO_4}$ *after hydration*

 Know: wt. Na_2SO_4= 3.50 g, $3.50\ g\ Na_2SO_4\left|\dfrac{1\ mole\ Na_2SO_4}{142.0\ g\ Na_2SO_4}\right| = 0.0246\ mole\ Na_2SO_4$

 wt. H_2O = 6.93 - 3.50 g = 3.43 g H_2O absorbed, $3.43\ g\ H_2O\left|\dfrac{1\ mole\ H_2O}{18.0\ g\ H_2O}\right| = 0.191\ mole\ H_2O$

 Have: $\dfrac{0.191\ mole\ H_2O}{0.0246\ mole\ Na_2SO_4} = \dfrac{\mathbf{7.75\ mole\ H_2O}}{\mathbf{1.0\ mole\ Na_2SO_4}}$ | **The ratio of H_2O to Na_2SO_4 less than 10/1, so salt not fully hydrated.** |

15. A)   **Plan:** Find moles of C, H,O in sample, divide by smallest moles to get subscripts.

$$0.2691\ g\ H_2O\left|\frac{1\ mole\ H_2O}{18.0\ g\ H_2O}\right|\left|\frac{2\ mole\ H}{1\ mole\ H_2O}\right| = 0.0299\ mol\ H\ /\ \mathbf{0.00372 = 8\ H}$$

$$1.152\ g\ CO_2\left|\frac{1\ mole\ CO_2}{44.0\ g\ CO_2}\right|\left|\frac{1\ mole\ C}{1\ mole\ CO_2}\right| = 0.0262\ mol\ C\ /\ \mathbf{0.00372 = 7\ C}$$

 wt. O = 0.4039 g − (0.3144 g + 0.0299 g) = 0.0596 g O

$$0.0596\ g\ O\left|\frac{1\ mole\ O}{16.0\ g\ O}\right| = 0.00372\ mol\ O\ /\ \mathbf{0.00372 = 1\ O}$$ | **Empirical formula: C_7H_8O** |

 B) $\mathbf{2\ C_7H_8O + 17\ O_2 \to 14\ CO_2 + 8\ H_2O}$

16. A) $CaCN_2 + 3\ H_2O \rightarrow CaCO_3 + 2\ NH_3$ B) Answering this part will involve 2 steps:

Looking for: % yield NH_3 **Know**: $\% \ yield = \dfrac{actual\ yield}{theoretical\ yield} \times 100$

- *actual wt. NH_3 produced = 360 g , but theoretical yield = ? g NH_3*
- *limiting reactant situation since we have starting amounts of each reactant , so:*

$$1000\ g\ CaCN_2 \left[\dfrac{1\ mole\ CaCN_2}{80.0\ g\ CaCN_2}\right]\left[\dfrac{2\ mole\ NH_3}{1\ mole\ CaCN_2}\right] = 25.0\ mol\ NH_3 \left[\dfrac{17.0\ g\ NH_3}{1\ mole\ NH_3}\right] = 425\ g\ NH_3$$

$\boxed{CaCN_2\ is\ limiting}$

$$1000\ g\ H_2O \left[\dfrac{1\ mole\ H_2O}{18.0\ g\ H_2O}\right]\left[\dfrac{2\ mole\ NH_3}{3\ mole\ H_2O}\right] = 37.1\ mol\ NH_3$$

Substitute theoretical yield into formula: $\% \ yield = \dfrac{360\ g\ NH_3}{425\ g\ NH_3} \times 100 = \underline{\textbf{84.7\%}}$

17. A) Both compounds have different C/H mole ratios. Comparing the mole ratio of C/H in the sample from the combustion analysis data to that from the formula of each compound will tell if sample either compound or mixture.

B) **Plan:** Find C/H mole ratio in sample, compare to ratios in compounds

$$51.31\ mg\ CO_2 \left[\dfrac{1\ g\ CO_2}{1000\ mg\ CO_2}\right]\left[\dfrac{1\ mole\ CO_2}{44.00\ g\ CO_2}\right]\left[\dfrac{1\ mole\ C}{1\ mole\ CO_2}\right] = 0.001166\ mol\ C$$

$$12.99\ mg\ H_2O \left[\dfrac{1\ g\ H_2O}{1000\ mg\ H_2O}\right]\left[\dfrac{1\ mole\ H_2O}{18.01\ g\ H_2O}\right]\left[\dfrac{2\ mole\ H}{1\ mole\ H_2O}\right] = 0.001442\ mol\ H$$

- Comparison of experimental mole ratio C/ H to formula C/H ratios:

From combustion data: Cocaine: Sucrose:

$\dfrac{0.001166\ mol\ C}{0.001442\ mol\ H} = \dfrac{\textbf{0.809\ mole\ C}}{\textbf{1.0\ mole\ H}}$ $\dfrac{17\ mole\ C}{21\ mole\ H} = \dfrac{0.810\ mole\ C}{1.0\ mole\ H}$ $\dfrac{12\ mole\ C}{22\ mole\ H} = \dfrac{0.545\ mole\ C}{1.0\ mole\ H}$

So **compound most likely pure cocaine**, since very close match to formula ratio.

18. A) $4\ Fe + 3\ O_2 \rightarrow 2\ Fe_2O_3$

B) **Looking for**: wt Fe_2O_3 produced
Know: wt. of O_2 reacted = 2.381 g, wt of Fe= 13.263 g but NOT told all Fe reacts.
Plan: *Can use Method II. mole ratios, to see if Fe limiting or in excess.*

Need: $\dfrac{4\ mole\ Fe}{3\ mole\ O_2} = \dfrac{1.33\ mole\ Fe}{1\ mole\ O_2}$ *Have:* $\dfrac{13.263\ g\ Fe}{2.381\ g\ O_2} = \dfrac{0.2375\ mole\ Fe}{0.07443\ mole\ O_2} = \dfrac{3.191\ mole\ Fe}{1\ mole\ O_2}$

- *So Fe in excess, O_2 limiting the amount of product*

$$2.381\ g\ O_2 \left[\dfrac{1\ mole\ O_2}{32.0\ g\ O_2}\right]\left[\dfrac{2\ mole\ Fe_2O_3}{3\ mole\ O_2}\right]\left[\dfrac{159.8\ g\ Fe_2O_3}{1\ mole\ Fe_2O_3}\right] = \textbf{7.93\ } \boldsymbol{g}\ \textbf{Fe}_2\textbf{O}_3$$

C) **Looking for**: $\% \ Fe = \dfrac{wt.\ Fe\ reacted}{wt.\ Fe\ at\ start} \times 100$ **Know**: wt Fe at start = 13.263 g
Need: wt. Fe reacted   **Plan**: *Calculate wt. Fe reacted from O_2 reacted*

$$2.381\ g\ O_2 \left[\dfrac{1\ mole\ O_2}{32.00\ g\ O_2}\right]\left[\dfrac{4\ mole\ Fe}{3\ mole\ O_2}\right]\left[\dfrac{55.85\ g\ Fe}{1\ mole\ Fe}\right] = 5.541\ g\ Fe\ reacted$$

then $\% \ Fe\ converted = \dfrac{5.541\ g\ Fe}{13.263\ g\ Fe} \times 100 = \underline{\textbf{41.8\%}}$

Chapter 5

Answers to Key Terms Exercise:

Concerning Exchange Reactions:

1. precipitate 2. net ionic reaction 3. spectator ions 4. acid 5. base
6. salt 7. weak electrolyte 8. Acid 9. hydronium ion 10. strong acids
11. weak acid 12. hydronium ion 13. hydroxide ion 14. strong base
15. hydroxide ion 16. base 17. weak base 18. strong acid 19. strong base
20. net ionic reaction 21. weak acid 22. strong base 23. strong base

Concerning Oxidation and Reduction Reactions:

24. oxidation-reduction reaction 25. redox reaction 26. oxidation
27. reduction 28. reduction 29. oxidizing agent 30. oxidation
31. oxidized 32. reducing agent 33. reduction 34. oxidation number
35. oxidation number 36. oxidation 37. oxidation number 38. reduced
39. oxidized 40. oxidation number 41. Reduction 42. oxidation number
43. redox reaction 44. redox reaction 45. oxidizing agent 46. reducing agent
47. acid 48. redox reaction 49. acid 50. oxidizing agent
51. metal activity series 52. metal activity series 53. reducing agent 54. oxidation
55. metal activity series 56. oxidized 57. acid

Concerning Concentration Units

58. solute 59. solvent 60. solute 61. solvent 62. solute
63. solvent 64. concentration 65. molarity 66. solute 67. molarity
68. solute 69. molarity 70. titration 71. titration 72. molarity
73. standard solution 74. titration 75. equivalence point 76. titration
77. equivalence point 78. titration 79. molarity

Answers to True/False:

1. A). **False**   Dissociation into ions is not a decomposition reaction.

   B) **True**   They are chemical opposites and will neutralize each other.

   C) **True**   As long as the concentration of weak acid is much higher than that of the strong acid, this is possible.

   D) **True**   Dilution adds solvent, not solute, so the moles of solute always stays unchanged.

   E) **False**   The term anion incorrect, sodium and potassium ions form cations and their salts are always soluble in water.

   F) **False**   The cation comes from the base and the anion from the acid, for example HCl + NaOH produces the salt, NaCl

   G) **False**   Both the balanced equation and the standard solution molarity are needed, but so is the volume needed to reach the equivalence point.

   H) **True**   Combustion or burning is the process of adding oxygen to the elements in a compound. In the procees, oxygen always reduced to O^{-2} in oxides.

   I) **True**   If no spectator ions, the overall reaction need not be rewritten.

   J) **False**   Strong does not mean a high concentration of solute, but only means completely ionized.

   K) **False**   Elemental metals can only act as reducing agents and must be oxidized. The noble metals are always poor reducing agents, not oxidzing agents.

2. A) **(c)** since a metal hydroxide   B) **(e)** molecular, but not an acid or base
   C) **(b)** on strong acid list   D) **(a)** weak acid, not listed as strong
   E) **(a)** weak acid, not listed as strong   F) **(a)** carboxylic acid
   G) **(d)** amine group   H) **(e)** salt, so neither
   I) **(b)** on strong acid list

3. A) (a) +5   (b) -3   (c) +3   (d) +2   (e) -2   (f) +4

   B) **(a), (c), (d), (f)**   To be reduced, the oxidation number of N must be greater than zero so it can be reduced to zero in N_2.

   C) **(b), (e)**   To be oxidized, the oxidation number of N must be negative (less than than zero) so it can be raised to zero in N_2 when oxidized.

   D) **(a)**   The group number for N is 5A, so the +5 oxidation state is the maxumum positive state for N.

   E) **(b)**   The -3 oxidation state is when N is completely reduced.

4. A) (b)   B) (c)   C) (a)   D) (d)   E) (b)   F) (c)

5. A) **Answer = (a)** H_3PO_4.   *In a combination reaction, there is one product, so only (a) or (b) is possible, but only in (a) does P have a +5 oxidation number.*

   B) **phosphoric acid**   C) H_3PO_4 is a **weak acid**

   D) 1 P_4O_{10} + 6 H_2O → 4 H_3PO_4 (aq)

   E) **Neither**   *This is not a redox reaction, since P stays at +5 , also not acid-base since not a neutralization, but is a combination reaction.*

6. (a) *Overall:* $Zn(OH)_2$ (aq) + 2 HCl(aq) → 2 H_2O(l) + $ZnCl_2$(aq)

   Ne. Ionic: OH^-(aq) + H^+(aq) → H_2O(l)

   • *For a neutralization reaction, water and $ZnCl_2$ will be the products*

   (b) *Overall:* $ZnSO_4$(aq) + $CaCl_2$ (aq) → $CaSO_4$(s) + $ZnCl_2$(aq)

   Net Ionic: Ca^{2+} (aq) + $SO4^{2-}$(aq) → $CaSO_4$(s)

   • *For an exchange reaction, there must be two products, so $CaCl_2$ + $ZnSO_4$ are used to have one insoluble product.*

   (c) *Overall:* : Zn(s) + $CuCl_2$(aq) → Cu(s) + $ZnCl_2$ (aq)   or   Zn(s)+ 2 HCl(aq) → H_2(g) + $ZnCl_2$(aq)

   Net Ionic: Zn(s) + Cu^{2+} (aq) → Cu(s) + Cu^{2+} (aq)   Zn(s) + Cu^{2+} (aq) → Cu(s) + Cu^{2+} (aq)

   • *Both are redox reactions that can occur, based on the activity series*

7. **(a) Yes**   *Co in $CoCl_3$ is reduced from +3 oxidation state(its charge in the compound) to 0 as an elemental metal. Zn is oxidized from zero to +2.*

   (b) No   *Following the guidelines in the text, Cl has a -1 oxidation number in PCl_3, and P has +3. P also has +3 in H_3PO_3 so net change for P in the reaction.*

   (c) Yes   *Letting H stay as +1 throughout, C is oxidized from -8/3 to + 4 and O is reduced from 0 to -2 in this reaction. (Atoms can have fractional oxidation numbers, but the number of electrons lost or gained must be a whole number. Total loss or gain = no. atoms changing × oxidation number).*

8. **(a) No**   **Fe is not above Mg in the activity series** which so Fe(s) cannot be oxidized by Mg^{2+}. The metal ion must be from a metal listed below Fe in the series.

   (b) No   **Mg is above H_2 in the activity series, but not reactive enough** to be oxidized by H_2O(l), would need to have H_2O(g) or acid.

   (c) Yes   **Ca is above H_2 in activity series**, which means Ca is oxidized by acids.

9. A) *Overall :* $CuBr_2$ (aq) + Na_2S(aq) → CuS (s) + 2 NaBr(aq)

   Full ionic equation: Cu^{2+} + 2 Br^- + 2 Na^+ + S^{2-} → CuS (s) + 2 Na^++ 2 Br^-

   Net Ionic: Cu^{2+} + S^{2-} → CuS (s)

   • *Precipitates must stay intact in full and net ionic equations.*

   B) *Overall :* $Ba(OH)_2$(aq) + 2 HNO_3(aq) → $Ba(NO_3)_2$(aq) + 2 H_2O(l)

   Full ionic equation: Ba^{2+} + 2 OH^- + 2 H^+ + 2 NO_3^- → Ba^{2+}+ 2 NO_3^-+ 2 H_2O(l)

9. B) ***Net Ionic:*** $OH^- + H^+ \rightarrow H_2O(l)$
- *Water molecules stay intact as molecular compound*
- *Nitric acid is a strong acid in water.*

C) ***Overall :*** $AgNO_3(aq) + HCl(aq) \rightarrow AgCl(s) + HNO_3(aq)$
Full ionic equation: $Ag^+ + NO_3^- + H^+ + Cl^- \rightarrow AgCl(s) + H^+ + NO_3^-$
Net Ionic: $Ag^+ + Cl^- \rightarrow AgCl(s)$
- *Both hydrochloric and nitric acids are strong acids in water.*

D) ***Overall :*** $2\ NaHCO_3(aq) + 2\ KOH\ (aq) \rightarrow 2\ H_2O(l) + Na_2CO_3 + K_2CO_3$
Full ionic equation: $2\ Na^+ + 2\ HCO_3^- + 2\ K^+ + 2\ OH^- \rightarrow 2\ H_2O(l) + 2\ Na^+ + 2\ K^+ + 2\ CO_3^{2-}$
Net Ionic: $HCO_3^- + OH^- \rightarrow H_2O(l) + CO_3^{2-}$
- *Sodium hydrogen carbonate, $NaHCO_3$ acts as an acid and loses H^+*

10. A) **Looking for:** Molarity H_2SO_4

Need: $Molarity = \dfrac{No.\ mole\ solute}{Vol.\ Soln\ ,\ L}$ **Plan:** $\dfrac{wt.\ H_2SO_4}{800\ mL} \rightarrow \dfrac{?\ mole\ H_2SO_4}{?\ L\ soln}$

$$\dfrac{245.0\ g\ H_2SO_4 \left| \dfrac{1\ mole\ H_2SO_4}{98.1\ g\ H_2SO_4} \right|}{800\ mL \left| \dfrac{1.0\ L\ soln}{1000\ mL\ soln} \right|} = \dfrac{2.50\ mole\ H_2SO_4}{0.800\ L\ soln} = \textbf{3.12 M } H_2SO_4$$

B) **Looking for:** Volume Solution   **Know:** Solution contains 1.5 mole H_2SO_4
Plan: *1.50 mol H_2SO_4 $\rightarrow$? vol. soln*   **Need:** Molarity H_2SO_4 from (A) as conversion factor

$$1.50\ mole\ H_2SO_4 \left| \dfrac{1.0\ L\ solution}{3.12\ mole\ H_2SO_4} \right| = 0.481\ L \left| \dfrac{1000\ mL\ soln}{1.0\ L\ soln} \right| = \textbf{481 mL}$$

C) *Titration Reaction:* $H_2SO_4\ (aq) + 2\ KOH(aq) \rightarrow 2\ H_2O + K_2SO_4(aq)$
Plan: 20.00 mL, 3.12M H_2SO_4 $\rightarrow$? mol H_2SO_4 $\rightarrow$? mole KOH

$$20.00\ mL\ H_2SO_4 \left| \dfrac{1.0\ L\ soln}{1000\ mL\ soln} \right| \left| \dfrac{3.12\ mole\ H_2SO_4}{1.0\ L\ solution} \right| \left| \dfrac{2\ mole\ KOH\ soln}{1\ mole\ H_2SO_4} \right| = 0.125\ mole\ KOH$$

$$Molarity\ KOH = \left| \dfrac{0.125\ moles\ KOH}{0.03015\ L} \right| = \textbf{4.14 M KOH}$$

D) **Looking for:** Molarity H_2SO_4 after dilution
Know: Dilution, $V_{conc} \times M_{conc} = V_{dil} \times M_{dil}$ **Plan:** Solve equation for M_{dil}
$V_{conc} = 50.0$ mL, $M_{conc} = 3.12$ M , $V_{dil} = 100 + 50 = 150$ mL

$$M_{dil} = \dfrac{V_{conc}(M_{dil})}{V_{dil}} = \left| \dfrac{(50.00\ mL)\ (3.12\ M)}{(150\ mL)} \right| = \textbf{1.04 M } H_2SO_4$$

11. A) $Pb(OH)_2\ (aq) + 2\ HBr(aq) \rightarrow PbBr_2\ (s) + 2\ H_2O(l)$
B) **Looking for:** Volume HBr soln needed for complete reaction
Know: wt. $Pb(OH)_2 = 80.0$ g , M HBr = 3.40 M   **Plan:** *Use route from Figure 5.17*
80.0 g $Pb(OH)_2 \rightarrow$? moles $Pb(OH)_2 \rightarrow$? mol HBr $\rightarrow$? vol. soln HBr

$$80.0\ g\ Pb(OH)_2 \left| \dfrac{1\ mole\ Pb(OH)_2}{241.2\ g\ Pb(OH)_2} \right| \left| \dfrac{2\ mole\ HBr}{1\ mole\ Pb(OH)_2} \right| \left| \dfrac{1.0\ L\ solution}{3.50\ mole\ HBr} \right| = 0.190\ L = \textbf{190 mL}$$

12. A) **Physical state = aqueous**, since it is a soluble compound; **calcium chloride**.
 - *The naming follows ionic naming rules (Chapter 3).*

 B) **Looking for:** milligrams $CaCO_3$ reacted (from tablet)
 Know: Vol. HCl used = 29.47 mL , M HCl = 0.430 M
 Plan: 29.47 mL soln → ? mol HCl → ? mol $CaCO_3$ → ? g $CaCO_3$ → ? mg $CaCO_3$

$$29.47 \text{ mL} \left| \frac{1 \text{ L}}{1000 \text{ mL}} \right| \left| \frac{0.430 \text{ mol HCl}}{1.0 \text{ L solution}} \right| \left| \frac{1 \text{ mol } CaCO_3}{2 \text{ mol HCl}} \right| \left| \frac{100.0 \text{ g } CaCO_3}{1 \text{ mol } CaCO_3} \right| \left| \frac{1000 \text{ mg}}{1 \text{ g}} \right| = \textbf{634 mg}$$

13. A) Reaction is an oxidation-reduction reaction. Te is oxidized and N in HNO_3 reduced.
 - *Te going from 0 in Te(s) to +6 in TeO_3 (also adding O)*
 - *N going from +5 in HNO_3 to +2 in NO (also losing O)*

 B) $Te(s) + \underline{2} \ HNO_3(aq) \rightarrow H_2O(l) + \underline{2} \ NO(g) + TeO_3(s)$

 C) **Net ionic reaction is the same as overall**
 - *Full ionic equation:* $Te(s) + 2 \ H^+ + 2 \ NO_3^- \rightarrow H_2O(l) + 2 \ NO(g) + TeO_3(s)$
 - *Only HNO_3 as strong acid can be represented as ions, so no spectator ions.*

 D) **Looking for:** % yield $= \dfrac{\text{actual yield } TeO_3}{\text{theoretical yield } TeO_3} \times 100$ [Refer to Objective 4.7]

 Know: Actual yield TeO_3(s) = 50.0 g , theoretical yield = ? g TeO_3(s)
 - *limiting reactant situation since you have starting amounts for each reactant*

$$0.100 \ L \ soln \left| \frac{6.12 \ mole \ HNO_3}{1.0 \ L \ soln} \right| \left| \frac{1 \ mole \ TeO_3}{2 \ mole \ HNO_3} \right| = 0.306 \ mol \ TeO_3 \left| \frac{175.6 \ g \ TeO_3}{1 \ mole \ TeO_3} \right| = 53.7 \ g \ TeO_3$$

$$\boxed{HNO_3 \ is \ limiting}$$

$$50.0 \ g \ Te \left| \frac{1.0 \ mole \ Te}{127.6 \ g \ Te} \right| \left| \frac{1 \ mole \ TeO_3}{1 \ mole \ Te} \right| = 0.392 \ mol \ TeO_3$$

$$\% \ yield = \frac{50.0 \ g \ TeO_3}{53.7 \ g \ TeO_3} \times 100 = \textbf{93.1 \%}$$

14. A) Coefficients = 10 CO_2 , 8 H_2O *Balanced reaction*

 $2 \ KMnO_4 + 5 \ H_2C_2O_4 + 3 \ H_2SO_4 \rightarrow 2 \ MnSO_4 + \underline{10} \ CO_2 + \underline{8} \ H_2O + K_2SO_4$

 B) $KMnO_4$ is the oxidizing agent, $H_2C_2O_4$ is the reducing agent
 - *since C is oxidized, Mn is reduced and H_2SO_4 is unchanged throughout the reaction.*

 C) **Looking for:** Molarity of $H_2C_2O_4$ (acid) from titration with $KMnO_4$ solution
 Plan: 26.50 mL, 0.203 M, → ? mol KMnO4 → ? mol $H_2C_2O_4$ in 50.00 mL →? M

$$0.02650 \ L \ KMnO_4 \left| \frac{0.203 \ mol \ KMnO_4}{1.0 \ L \ soln} \right| \left| \frac{5 \ mole \ H_2C_2O_4}{2 \ mole \ KMnO_4} \right| = 0.0134 \ mole \ H_2C_2O_4$$

$$Molarity \ H_2C_2O_4 = \left| \frac{0.0134 \ mole \ H_2C_2O_4}{0.0500 \ L} \right| = \textbf{0.269 M}$$

15. A) Yes, a redox reaction would occur since Al is above H_2 in the metal activity series.
 Balanced reaction: $\textbf{2 Al(s) + 6 HCl(aq)} \rightarrow \textbf{2 AlCl}_3\textbf{(aq) + 3 H}_2\textbf{(g)}$
 - *Al(s) will be oxidized to Al^{3+} and the H^+ ions from the acid reduced to H_2(g).*

15.   B) **Looking for:** Volume Al(s) reacted

Know: Vol. HCl = 0.05 mL , M HBr = 3.40 M , density Al = 2.70 g/cm^3

Plan: 0.05 mL HCl soln → ? L soln→ ? mol HCl→ ? mol Al→ ? g Al→ ? cm^3 Al

• *Use route Figure 5.17, add last step to convert mass of Al to volume using the density.*

$$0.05 \text{ mL soln} \left| \frac{1.0 \text{ L}}{1000 \text{ mL}} \right\| \frac{12.0 \text{ mol HCl}}{1.0 \text{ L soln}} \right\| \frac{2 \text{ mol Al}}{6 \text{ mol HCl}} \right\| \frac{27.0 \text{ g Al}}{1 \text{ mol Al}} \right\| \frac{1.00 \text{ cm}^3}{2.70 \text{ g Al}} \right| = \underline{\underline{2.0 \times 10^{-3} \text{ cm}^3 \text{ Al}}}$$

C) **Looking for:** Number of drops HCl needed to react certain area of Al(s)

Conversion needed: Area Al = ? drops HCl

Know: Vol. HCl = 0.05 mL per drop, Area Al(s) reacted = 1.8 cm^2,
        thickness foil = 0.10mm

• *From Chapter 2 recall: volume Al = area × height (or thickness)*

• *From part (B): 1 drop HCl = 4.00×10^{-3} cm^3 Al, can use to convert vol. Al → drops HCl*

Vol. Al reacted, cm^3 = 1.8 $cm^2 \left[0.10 \text{ mm} \times \left(1.00 \text{ cm}/10 \text{ mm} \right) \right]$ = 1.80×10^{-2} cm^3 Al reacted

$$1.80 \times 10^{-2} cm^3 \text{ Al} \times \left| \frac{1.0 \text{ drop HCl soln}}{2.0 \times 10^{-3} cm^3 \text{ Al}} \right| = \underline{\underline{\textbf{9.0 drops HCl solution needed}}}$$

Therefore adding between 9 drops of 12.0 M HCl makes a hole the size of a penny in the foil.

16. **Looking for:** Idenitity of compound

Know: Molar mass compound = 171.3, compound is hydroxide that is soluble in water,
Molarity OH^- = 0.160 M when wt. compound dissolved= 6.84g, volume solution = 500 mL

For plan, assess information given about compound:

• Soluble hydroxide so must be metal hydroxide, with general formula $M(OH)_x$.

• Solution data lets us calculate x, $x = \dfrac{\text{moles } OH^-}{1.0 \text{ mole compound}} = \dfrac{\text{moles } OH^- \text{in 500 mL}}{\text{moles compound dissolved}}$

$$x = \frac{0.500 \text{ L} \times (0.160 \text{ mol } OH^-/1.0 \text{ L})}{6.84 \text{ g compd} \times (1.0 \text{ mol compd}/171.3 \text{ g compd})} = \frac{0.080 \text{ mol } OH^-}{0.040 \text{ mol compd}} = 2$$

• *So formula is $M(OH)_2$, then can use molar mass data to identify M:*
    Molar mass $M(OH)_x$ = 171.3 = [(atomic mass M) + 2 (16.0 + 1.00)]
    Atomic mass of M = 171.3 - 2(17.0) = 137.3 g/mol, *so* **M= Ba**

• *$Ba(OH)_2$ is soluble and would react with H_2SO_4 to produce $BaSO_4(s)$,*

Compound that fits all the data = $Ba(OH)_2$.

Chapter 6

Describing Energy Changes:

1. thermodynamics   2. kinetic energy   3. potential energy   4. kinetic energy
5. Law of Conservation of Energy   6. system   7. surroundings   8. system
9. surroundings   10. internal energy   11- 12. potential and kinetic energy
13. first law of thermodynamics   14. internal energy   15. work/working
16. potential energy   17. heat/heating   18. thermal equilibrium   19. internal energy
20. system   21. surroundings   22. heat capacity   23. heat capacity
24. heat/heating   25. specific heat capacity   26. molar heat capacity
27. heat capacity   28. molar heat capacity   29. specific heat capacities

Concerning Enthalpy Changes:

30. phase change   31. change in state (or phase change)   32. enthalpy change
33. potential energy   34. kinetic energy   35. heat of fusion   36. heat of vaporization
37. heat of fusion   38. enthalpy change   39. exothermic   40. enthalpy change
41. endothermic   42. enthalpy change   43. state function   44. exothermic
45. endothermic   46. change in state   47. enthalpy change   48. thermochemical equation
49. thermostoichiometric factor   50. thermostoichiometric factor   51. enthalpy change
52. bond enthalpy(energy)   53. bond enthalpy(energy)   54. endothermic   55. exothermic

Measuring and Classifying Enthalpy Changes:

56. calorimeter   57. enthalpy change   58. heat capacity
59. enthalpy change   60. standard state   61. standard enthalpy change
62. standard enthalpy change   63. Hess's law   64. standard molar enthalpy of
formation   65. standard enthalpy change   66. standard enthalpy change
67. standard molar enthalpy of formation   68. chemical fuel   69. exothermic
70. chemical fuel   71. energy density   72. fuel value   73. chemical fuels
74. basal metabolic rate   75. chemical fuel   76. caloric value   77. fuel value
78. energy density

Answers to True/False:

1. A). **False**   The temperature is 25°C for the thermodynamic standard state.

   B). **False**   Only the most common form or allotrope is given the zero value.

   C). **True**   The direction of change must be either a loss or gain for the system being considered, so the sign is always included.

   D). **False**   The molar heat capacities can be nearly the same, as for metals, and do not function well as chemical identifiers.

   E). **True**   A calorie is about four times larger, since 1 calorie= 4.184 joule.

   F). **False**   Bonds are not broken when a gas is formed. Tthe energy used to overcome the forces of attraction between the particles in the liquid state.

   G). **False**   The Δ symbol not used for heat (q) and work (w) by convention.

   H). **False**   These are the conditions for an exothermic reaction, not endothermic.

   I). **True**   Hess's law only defined for chemical changes, which have ΔH's only.

   J). **False**   Used for combustion reactions, which are not decomposition reactions.

   K). **True**   The standard state is the most common form of element at 25°,1 bar.

   L). **False**   The energy density requires the volume of fuel per mole, at 25°C and 1 bar, to also be known, which cannot be determined from the formula.

   M). **False**   The units are in kilojoules only for the reactions as written.

   N). **True**   Energy flows into system, so internal energy of the system must increase.

   O). **True**   1750 Cal/day = 1750 Cal × 4.184 kJ/Cal = 7,000,000 joules.

2.   A) **False**   Only true when an element reacts with $O_2(g)$ at 25°C and 1 bar.

B) **True**   You can have a shortage of a <u>useful</u> form of energy, starving because of lack of food, or running out of gasoline to power a car, but not of energy itself.

C) **True**   Being colder just means having less kinetic energy, or heat, in a substance compared to something else. Temperature measures heat content.

D) **False**   The total heat will be greater in the larger sample, but the heat capacity per mole is the same for both samples.

E) **True**   The ΔH° for reaction is more negative (exothermic) when CO_2 formed, instead of CO (low O_2 conditions), since ΔH° formation of CO(g), -110 kJ/mol), is lower than that for $CO_2(g)$, -393 kJ/mol which makes it more likely.

3. A) Compare heat needed to reach the melting point for each substance using

$q = c_{benzene} \times (mass\ sample) \times \Delta T_{benzene}$ $C_6H_6 : q = 1.75\ J/g\text{-°}C \times (50.0\ g) \times (5-(-10)\text{°}C) = 1312\ J$
$H_2O : q = 4.18\ J/g\text{-°}C \times (50.0\ g) \times (0-(-10)\text{°}C) = 2092\ J$

The benzene, C_6H_6, would the first to begin to melt.

B) Compare the heat needed to completely melt the sample:

$q = \Delta H_{fus} \times (no.\ moles\ in\ sample)$ $C_6H_6\ q = 50.0\ g\ C_6H_6 \times \dfrac{1\ mol\ C_6H_6}{78.0\ g\ C_6H_6} \times \dfrac{10.56\ kJ}{1\ mol\ C_6H_6} = 6.77\ kJ$

$H_2O: q = 50.0\ g \times \dfrac{1\ mol\ H_2O}{18.0\ g\ H_2O} \times \dfrac{40.7\ kJ}{1\ mol\ H_2O} = 16.7\ kJ$

The benzene, C_6H_6, sample needs less heat, it will be completely melted first.

4. A)

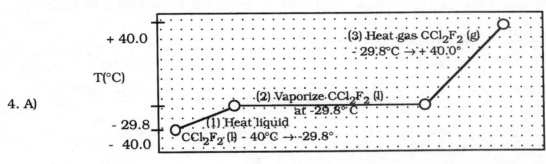

4. B) (1) $q_1 = c \times mass\ CCl_2F_2 \times \Delta T = 0.598\ J/g\text{-°}C \times 2000\ g \times (10.2\text{°}C) \times \dfrac{1.0\ kJ}{1000\ J} = 12.2\ kJ$

(2) $q_2 = \Delta H_{vap} \times (no.\ moles\ CCl_2F_2) = \dfrac{20.11\ kJ}{1\ mole\ CCl_2F_2} \times 2000\ g\ CCl_2F_2\ \dfrac{1\ mole\ CCl_2F_2}{121.0\ g\ CCl_2F_2} = 332.8\ kJ$

(3) $q_3 = c \times mass\ CCl_2F_2 \times \Delta T = 0.969\ J/g\text{-°}C \times 2000\ g \times (69.8\text{°}C) \times \dfrac{1.0\ kJ}{1000\ J} = 135.3\ kJ$

Total heat absorbed = 12.2 + 332.4 + 135.3 kJ = <u>480.3 kJ</u>

5. Sample 1, $T_1 = 24.0$°C gains heat, Sample 2, T_1 loses **Plan:** $q_{sample1} = -q_{sample2}$

$c_{ethanol} \times (mass\ sample\ 1) \times \Delta T_{sample1} = - (c_{ethanol} \times (mass\ sample\ 2) \times \Delta T_{sample2})$

$\Delta T_{sample2} = -\dfrac{(mass\ sample\ 1) \times \Delta T_{sample1}}{(mass\ sample\ 2)} = -\dfrac{(100\ g) \times (28-24\text{°}C)}{(75.0\ g)} = -5.33\text{°}C = (28.0 - T_1)$

$T_{1,\ sample\ 2} = 28.0 + 5.33\ \text{°}C = \textbf{33.3°C}$

6. Water gains heat (with ΔT) and metal cup, loses heat (with ΔT)   **Plan:** $q_{water} = -q_{metal}$

$$c_{water} \times (mass\ water) \times \Delta T_{water} = -(c_{metal} \times (mass\ metal) \times \Delta T_{metal})$$

$$c_{metal} = -\frac{(4.184\ J/g-°C)\ (44.0\ mL(1.0\ g/mL))\ (4.53°C)}{(48.9\ g)\ (-70.15\ °C)} = \underline{0.245\ J/g\text{-}°C}$$

7. Reaction gains: KI (s) $\rightarrow$ K$^+$ + I$^-$ (with ΔH), Solution loses (with ΔT) **Plan:** $q_{reaction} = -q_{soln}$

$$\frac{\Delta H°\ (kJ)}{1.0\ mole\ KI} \times (moles\ KI\ reacted) \times \frac{1000\ J}{1.0\ kJ} = -(c_{soln} \times (mass\ solution) \times \Delta T_{soln})$$

$$moles\ KI\ reacted = -\frac{(4.184\ J/g-°C)\ (200\ mL\ (1.21\ g/mL)\ (-4.20°C)}{(21.3)\ (1000)\ J/mol} = 0.199\ mol$$

$$mass\ KI = 0.199\ mol \times \frac{166.0\ g\ KI}{1.0\ mol\ KI} = \underline{\textbf{33.1 g KI}}$$

8. A) $2\ Ag\ (s) + Zn(NO_3)_2(aq) \rightarrow 2\ AgNO_3\ (aq) + Zn(s)$ B) exothermic

   C) Solution gains (with ΔT), reaction loses (with ΔH), **Plan:** $q_{soln} = -q_{reaction}$

$$c_{soln} \times (mass\ solution) \times \Delta T_{soln} = -(\Delta H) \times (no.\ mole\ Zn(NO_3)_2\ reacted)$$

$$\Delta H = -\frac{(4.184\ J/g-°C)\ (100\ mL\ (1.00\ g/mL)\ (23.85°C)}{0.100\ L\ soln\ (0.274\ mol\ Zn(NO_3)2/1.0\ L\ soln} = -\frac{18,300\ J}{0.0274\ moles} = \underline{\textbf{-668 kJ/mol}}$$

   D) $2\ Ag\ (s) + Zn^{+2} \rightarrow 2\ Ag^+ + Zn(s)$ The ΔH for the net ionic reaction should be the same as for the overall reaction since the net ionic reaction describes the same chemical change.

9. A) $\dfrac{\Delta H_{comb}}{1.0\ g} = \dfrac{C_{cal} \times \Delta T}{mass\ of\ asparagus} = \dfrac{(5.24\ kJ/°C)\ (23.17 - 22.45°C)}{0.50\ g} = \underline{\textbf{7.55 kJ/g}}$

   B) $No.\ Calories = 4.0\ oz. \times \dfrac{1.0\ lb}{16.0\ oz.} \times \dfrac{454\ g}{1.0\ lb} \times \dfrac{7.54\ kJ}{1.0\ g} \times \dfrac{1.0\ Cal}{4.184\ kJ} = \underline{\textbf{204 Cal}}$

10. A) $N_2(g) + O_2(g) \rightarrow 2\ NO(g)$ $\Delta H = +180.6$ kJ   B) endothermic

   C) $q = 1.50\ g\ NO \times \dfrac{1\ mol\ NO}{30.0\ g} \times \dfrac{180.6\ kJ}{2\ mole\ NO} = \underline{\textbf{4.51 kJ}}$

11. A) exothermic   B) $q = 18.0\ g\ C_6H_{12}O_6 \times \dfrac{1\ mol\ C_6H_{12}O_6}{180.0\ g} \times \dfrac{-68.4\ kJ}{1\ mole\ C_6H_{12}O_6} = \underline{\textbf{-6.84 kJ}}$

   C) Comparing the heats: $\% = \dfrac{-6.84\ kJ}{-280.2\ kJ} \times 100 = \underline{2.4\%}$

   The fermentaion produces only about 2.4% of the heat produced the combustion of one mole of glucose.

   D) The fermentation reaction produces fewer CO_2 molecules, which contain strong C-O bonds, and fewer bonds overall (20 versus 24 bonds) than the combustion reaction, so that the total heat released is less in the fermentation.

12. A) - 19.3 kJ   Precipitation reverses the reaction, so the sign changes to exothermic.

   B) $q = 0.050\ L\ soln \times \dfrac{0.102\ mole\ BaSO_4}{1.0\ L} \times \dfrac{-19.3\ kJ}{1\ mole\ BaSO_4} = -0.101\ kJ = \underline{\textbf{-101 J}}$

   C) No, it would be different since both neutralization and precipitation would occur. and the heat of neutralization would be added. The net ionic reaction would change from: $Ba^{+2} + SO_4^{-2} \rightarrow BaSO_4(s)$ to $Ba(OH)_2 + H_2SO_4 \rightarrow BaSO_4(s) + H_2O(l)$

13. Correct sequence for adding reactions:

$2 OF_2 + 2 H_2O \rightarrow 2 O_2 + 4 HF$ $2 \times (-276.6$ kJ$)$ • *Need OF_2 as reactant and need 2*

$+ \quad SO_2 + 4 HF \rightarrow SF_4 + 2 H_2O$ -1×-827.5 kJ$)$ • *Need SF_4 as product, reverse reaction*

$+ \quad 2 S + 2 O_2 \rightarrow 2 SO_2$ $2 \times (-296.9$ kJ$)$ • *Need S as reactant and need 2*

$2 OF_2 + \quad 2 S \quad \rightarrow \quad SO_2 + SF_4$ $\Delta H = -553.2 + 827.5 - 593.8 = -$ **319.5 kJ**

14. A) $\Delta H° = 321$ kJ $= [2 \ \Delta H°_f \ Fe(OH)_3(s) + 3 \ \Delta H°_f \ H_2(g)] - [2 \ \Delta H°_f \ Fe(s) + 6 \ \Delta H°_f \ H_2O(l)]$

         321 kJ $= [2(x) + 3(0)] - [2(0) + 6 (-285.8)]$ kJ $\rightarrow$ 321 kJ $= 2x + 1710$ kJ

       $\Delta H°_f \ Fe(OH)_3(s) = x = -$ **697.0 kJ**

B) $\Delta H° = [\ \Delta H°_f \ Fe_2O_3(s) + 3 \ \Delta H°_f \ H_2O(l)] - [2 \ \Delta H°_f \ Fe(OH)_3(s)]$

     $= [(-824.2) - 3 (-285.8)] - [2 (-694.5)]$ kJ $= -$ **286.1 kJ**

C) Formation equation: $Fe(s) + \frac{3}{2} O_2(g) = Fe_2O_3(s)$

     $(2 Fe(s) + 6 H_2O(l) \rightarrow 2 Fe(OH)_3 + 3 H_2(g))$ $+ 321.0$ kJ

     $(2 Fe(OH)_3 \rightarrow Fe_2O_3(s) + 3 H_2O(l))$ $- 290.2$ kJ

$+ \frac{3}{2} \ (2 H_2(g) + O_2(g) \rightarrow 2 H_2O(l))$ $+ \frac{3}{2} (- 571.6$ kJ $)$

$2 Fe(s) + \cancel{6 H_2O(l)} + \cancel{2 Fe(OH)_3} + \cancel{3 H_2(g)} + \frac{3}{2} O_2(g) \rightarrow \cancel{2 Fe(OH)_3} + \cancel{3 H_2(g)} + Fe_2O_3(s) + \cancel{6 H_2O(l)}$

reduces to : $Fe(s) + \frac{3}{2} O_2(g) = Fe_2O_3(s)$ with $\Delta H = (321) + (-290.2) + (-857.4) = -$**826.6 kJ**

• *The tabled value for $\Delta H°_f \ Fe_2O_3(s)$ is 824.4 kJ/ mol , so this value is in good agreement.*

15. A) $\underline{4} \ NH_3(g) + 6 \ NO(g) \rightarrow \underline{5} \ N_2(g) + 6 \ H_2O(l)$

B) $\Delta H° = [5 \ \Delta H°_f \ N_2(g) + 6 \ \Delta H°_f \ H_2O(l)] - [4 \ \Delta H°_f \ NH_3(g) + 6 \ \Delta H°_f \ NO(g)]$

    $= [5 \ (0) + 6(-285.8)] - [4(- 46.11) + 6(90.25)] = - 1714.8 - 357.1 = -$ **2072 kJ**

C) **exothermic**   D) $q = 60.0$ g NO $\times \dfrac{1 \text{ mol NO}}{30.0 \text{ g NO}} \times \dfrac{- 2072 \text{ kJ}}{6 \text{ mole NO}} = -$ **691 kJ**

16. A) Both have same reaction for combustion: $2 C_3H_6 + 9 O_2(g) \rightarrow 6 CO_2(g) + 6 H_2O(g)$

     and $\Delta H°_{comb} = [6 \ \Delta H°_f \ CO_2(g) + 6 \ \Delta H°_f \ H_2O(l)] - [2 \ \Delta H°_f \ C_3H_6]$

     • **Propene has the largest fuel value** since the smallest value of $\Delta H°_f \ C_3H_6$ will result in the largest $\Delta H°_{comb}$ and both compounds have the same molecular weight.

B) Since propene has the more exothermic reaction for the combustion, there are weaker bonds in propene than in cyclopropane. The lower heat of formation of propene also indicates that it takes less energy to break all the bonds in propene to form gaseous atoms than in cyclopropane.

17. A) $2 H_2S(g) + 3 O_2(g) \rightarrow 2 SO_2(g) + 2 H_2O(g)$ $\Delta H° = -$ **1036 kJ**

B) $100 \ g \ H_2S \times \dfrac{1 \text{ mol } H_2S}{34.0 \text{ g } H_2S} = 2.94 \ mol \ H_2S \times \dfrac{- 1036 \ kJ}{2 \text{ mol } H_2S} = - 1523 \ kJ$

    $100 \ g \ O_2 \times \dfrac{1 \text{ mol } O_2}{32.0 \text{ g } O_2} = 3.12 \ mol \ O_2 \times \dfrac{- 1036 \ kJ}{3 \text{ mol } O_2} = -$ **1079 kJ** $\boxed{O_2 \text{ limiting}}$

C) **Fuel value $H_2S(g)$:**   **Energy density $H_2S(g)$:**

$\dfrac{1 \text{ mol } H_2S}{34.0 \text{ g } H_2S} \times \dfrac{1036 \text{ kJ}}{1 \text{ mol } H_2S} = 15.2$ kJ/g $\dfrac{15.2 \ kJ}{1.0 \text{ g } H_2S} \times \dfrac{1.39 \text{ g } H_2S}{1.0 \text{ L } H_2S} = 21.1$ kJ/L

     The fuel value of H_2S is less than H_2 or CH_4, but the energy density of H_2S is higher than H_2 and about one-half that of CH_4. Based on its energy density, it could be used as a chemical fuel, but as a poisonous, as well as bad-smelling, gas it could be dangerous to use.

Chapter 7

Answers to Key Terms Exercise:

Defining Light Energy and Connection to Atoms:

1. Electromagnetic radiation 2. wavelength 3. frequency 4. spectrum
5. frequency(ies) 6. wavelength 7. wavelength 8. frequency 9. wavelength
10. spectrum 11. quantum 12. Planck's constant 13. frequency
14. quantum 15. quantum theory 16. photoelectric effect 17. frequency
18. wavelength 19. quantum 20. photon 21. photon (or quantum)
22. quantum theory 23. line emission spectrum 24. continuous spectrum
25. wavelength 26. spectrum 27. ground state 28. excited state
29. ground states 30. principal quantum number

The Quantum Mechanical Model for Electrons:

31. momentum 32. Heisenberg uncertainty principle 33. momentum
34. boundary surface 35. boundary surface 36. orbital 37. principal quantum number 38. principal energy level 39. orbital 40. shell 41. orbital
42. shell 43. subshell 44. orbital 45. subshell 46. orbital 47. orbital
48. Pauli exclusion principle 49. orbital 50. Nuclear magnetic resonance (NMR)

Describing the Electrons on Atoms:

51. Electron configuration 52. orbital 53. shell 54. subshell (orbital)
55. shell 56. subshell (orbital) 57. subshell 58. Hund's rule 59. orbital
60. subshell 61. subshell 62. electron configuration 63. electron configuration
64. shell 65. core electrons 66. shell 67. valence electrons 68. noble gas notation 69. core electrons 70. electron configuration 71. subshell 72. valence electron 73. valence electron 74. s-block elements 75. p-block elements
76. transition elements 77. valence electrons 78. subshell (orbital) 79. subshell
80. transition metals 81. valence electron 82. subshell (orbital) 83. subshell

Properties Related to Configurations:

84. subshell 85. s-block elements 86. isoelectronic 87. p -block elements
88. subshell 89. p-block elements 90. isoelectronic 91. Lewis dot symbol
92. valence electrons 93. atomic radii 94. Ionic radii 95. ionic radii
96. atomic radii 97. diamagnetic 98. paramagnetic 99. paramagnetic
100. ferromagnetic 101. ionization energy 102. core electron 103. ionization energy 104. valence electrons 105. ionization energy 106. electron affinity
107. ionization energy 108. electron affinity 109. p-block elements

Answers to True/False:

1. A). **False** The effective nuclear attraction increases going across a row.

   B). **False** Sc in group 3B, will have one valence electron left in +3 ion.

   C). **False** The second part not true, the atomic radius increases going down a group.

   D). **True** Have 5 electrons in the p subshell, so one must be unpaired.

   E). **False** Electron removal requires absorption, not emission.

   F). **False** Electrons fill the 4s before the 3d is filled.

   G). **False** The value of m_ℓ determines orientation.

   H). **True** A shorter wavelength will mean a higher frequency.

   I). **True** Removing the second electron from Na requires removing a core electron.

   J). **False** Consult Figure 7.17 in text.

   K). **True** Ionization is require to produce the electrical current.

2. A) The electrons on the sodium atom are quantized and only certain frequencies can appear in the line spectrum. The intensity of the photons emitted is a function of how many photons are emitted, as joules per second, not the frequency or wavelength.

2. B) The electrons are the part that respond to visible light by undergoing transitions between the ground state and excited states.

C) (a) An atom needs to have unpaired spins to respond and be paramagnetic.
(b) A nucleus needs to have different spin states, such as in ^1H.

D) The number of types of subshells will equal n itself, while the number of orbitals total will be equal to n^2.

E) The core electrons are tightly bound in filled shells, while the valence electrons are in unfilled subshells.

F) The metals in the p-block form cations by losing electrons. The most stable positive ion will be isoelectronic with the noble gas from the row before the element. The nonmetals gain electrons to form anions that are isoelectronic with the noble gas at the end of the row in which they appear.

G) Ga^{+4} is not stable. Ga has 3 valence electrons and an electron from a filled d subshell would have to be removed to make the ion. Mn has 5 valence electrons so Mn^{+4} is stable.

H) (a) After the first electron removed the remaining electrons feel a much stronger pull from the nucleus since the number of protons doesn't change.

(b) The first ionization energy for sulfur would be lower than that of oxygen, since the same number of valence electrons are further out from the nucleus.

3. A) $\Delta E = \frac{hc}{\lambda} = \frac{6.63 \times 10^{-34} J\text{-sec}\ (3.00 \times 10^{8}) m/\text{sec}}{550\ nm \times (\ 1.0\ m\ /\ 1.00 \times 10^{9} nm)} = \mathbf{3.61 \times 10^{-19} J}$

B) **Answer = Ultraviolet light** *Light of smaller frequency must be higher energy*

4. A) $\Delta E = h\nu$, so $\nu = \frac{\Delta E}{h} = \frac{7.0 \times 10^{-19} J}{6.63 \times 10^{-34} J\text{-sec}} = \mathbf{1.06 \times 10^{15} s^{-1}}$

B) λ = **220 nm** $\lambda = \frac{c}{\nu} = \frac{3.00 \times 10^{8} m/\text{sec}}{1.06 \times 10^{15} 1/\text{sec}} = 2.83 \times 10^{-7} m \times \frac{1.00 \times 10^{9} nm}{1.0\ m} = 284\ nm$

The photon must have a λ equal to or less than 284 nm to have sufficient energy.

5. A) Answer = **3 photons**

B) For the three transitions: $\Delta E = h\nu = 2.179 \times 10^{-18} J \left[\frac{1}{n^2_{lower}} - \frac{1}{n^2_{upper}} \right]$

(a) $n = 7 \rightarrow n = 5$ $\Delta E = 2.179 \times 10^{-18}$ J [0.0400 - 0.0204] = 4.27×10^{-20} J
(b) $n = 5 \rightarrow n = 4$ $\Delta E = 2.179 \times 10^{-18}$ J [0.0625 - 0.0400] = 4.90×10^{-20} J
(c) $n = 4 \rightarrow n = 1$ $\Delta E = 2.179 \times 10^{-18}$ J [1.000 - 0.0625] = 2.04×10^{-18} J

so the **transition in (c) has the highest energy**

C) Calculate λ(or ν) and consult Figure 7.1: $\lambda = \frac{hc}{\Delta E} = \frac{1.986 \times 10^{-25} J - m}{\Delta E,\ J}$ then:

(a) $\lambda = 4.65 \times 10^{-6}$ m (b) $\lambda = 4.05 \times 10^{-6}$ m (c) $\lambda = 9.70 \times 10^{-8}$ m
Infrared region **Infrared region** **X-ray region**

6. A) (a)  (b)  (c) 

B) All three would appear first in **n= 3**
C) (a) **2 electrons** (b) **10 electrons** (c) **6 electrons**

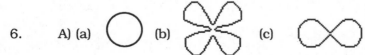

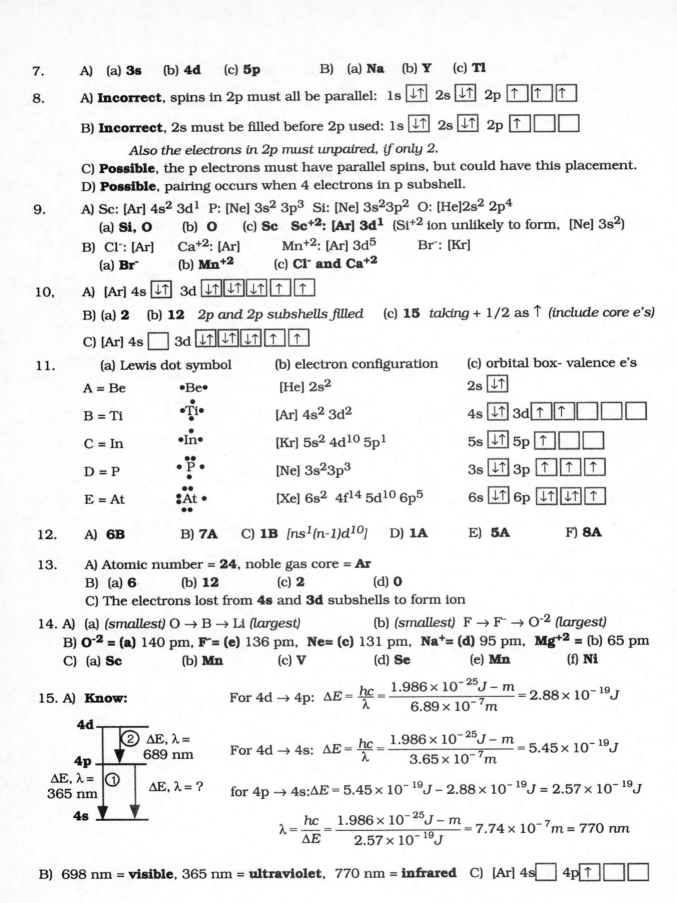

7.   A) (a) **3s**   (b) **4d**   (c) **5p**   B) (a) **Na**   (b) **Y**   (c) **Tl**

8.   A) **Incorrect**, spins in 2p must all be parallel: 1s ⟨↓↑⟩ 2s ⟨↓↑⟩ 2p ⟨↑⟩⟨↑⟩⟨↑⟩

 B) **Incorrect**, 2s must be filled before 2p used: 1s ⟨↓↑⟩ 2s ⟨↓↑⟩ 2p ⟨↑⟩⟨ ⟩⟨ ⟩
     Also the electrons in 2p must unpaired, if only 2.
 C) **Possible**, the p electrons must have parallel spins, but could have this placement.
 D) **Possible**, pairing occurs when 4 electrons in p subshell.

9.   A) Sc: [Ar] $4s^2\,3d^1$ P: [Ne] $3s^2\,3p^3$ Si: [Ne] $3s^23p^2$ O: [He]$2s^2\,2p^4$
      (a) **Si, O**   (b) **O**   (c) **Sc** Sc^{+2}: **[Ar] $3d^1$** (Si^{+2} ion unlikely to form, [Ne] $3s^2$)
   B) Cl⁻: [Ar]   Ca^{+2}: [Ar]   Mn^{+2}: [Ar] $3d^5$ Br⁻: [Kr]
      (a) **Br⁻**   (b) **Mn^{+2}**   (c) **Cl⁻ and Ca^{+2}**

10,   A) [Ar] 4s ⟨↓↑⟩ 3d ⟨↓↑⟩⟨↓↑⟩⟨↓↑⟩⟨↑⟩⟨↑⟩

 B) (a) **2**   (b) **12**   *2p and 2p subshells filled*   (c) **15** *taking + 1/2 as ↑ (include core e's)*

 C) [Ar] 4s ⟨ ⟩ 3d ⟨↓↑⟩⟨↓↑⟩⟨↓↑⟩⟨↑⟩⟨↑⟩

11.   (a) Lewis dot symbol   (b) electron configuration   (c) orbital box- valence e's

 A = Be   •Be•   [He] $2s^2$ 2s ⟨↓↑⟩

 B = Ti   •T̈i•   [Ar] $4s^2\,3d^2$ 4s ⟨↓↑⟩ 3d ⟨↑⟩⟨↑⟩⟨ ⟩⟨ ⟩⟨ ⟩

 C = In   •Ïn•   [Kr] $5s^2\,4d^{10}\,5p^1$ 5s ⟨↓↑⟩ 5p ⟨↑⟩⟨ ⟩⟨ ⟩

 D = P   •P̈•   [Ne] $3s^23p^3$ 3s ⟨↓↑⟩ 3p ⟨↑⟩⟨↑⟩⟨↑⟩

 E = At   :Ät•   [Xe] $6s^2\,4f^{14}\,5d^{10}\,6p^5$ 6s ⟨↓↑⟩ 6p ⟨↓↑⟩⟨↓↑⟩⟨↑⟩

12.   A) **6B**   B) **7A**   C) **1B** $[ns^1(n\text{-}1)d^{10}]$ D) **1A**   E) **5A**   F) **8A**

13.   A) Atomic number = **24**, noble gas core = **Ar**
   B) (a) **6**   (b) **12**   (c) **2**   (d) **0**
   C) The electrons lost from **4s** and **3d** subshells to form ion

14. A) (a) *(smallest)* O → B → Li *(largest)*   (b) *(smallest)* F → F⁻ → O^{-2} *(largest)*
   B) O^{-2} = **(a) 140 pm**, F⁻= **(e) 136 pm**, Ne= **(c) 131 pm**, Na$^+$= **(d) 95 pm**, Mg^{+2} = **(b) 65 pm**
   C) (a) **Sc**   (b) **Mn**   (c) **V**   (d) **Se**   (e) **Mn**   (f) **Ni**

15. A) **Know:**

For 4d → 4p: $\Delta E = \dfrac{hc}{\lambda} = \dfrac{1.986 \times 10^{-25} J-m}{6.89 \times 10^{-7} m} = 2.88 \times 10^{-19} J$

For 4d → 4s: $\Delta E = \dfrac{hc}{\lambda} = \dfrac{1.986 \times 10^{-25} J-m}{3.65 \times 10^{-7} m} = 5.45 \times 10^{-19} J$

for 4p → 4s: $\Delta E = 5.45 \times 10^{-19} J - 2.88 \times 10^{-19} J = 2.57 \times 10^{-19} J$

$\lambda = \dfrac{hc}{\Delta E} = \dfrac{1.986 \times 10^{-25} J-m}{2.57 \times 10^{-19} J} = 7.74 \times 10^{-7} m = 770\ nm$

 B) 698 nm = **visible**, 365 nm = **ultraviolet**, 770 nm = **infrared**   C) [Ar] 4s ⟨ ⟩ 4p ⟨↑⟩⟨ ⟩⟨ ⟩

Chapter 8

Answers to Key Terms Exercise:
Concerning Covalent Bonding:
1. covalent bond   2. bonding electrons   3. lone pair electrons   4. covalent bonds
5. octet rule   6. multiple covalent bonds   7. double bond   8. triple bond
9. Multiple covalent bonding   10. single covalent bond   11. Lewis structure
12. covalent bonds   13. lone pair electrons   14. octet rule   15. Lewis structure
16. octet rule   17. free radicals   18. Lewis structure   19. octet rule
20. multiple covalent bonding   21. single covalent bonds   22. saturated hydrocarbons
23. multiple covalent bonds   24. unsaturated hydrocarbons   25. covalent bond
26. Alkenes   27. double bond   28. alkynes   29. triple bond
30. aromatic compounds   31. single covalent bonds   32. double bonds

Different Characteristics of Bonding:
33. single covalent bond   34. double bond   35. alkenes   36.cis-trans isomerism
37. cis isomer   38. double bond   39. trans isomer   40. double bond   41. Cis
trans isomerism   42.Saturated fats   43. single covalent bonds   44. unsaturated fats
45. double bond   46. saturated fats   47. unsaturated fats   48. unsaturated fats
49. cis isomers   50. trans isomers   51. saturated fats   52. unsaturated fats
53. saturated fats   54. double bond(s)   55. electronegativity   56. covalent bond
57. electronegativity   58. nonpolar covalent bond   59. electronegativity
60. polar covalent bond   61. electronegativity   62. bond length   63. triple bond
64. bond length   65. double bond   66. bond length   67. single covalent bond

Properties from Lewis Structures:
68. formal charge   69. Lewis structure   70. formal charge
71. lone pair electrons   72. bonding electrons   73. Lewis structure
74. resonance structures   75. single covalent bonds   76. double bond
77. triple bond   78. lone pair electrons   79. resonance hybrid
80. free radicals   81. covalent bond   82. lone pair electrons
83. octet   84. free radical   85. antioxidants   86. antioxidants

Answers to True/False:

1.   A). **True**   Covalent always means sharing bonded electrons.

   B). **True**   H can only form one bond, so never can be in the center.

   C). **True**   D). **True**   E). **False**   May have just an octet, not always expanded.

   F). **True**   G). **True**

   H). **False**   Need at least one multiple bond, as well as lone pairs, for resonance.

   I). **False**   Atoms can be different, but have the same electronegativity which produces
   nonpolar covalent bonds. Also a large difference in electronegativity produces
   ionic bonds rather than polar covalent bonds.

   J). **True**   K). **False**   Alkenes, not alkanes, can have cis-trans isomers.

   L). **True**

2. A) Structures **(a), (b) and (d)** are incorrect

   B) (a) OFN has 18 valence electrons and 20 are shown in the structure and F can never be
   the central atom.   Using formal charges as guide, the F must be
   bonded to the N, producing a structure where all formal charges are
   zero.

   (b) ICl_4^- has 36 valence electrons and only 32 are shown. I has an
   expanded octet in the molecule.

   (d) Cannot multiple bond to Cl, since formal charges indicate this is
   unstable. The multiple bond appears between P and O.

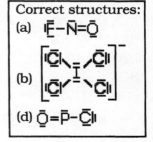

Correct structures:

3. A)

(a) (b) (c) (d) (e)

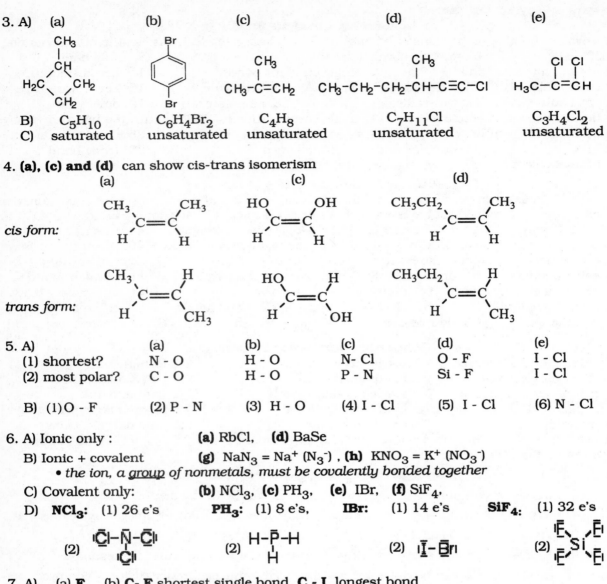

B) C_5H_{10} $C_6H_4Br_2$ C_4H_8 $C_7H_{11}Cl$ $C_3H_4Cl_2$

C)   saturated   unsaturated   unsaturated   unsaturated   unsaturated

4. (a), (c) and (d) can show cis-trans isomerism

           (a)   (c)   (d)

cis form:

trans form:

5. A)

	(a)	(b)	(c)	(d)	(e)
(1) shortest?	N - O	H - O	N - Cl	O - F	I - Cl
(2) most polar?	C - O	H - O	P - N	Si - F	I - Cl

B) (1) O - F   (2) P - N   (3) H - O   (4) I - Cl   (5) I - Cl   (6) N - Cl

6.
A) Ionic only :   **(a)** RbCl,   **(d)** BaSe

B) Ionic + covalent   **(g)** $NaN_3 = Na^+ (N_3^-)$, **(h)** $KNO_3 = K^+ (NO_3^-)$
* the ion, a <u>group</u> of nonmetals, must be covalently bonded together

C) Covalent only: **(b)** NCl_3, **(c)** PH_3,   **(e)** IBr,   **(f)** SiF_4,

D) **NCl_3:** (1) 26 e's   **PH_3:** (1) 8 e's,   **IBr:** (1) 14 e's   **SiF_4:** (1) 32 e's

(2)   (2)   (2)   (2)

7.
A)   (a) **F**   (b) **C - F** shortest single bond, **C - I** longest bond
       (c) **C - F** most polar, both **C= C** and **C - I** are nonpolar bonds   (d) **C - F**

B) No, it could not have cis and trans isomers because all four hydrogens have been replaced and there are no two like atoms are bonded to the carbons. Structural isomers could exist however.

8.
A)

	(a) central atom?	(b) total valence e's	(c) No. single bonds	(d) no. lone pairs
CS_2	C	16	2	4
NH_3	N	8	3	1
$HCCl_3$	C	26	4	9
SF_6	S	48	6	18
BrO_2^-	Br	20	2	8
SO_2	S	18	2	6

B)   (1) SF_6 (2) CS_2 and SO_2 (3) CS_2, NH_3 , $HCCl_3$ and SF_6

       (4) SO_2, (a resonance structure for CS_2 requires that formal charges are not minimized)

9. A) Formamide: B) resonance structure: C) formal charges
                                                    O   C   N   H

(I) $H-C-N-H$ (II) $H-C=N-H$ (I)   6-2-4= 0   4-4-0= 0   5-3-2= 0   1-1-0= 0
      with $\ddot{O}$, H   with $\ddot{O}$, H   (II) 6-1-6= -1 4-4-0= 0   5-4-0= +1 1-1-0= 0

Structure (I) is the most likely structure.   *No resonance, so no hybrid is possible.*

A) Hydrazoic acid: B) resonance structures: C) formal charges:
                                                        N(a)   N(b)   N(c)   H

(I) $\ddot{N}=N=\ddot{N}-H$ (II) $N\equiv N-\ddot{N}-H$ (I)   5-3-2= 0   5-4-0= +1 5-2-4= -1 1-1-0= 0
      (a) (b) (c)   (II) 5-2-4= -1 5-4-0= +1 5-3-2= 0   1-1-0= 0
                                 (III) $\ddot{N}-N\equiv N-H$ (II) 5-4-0= +1 5-4-0= +1 5-1-6= -2 1-1-0= 0

Structures (I) and (II) are equally likely
The resonance hybrid should look like (I)

10. A) $H-C\equiv N + 2H-H \rightarrow H-C-N-H$ (with H, H on C, H on N)   B) $\ddot{O}=C=\ddot{O} + H-\ddot{O}-H \rightarrow H-\ddot{O}-C-\ddot{O}-H$ (with $\ddot{O}$ above C)

B) Bonds broken, B.E.   Bonds formed, B.E.   Bonds broken, B.E.   Bonds formed, B.E.
   1 C-H   416 kJ   3 C-H   3 (416 kJ)   2 C=O   2 (803 kJ)*   2 O-H   2 (467 kJ)
   1 C≡N   866 kJ   2 N-H   2 (391 kJ)   * B. E. in CO_2 only   2 C-O   2 (336 kJ)
   2 H-H   2 (436 kJ)   C-N   285 kJ   2 O-H   2 (467 kJ)   1 C=O   695 kJ
   ─────────────   ─────────────   ─────────────   ─────────────
       2154 kJ   2315 kJ   2540 kJ   2301 kJ

C)   ΔH = 2154 - 2315 kJ = **- 161 kJ**   ΔH = 2540 - 2301 kJ = **+ 239 kJ**
D)   **exothermic**   **endothermic**

11. A) $H-C-C-C-\ddot{O}-H$ (with H, H on first C; H, H on second C; $\ddot{O}$ on third C)   B) The two types are the C = O and C-O and the shortest is C = O.

All zeros appear for formal charges on the atoms when the double bond is between the C and the O not bonded to H and this is the most likely structure.

C)   In the anion, the same formal charges result when either oxygen has the double bond, so a resonance hybrid exists and the bonds have the same length.

D)   A resonanace with a C= C is impossible since a H would have to be removed or moved. Atoms cannot be moved in a resonance structure.

12. A) NO_2 17 e's N_2O 16 e's NO_3^- 24 e's NO^+ 10 e's N_2O_3 28 e's

     $\ddot{O}-\ddot{N}=\ddot{O}$ $\ddot{N}=O=\ddot{N}$ $\ddot{O}-N=\ddot{O}$ (with $\ddot{O}$ above N, ⊖) $N\equiv O^+$ $\ddot{O}-N-N=\ddot{O}$ (with $\ddot{O}$ above N)

     B) NO^+ C) N_2O_3 D) NO_2 E) N_2O F) NO^+

13. A) They all aromatic compounds   OH   OH   HO
    B) resorcinol   C) resorcinol   (structure) (structure) (structure)
                                            meta-   ortho-   para - isomers

14. A) X = **C or Si**   B) X = **Br**   C) X = **Xe**   D) X = **P**

15. A) $3\, XO_2 + H_2O \rightarrow XO + 2\, HXO_3$
    B)   To identify X need atomic weight, so find molecular weight of XO_2

$$0.0621\ L\ soln \times \frac{0.1225\ mol\ OH^-}{1.0\ L\ soln} \times \frac{1.0\ mol\ HXO_3}{1.0\ mol\ OH^-} \times \frac{3.0\ mol\ XO_2}{2.0\ mol\ HXO_3} = 1.14 \times 10^{-2}\ mol\ XO_2$$

$$MW\ XO_2 = \frac{0.523\ g}{0.00114\ mol} = 45.9\ g/mol$$ **Lewis structure, $HXO_3 = HNO_3$**
                                                                        $\ddot{O}$ (above N)
     AW X = 45.9 – 3(16.0) = **13.9 g/mol** and **X = N** $\ddot{O}-N-\ddot{O}-H$

Chapter 9

Geometry and Shapes of Molecules:

1. valence shell electron repulsion model   2. electron pair geometry   3. bond angle
4. electron pair geometry   5. molecular geometry   6. electron pair geometry   7. axial
8. equatorial   9. axial   10. bond angles   11. equatorial   12. equatorial
13. spectroscopy   14. valence bond theory   15. sigma bond   16. pi-bond
17. sigma bond   18. sigma bond   19. pi bond   20. pi-bond
21. sigma bond   22. hybridized   23. hybridized   24. sigma
25. Pi-bonds   26. hybridized   27. sp hybrid orbitals   28. sigma bonds
29. sp^2 hybrid orbitals   30. sp^3 hybrid orbitals   31. sp^3d hybrid orbitals
32. sp^3d^2 hybrid orbitals   33. hybridized   34. bond angles
35. electron pair geometry   36. valence electron pair repulsion model

Polarity and Noncovalent Interactions:

37. polar molecule   38. nonpolar molecule   39. Polar molecule
40. molecular geometry   41. Nonpolar molecule   42. molecular geometry
43. noncovalent interactions   44. intermolecular forces   45. London forces
46. London forces   47. induced dipoles   48. induced dipoles   49. polarization
50. polarization   51. London forces   52. Dipole-dipole attractions
53. polar molecule   54. dipole moment   57. noncovalent interactions
56. Hydrogen bonding   57. noncovalent interactions (or intermolecular forces)
58. polarization   59. London forces   60. dipole- dipole attractions
61. hydrogen-bonding   62. polar molecule   63. Molecular geometry
64. noncovalent interactions 65. intermolecular forces   66. noncovalent interactions
67. phospholipid molecules 68. dipole-dipole attractions 69. London forces
70. lipid bilayer   71. hydrophobic   72. hydrophilic

Superimposable Isomers and Structure of Biological Substances:

73. deoxyribonucleic acid (DNA)   74. noncovalent interactions   75. polymer
76. nucleotide   77. nucleotide   78. hydrogen-bonding   79. complementary
base pairs   80. complementary base pairs   81. molecular geometry
82. hydrogen-bonding   83. Chiral   84. enantiomers   85. chiral
86. enantiomers   87. molecular geometry   88. sp3 hybrid orbitals
89. asymmetric   90. symmetrical   91. achiral

Answers to True/False:

1.   A). **True**   B). **False** Lone pairs require more space than bonded electrons.

C). **False**   A square planar shape requires 2 lone pairs and 4 bonds.

D). **True**   E). **True**

F). **False**   The geometry is different. NCl_3 has angles near 109°, while BCl_3 of 120°.

G). **False** CO_2 is a nonpolar molecule, and has only London forces.

H). **False**   Both are polar molecules and would have dipole-dipole attractions.

I). **True**   Both have the same hybridization, but N has a lone pair of electrons

J). **False**   Either half-filled, filled or empty atomic orbitals can be mixed together.

K). **True**

L). **True**   Hybrid orbitals cannot be used for pi-bonding.

M). **False**   Only the sigma bonds, not pi bonds, and lone pairs can be counted.

N). **True**   A larger molecule can have both polar and nonpolar regions.

O). **True**   The second carbon has 4 different groups bonded to it.

P). **False**   Only is all the terminal atoms are the same.

Q). **False**   IR energy matches the energy of motion, not the bond energy.

R). **False**   The nitrogen bases are bonded to the sugar units.

2. A)   Noncovalent interactions are attractions that exist between atoms or molecules that are not ionic, covalent or metallic bonds. They can be intermolecular or intramolecular and are much weaker than covalent bonds with energies from 0- 40 kJ/mol whereas covalent bonds have energies that range from 150-1000 kJ/mol.

    B)   Dipole-dipole attractions result from permanent dipole moments in molecules, London forces result from temporary induced dipoles.

    C)   In a resonance structure only pi-bonds and lone pairs are moved. The electron pair geometry is the same and if the lone pairs shift on terminal atoms, instead of the central atom, the shape will be exactly the same. In constitutional isomers, sigma bonds are arranged differently, as in straight chain versus branched alkanes, and this is likely to change the electron pair geometry or shape+.

    D) CH_4 has only weak London forces since it is too small to be easily polarizable and both alcohols are polar and also have H-bonding. However, the polar -OH group in CH_3OH contributes more than half of the molar mass or number of electrons in the molecule. Therefore, the influence on intermolecular attractions is great. In decanol, the polar -OH group contributes only about one-tenth of the molar mass or number of electrons, so its influence on the intermolecular forces is minor.

    E)   The strongest intermolecular force is hydrogen bonding. When any solid melts, the particles have enough K.E. to overcome some of the intermolecular forces. Since there are fewer hydrogen bonds in the liquid to hold the water molecules, the structure collapses on itself and liquid water becomes more dense than the solid.

3. A) **Hydrogen- bonding**   B) **hydrophilic**   C) As x increases, the water solubility will decrease, as the -COOH region has less influence on the overall properties.
    D) The solubility in fats would increase as x increases since influence of the London forces increases.

4. A) **London forces**   B) **Dipole-dipole attractions**   C) The molar mass is larger, as is the total valence electrons, together with the dipole-dipole attractions produce stronger forces.

5. A)   B) **4 sigma bonds**   C) **zero pi-bonds**
    D) electron pair geometry = **trigonal bypyramid**, molecular geometry = **seesaw**

    E) (axial) Br- Se- (equatorial) Br ≤ 90°, (equatorial) Br-Se- (equatorial) Br ≤ 120°
    F) **nonpolar covalent**, since difference in electronegativity = 0.4

6. A)   (a) **Br$_2$** more polarizable   (b) **HCl** since greater polarity in bond
    B)   (a) **CH$_3$Br**   (b) **CH$_3$CH$_2$CH$_2$CH$_2$Cl**   (c) **butane**, as straight chain
    C)   (d) **London dispersion forces**

7. A) **a, e**   B) **b, d**   C) **c**   D) **e**   E) **b**

8. A) (a)   (b)   (c)   (d)

    B) (a) **sp^2**   (b) **sp^{3}d**   (c) **sp^3**   (d) **sp^2**

    C) (a)   (b)   (c)   (d)

    D) (a) **polar**   (b) **polar**   (c) **nonpolar**   (d) **polar**
    E) (b) **dipole-dipole**   (b) **dipole-dipole**   (c) **London forces**   (d) **dipole-dipole**

    F) **Only (d) could show resonance**. The shape of the resonance structures would be the same for both of the resonance structures.

9. A) **b**   B) **a**   C) **c**   D) **e**   E) **none**   F) **b**   G) **d**

10. A) **a, c**   B) **d**   C) **b**   D) **e**   E) **a, d**   F) **e**   G) **b**   H) **e**

11. (a) $CH_3CH_2CH_3$ (b) CH_3OCH_3 (c) $\underline{C}H_3Cl$ (d) $CH_3\underline{C}HO$ (e) $CH_3\underline{C}N$

   A) **nonpolar**   **polar**   **polar**   **polar**   **polar**

   B) **London forces**   **dipole-dipole**   **dipole-dipole**   **dipole-dipole**   **dipole-dipole**

   C) $\underline{C}$- sp^3, 109° $\underline{O}$- sp^3, 109° $\underline{C}$- sp^3, 109° $\underline{C}$- sp^2, 120° $\underline{C}$- sp, 180°

   D)   (5)   (1)   (3)   (4)   (2)

   E)   (2)   (1)

   F) **CH_3CN is the most polar** since it has a molecular geometry that maximizes the dipole-dipole attractions. **$CH_3CH_2CH_3$ will have the weakest intermolecular forces** of the molecules, since $CH_3CH_2CH_3$ is nonpolar, has only London forces and has about the same molar mass (and polarization) as the other molecules in the group.

12. A)

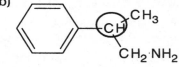

B) The hybridization for C in H_2NCN is sp, and it changes to sp^2 in melamine. There are two types of hybridization for N in H_2NCN, both sp and sp^3. In melamine, the N's in the ring are sp^2, but the N in the - NH_2 groups are still sp^3.

C) Because the atoms in the ring are all sp^2, the ring is a planar shape, with the NH_2 groups (tetrahedral shapes) projecting off of the flat ring. The intermolecular force will be hydrogen-bonding, between the H's on the NH_2 groups and N's in the ring or NH_2 groups (much like between the bases in DNA). The mostly flat shape will make it easier to align the molecules to form many hydrogen bonds.

13. A) **sp^3**, tetrahedral, **achiral**   B) **sp^3**, tetrahedral, **chiral**

   C) **sp^3**, tetrahedral, **chiral**   D) **sp^2**, triangular planar, **achiral**

14. Each of the molecules (a)- (c) will have enantiomers. The chiral carbons are:

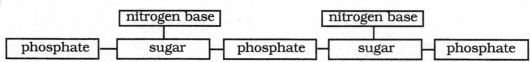

15. A)

| phosphate | | sugar | | phosphate | | sugar | | phosphate |

   B)   - T - C - G - T - A - G - A -

   C)   They are linked through covalent bonds. The two bonded groups serve as the structural backbone for each helical strand in DNA.

   D)   Hydrogen bonding between the nitrogen bases hold the strands together.

   E)

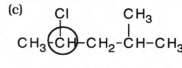

Chapter 10

Answers to Key Terms Exercise:

Describing Behavior of Gases:

1. pressure   2. pressure   3. Pascal   4. Newton   5. pressure
6. barometer   7. pressure   8. milliliters of mercury   9. torr
10. standard atmosphere   11. barometer   12. standard atmosphere   13. Pascal
14. standard atmosphere   15. bar   16. Pascal   17. bar   18. absolute (or Kelvin) temperature scale   19. absolute (or Kelvin) temperature scale   20. absolute (or Kelvin) temperature scale   21. pressure   22. Boyle's Law   23. pressure
24. pressure   25. Charles's Law   26. absolute (or Kelvin) temperature scale
27. pressure   28. Avogadro's law   29. pressure   30. ideal gas law
31. ideal gas constant   32. ideal gas constant   33. pressure   34. standard temperature and pressure (STP)   35. standard molar volume   36. ideal gas law

Modifications of the Ideal Gas law and Comparing Two gases:

37. combined gas law   38. ideal gas law   39. ideal gas law   40. ideal gas law   41. ideal gases   42. pressure   43. partial pressure   44. Dalton's law of partial pressures   45. pressure   46. partial pressure   47. partial pressure
48. mole fraction   49. ideal gas law   50. law of combining volumes   51. pressure
52. ideal gas   53. ideal gas law   54. van der Waals equation

Reaction of Gases in the Earth's Atmosphere:

55. troposphere   56. stratosphere   57. stratosphere   58. troposphere
59. photochemical reactions   60. photochemical reactions   61. photodissociation
62. electronically excited state   63. stratosphere   64. photodissociation
65. stratosphere   66. chlorofluorocarbons (CFC's)   67. Photodissociation
68. chlorofluorocarbons   69. stratosphere   70. ozone hole   71. stratosphere
72. troposphere   73. air pollutant   74. primary pollutant
75. secondary pollutant   76. primary pollutant   77. Particulates
78. secondary pollutant   79. air pollutant   80. Aerosols   81. particulates
82. primary pollutant   83. smog   84. aerosol   85. NO_x
86. NO_x 87. aerosol   88. photochemical smog
89. photochemical smog   90. secondary pollutant   91. photochemical reactions

Answers to True/False:

1.   A). **True**   B). **False**   Standard atmosphere is 760 mmHg or 1.0 bar.

     C). **False**   True at many low pressures and high temperatures.

     D). **False**   Only true if at same T and P

     E). **False**   Only true if V and n constant.

     F). **True**   G). **True**   H). **True**

     I). **False**   Aerosols contain smaller particles than those classified as particulates.

     J). **False**   Also need lots of continuous sunlight to produce photochemical smog.

     K). **True**   L). **True**   M). **True**

     N). **False**   Water evaporates producing a mixture of the gas with water vapor.

2.   A) **c**   B) **e**   C) **a, c**   D) **c**   E) **a, c**   F) (a) **decreases**, (b) **increases**

3.   A)   The temperature of - 273.15 °C which corresponds to absolute zero, 0 K, the temperature when all particles have zero kinetic energy or motion.

     B)   (a) CFC's increase the rate of loss of ozone. (b) ClO·, Cl·   (c) Sunlight dissociates Cl_2 to produce Cl· which initiates the destructive cycle from the inactive forms trapped by the ice crystals formed during the winter months.

     C)   Primary pollutants are introduced directly from a source into atmosphere. Secondary pollutants result from chemical conversion of primary pollutants in the atmosphere.

3.   D) The evidence that chlorofluorocarbons introduced into troposphere eventually appeared in stratosphere, indicates the layers can mix at interface.

    E) Ozone is produced late in the sequence of reactions and the winds can then carry the reaction products far away from the source of pollution.

    F) Sunlight, hydrocarbons and NOx compounds.

    G) In smog, the three major secondary pollutants are: (1) NO_2 from NO_x, (2) PAN's from hydrocarbons and (3) Aldehydes from hydrocarbons.

    H) Smog is chemically reducing, since it contains primarily SO_2 which can act as reducing agent. Photochemical smog has NO_x or O_3 which can act as oxidizing agents.

4. A) *Use ideal gas law to calculate moles, then use molar mass:*

$$n = \frac{PV}{RT} = \frac{113\ kPa\ (1\,atm/101.3\ kPa)\ (0.250\ L)}{0.0821\ \frac{L-atm}{K-mol}\ (40 + 273)\ K} = 0.0108\ \text{mol}\ N_2O \left[\frac{44.0\ \text{g}\ N_2O}{1\ \text{mole}\ N_2O}\right] = \mathbf{0.478\ g\ N_2O}$$

    B) $d = \dfrac{0.478\ \text{g}\ N_2O}{0.250\ L\ N_2O} = \mathbf{1.91\ g/L}$

    C) **You can change either P or T and not affect the density**. Since the volume and moles are fixed, changing either T or P will cause the other to change in proportion, but not mass per unit volume.

5. A) **Flask B:** *Flask A contains 0.125 mole Ar and Flask B, 0.132 moles CS_2*

    B) **Flask A:** *Comparing two gases, assume ideal gas law applies to both, then:*

$$\frac{P_A V_A}{n_A T_A} = \frac{P_B V_B}{n_B T_B} \quad \text{with equal volumes: } P_A = P_B \times \frac{n_A T_A}{n_B T_B} = P_B \times \frac{0.125\ \text{mol}\ (400\ K)}{0.131\ \text{mol}\ (200\ K)} = 1.91 P_B$$

    C) **Flask B:**   *Since lower temperature means lower K.E.*

    D) $n_A T_A = n_B T_B$ so $T_A = \dfrac{n_B T_B}{n_A} = \dfrac{0.132\ \text{mol}\ (200\ K)}{0.125\ \text{mol}} = \mathbf{211.2\ K}$,   T(°C) = $\underline{\mathbf{-62.0°C}}$

6. **Molar mass =** $M = d\left[\dfrac{RT}{P}\right] = \dfrac{0.855\ \text{g}}{1.0\ L}\left|\dfrac{0.0821\ \frac{L-atm}{K-mol}\ (273)\ K}{0.4355\ atm}\right| = \mathbf{44.0\ g/mol}$

7. *Start with:* $\dfrac{P_1 V_1}{n_1 T_1} = \dfrac{P_2 V_2}{n_2 T_2}$ *Constant n, P:* $\dfrac{V_1}{T_1} = \dfrac{V_2}{T_2}$ so $T_2 = \dfrac{V_2 T_1}{V_1} = \dfrac{2.50\ L\ (471\ K)}{5.10\ L} = 231\ K$

    T(°C) = 231 - 273 = $\underline{\mathbf{-42°C}}$

8. A) $P_4S_3 + 8\ O_2(g) \rightarrow P_4O_{10} + 3\ SO_2(g)$

    B) $0.800\ \text{g}\ P_4S_3 \left[\dfrac{1\ \text{mole}\ P_4S_3}{220.0\ \text{g}\ P_4S_3}\right]\left[\dfrac{3\ \text{mole}\ SO_2}{1\ \text{mole}\ P_4S_3}\right] = 0.0109\ \text{mole}\ SO_2$

$$V = \frac{nRT}{P} = \frac{(0.0109\ \text{mol})\ 0.08206\ \frac{L-atm}{K-mol}\ (305\ K)}{0.954\ atm} = 0.286\ L = \mathbf{286\ mL\ SO_2}$$

9. *Apply Law of combining volumes since same P, T*

    • Need balanced reaction: $C_3H_8 + 5\ O_2(g) \rightarrow 3\ CO_2(g) + 4\ H_2O(g)$

$$\frac{V_{oxygen}}{V_{propane}} = \frac{n_{oxygen}}{n_{propane}} = \frac{3\ \text{mol}}{1\ \text{mol}} \quad V_{oxygen} = \frac{30.0\ L\ (3)}{1} = 150\ L\ \text{needed} \quad \text{So need } \underline{\mathbf{10\ tanks\ O_2(g)}}$$

10. *Using ideal gas law:* $M = \dfrac{mRT}{PV} = \dfrac{(0.482\ \text{g})\ 0.08206\ \frac{L-atm}{K-mol}\ (374\ K)}{1.009\ atm\ (0.204\ L)} = 71.9\ \text{g/mol}$

    and 71.9 - 5(12.0) = 12 g H in 1 mole of compound so **formula = C_5H_{12}**

11. A) $20.3 \text{ g } (NH_4)_2Cr_2O_7 \left[\dfrac{1 \text{ mol } (NH_4)_2Cr_2O_7}{252 \text{ g } (NH_4)_2Cr_2O_7} \right]\left[\dfrac{1 \text{ mol } N_2}{1 \text{ mol } (NH_4)_2Cr_2O_7} \right] = 0.0806 \text{ mole } N_2$ produced

$0.0806 \text{ mole } N_2 \times \dfrac{4 \text{ mol } H_2O}{1 \text{ mol } N_2} = 0.322 \text{ mole } H_2O \text{ (g) produced}$

$P_{N_2} = \dfrac{nRT}{V} = \dfrac{(0.0806 \text{ mol}) \, 0.08206 \frac{L-atm}{K-mol} \, (293 \text{ K})}{5.0 \text{ L}} = \mathbf{0.388 \text{ atm}} \; ; \; P_{H_2O} = \mathbf{1.55 \text{ atm}}$

B) $P_{total} = 1.55 + 0.388 = \underline{\mathbf{1.94 \text{ atm}}}$

12. A) *Collected over water, calculate* $\mathbf{P_{C_2H_2}}$ *from* $\mathbf{P \, C_2H_2 = P_{total}} - \mathbf{P \, H_2O} = 738 - 21 = 717 \text{ mmHg}$

$n = \dfrac{PV}{RT} = \dfrac{717 \text{ mmHg } (1 \text{atm}/760 \text{ mmHg}) \, (0.528 \text{ L})}{0.08206 \frac{L-atm}{K-mol} \, (296 \text{ K})} = 0.0205 \text{ mol } C_2H_2 \left[\dfrac{26.0 \text{ g } C_2H_2}{1 \text{ mole } C_2H_2} \right] = \underline{\mathbf{0.533 \text{ g}}}$

B) % *yield* $= \dfrac{actual \ yield}{theoretical \ yield} \times 100$ *Need theoretical yield* C_2H_2 *from mass of* CaC_2 *given:*

$1.50 \text{ g } CaC_2 \left[\dfrac{1 \text{ mole } CaC_2}{64.0 \text{ } CaC_2} \right]\left[\dfrac{1 \text{ mole } C_2H_2}{1 \text{ mole } CaC_2} \right]\left[\dfrac{26.0 \text{ g } C_2H_2}{1 \text{ mole } C_2H_2} \right] = \mathbf{0.609 \text{ g } C_2H_2}$

% *yield* $= \dfrac{0.533 \text{ g } C_2H_2 \times (100)}{0.609 \text{ g } C_2H_2} = \underline{\mathbf{87.5\%}}$

13. A) **Need:** $mg \ NO_2 \rightarrow ? \ mol \ NO_2 \rightarrow ? \ \mu L$ at STP; $1 m^3$ air $= 1000 L$, 1 mol $NO_2 = 22.4$ L at STP

$\dfrac{10 \text{ mg } NO_2}{1000 \text{ L air}} \left[\dfrac{1 \text{ g } NO_2}{1000 \text{ mg}} \right]\left[\dfrac{1 \text{ mole } NO_2}{46.0 \text{ g } NO_2} \right]\left[\dfrac{22.4 \text{ L } NO_2}{1 \text{ mole } NO_2} \right]\left[\dfrac{1.0 \times 10^6 \, \mu L \, NO_2}{1 \text{ L } NO_2} \right] = \underline{\mathbf{4.87 \text{ ppm } NO_2}}$

B) *Convert ppm* ($\mu L/L$) *to mole fraction:* $\dfrac{n_{NO_2}}{n_{air}} = \dfrac{200 \text{ L } NO_2 \times (1 \text{ mol } NO_2/22.4 \text{ L})}{1.0 \times 10^6 \text{ L air} \times (1 \text{ mol air}/22.4 \text{ L})} = 0.0020$

Since $\dfrac{P_{NO_2}}{P_{air}} = \dfrac{n_{NO_2}}{n_{air}}$ then $P_{NO_2} = P_{air} \times \dfrac{n_{NO_2}}{n_{air}} = 0.0020 \times (1.00 \text{ atm}) = \underline{\mathbf{0.0020 \text{ atm } NO_2}}$

14. A) Have 0.361 mol CH_4, 0.108 mol Ne and 0.106 mol SO_3, so $n_{total} = 0.575$ moles, then:

$P_{total} = n_{total} \left[\dfrac{RT}{V} \right] = (0.575 \text{ mol}) \left[\dfrac{0.08206 \frac{L-atm}{K-mol} \, (358 \text{ K})}{75.0 \text{ L}} \right] = \underline{\mathbf{0.225 \text{ atm}}}$

B) **CH_4** , P $CH_4 = P_{total} \times (n \, CH_4 / n_{total}) = 0.245 \text{ atm} \times (0.361 \text{mol}/0.573 \text{ mol}) = \mathbf{0.141 \text{ atm}}$

C) **SO_2**, since greatest mass and d = 6.80 g/75.0 L = **0.0907 g/L**

15. A) $n_{N_2} = \dfrac{PV}{RT} = \dfrac{170 \text{ atm } (60.0 \text{ L})}{0.08206 \frac{L-atm}{K-mol} \, (294 \text{ K})} = 422.5 \text{ mol } N_2 \left[\dfrac{28.0 \text{ g } N_2}{1 \text{ mole } N_2} \right] = \underline{\mathbf{1.18 \times 10^4 \text{ g } N_2}}$

B) **Answer = density is 158 times greater than STP value**

$d_{N_2} \text{ tank} = \dfrac{1.18 \times 10^4 \text{ g}}{60.0 \text{ L}} = 197 \text{ g/L} \quad d_{N_2} \text{ STP} = \dfrac{1 \text{ mol } N_2 (28.0 \text{ g}/1 \text{ mol})}{22.4 \text{ L}} = 1.25 \text{ g/L}$

C) Because of high pressure, the free space very small, so likely to be acting as a **real gas.**

16. A) *Can calculate moles total from* $n_{total} = P_{total} \left[\dfrac{V}{RT} \right]$ *but need V of flask from first data:*

$V_{flask} = \dfrac{(1.9 \text{ mol}) \, 0.08206 \frac{L-atm}{K-mol} \, (294 \text{ K})}{0.917 \text{ atm}} = 50.0 \text{ L} \quad n_{total} = \left[\dfrac{(1.046 \text{ atm}) \, 50.0 \text{ L}}{0.08206 \frac{L-atm}{K-mol} \, (294 \text{ K})} \right] = \mathbf{2.14 \text{ mol}}$

B) 2.13 mol - 1.90 mol = **0.24 mol added**   C) M = 10.5 g / 0.23 mol = **44 g/mol = NO_2**

Chapter 11

Answers to Key Terms Exercise:
Concerning Liquid Properties, Phase Changes and Phase Diagrams:

1. viscosity   2. surface tension   3. surface tension   4. capillary action
5. meniscus   6. surface tension   7. capillary action   8. vaporization
9. evaporation   10. boiling point   11. evaporation   12. vapor pressure
13. volatility   14. volatility   15. vapor pressure   16. evaporation
17. condensation   18. vapor pressure   19. equilibrium vapor pressure
20. evaporation   21. equilibrium vapor pressure   22. vapor pressure   23. boiling
24. boiling point   25. (equilibrium) vapor pressure   26. normal boiling point
27. vapor pressure   28. sublimation   29. sublimation   30. deposition
31. Deposition   32. crystallization   33. heating curve   34. phase diagram
35. phase diagram   36. triple point   37. critical temperature (Tc)   38. critical pressure
39. critical temperature   40. critical temperature   41. supercritical fluid

Concerning Arrangements of Atoms in Solids and Types of Materials:

42. crystalline solids   43. amorphous solids   44. crystalline solids
45. crystal lattice   46. unit cell   47. closest packing   48. cubic close packing
49. cubic unit cell   50. hexagonal close packing   51. X-ray crystallography
52. closest packing   53. crystalline solids   54. unit cell   55. Network solids
56. Materials science   57. ceramics   58. Composites   59. Ceramics
60. network solids   61. ceramics   61. amorphous solid   62. unit cell
63. glass   64. ceramics   65. glass   66. glass
67. optical fiber   68. cement   69. concrete

Concerning the Conduction Properties of Solids:

70. metallic bonding   71. energy band   72. conduction band   73. valence band
74. valence band   75. conduction band   76. conductor   77. conduction band
78. valence band   79. conduction band   80. insulator   81. conduction band
82. semiconductor   83. conduction band   84. superconductor   85. Semiconductor
86. zone refining   87. semiconductor   88. doping   89. n-type semiconductor
90. p-type semiconductor   91. n-type semiconductor   92. p-type semiconductor
93. semiconductor   94. p-n junction   95. solar cell   96. conduction band

Answers to True/False:

1.   A). **False**   Molecular crystals have low melting and many crystals are soft.

     B). **True**   C). **False** Triple point must be higher so all three phases exist.

     D). **False**   Ionic solids have high ΔH_{vap}, but many solids such as network covalent solids or polymers have even values for ΔH_{vap}.

     E). **False**   Hexagonal close packing has 12 versus 6 nearest neighbors for cubic.

     F). **True**   G). **True**   H). True   I). **True**

     J). **False**   True of metals but not network covalent solids.

     K). **False**   Insulators have a large energy gap between bands.

     L). **False**   Need to use similar sized atom from Group V, N too small.

     M). **True**   N). **True**   O). **True**

     P). **True**   Q). **False** The boiling point changes with atmospheric pressure.

2.   A) Both are polar substances, but SeO_2 has higher ΔH_{vap} and ΔH_{fus} because of the stronger intermolecular forces due to the higher number of electrons in the molecule.

     B) Melting points do not vary significantly with atmospheric pressure, so the measured values are accurate.

     C) (a) X-ray crystallography is used to measure the crystal structure of solids.
     (b) The wavelength used must be a multiple of the spacing between the layers of atoms.
     (c) Only X-rays have wavelengths in this region (1.0- 0.1 nm).
     (d) In amorphous materials the lack of long range order scatters the X-rays into a random pattern.

2.   D)   (a) Electrical conduction requires that electrons be excited into the conduction band from the valence band, so the energy required to move through a gap strongly influences conduction properties. A metal has no gap and is the best conductor, while a semiconductor has a small energy gap and an insulator has a large gap between the valence and conduction bands.

(b) Doping doesn't change the size of the gap, but creates new states within the gap.

E)   Cement is made from calcium compounds and sand (source of silicates) while ceramics are formed from clay which acts as the source of silicates. Cement does not require high temperatures to to form the network like ceramics, but does require water.

3.   A) **Group 4A**

B) **dopant for p-type = Ga**, since similar size and 3 valence electrons instead of 4;
   dopant for n-type = As, since has 5 e's and similar size.

4.   A) **Ag**   Although I_2 has higher molar mass, metallic forces are stronger than the London forces in I_2.

B) **NaI_3** NaI_3 is ionic while NI_3 is a polar molecule.

C) **CCl_4**   Both are nonpolar but CCl_4 has greater number electrons, so stronger forces.

D) **H_2O**   The H-bonding in H_2O will result in stronger forces to overcome.

5.   In closed container, condensation occurs in the same container and the rate of evaporation and condensation can become equal and the vapor pressure become constant. Condensation cannot occur as readily when the liquid is in an open container, so that the rate of evaporation is generally greater than the condensation.

6.   A) Capillary action and surface tension.

B) Surface tension > capillary action for <u>upward curve</u> and surface tension < capillary action for <u>downward curve</u>.

C) In the plastic tube the surface tension and capillary action must be equal strengths.

7.   (a) Since the density of water increases as the temperature decreases, the cool water (coffee) sinks to the bottom of the cup, forcing the less dense hotter water to the top.

(b) Possible experiment could be constructed by taking a tall cup (well-insulated) and fixing a high precision temperature probe near the bottom of the cup and one near the top of the cup. Then cup could filled with hot coffee until the level of the liquid covered the upper probe and the temperatures monitored over time.

8.   Lowest (most exothermic) ----------------------> Highest (most endothermic)
   $\Delta H_{deposition} < \Delta H_{condensation} < \Delta H_{crystallization} < \Delta H_{fusion} < \Delta H_{vaporization} < \Delta H_{sublimation}$

9.   The unit cell must contain the empirical formula so need $2Cl^-$ for each Ca^{+2}. In the CsCl unit cell there is only one Cl^- for each Cs^+ in the unit cell.

10.   (a) $SnCl_4$ **molecular nonpolar compound** - London forces
        (b) B(s)   **network covalent** - all atoms covalently bonded to each other in solid.
        (c) Ga(s)   **metallic** - metallic bonding occurs in the solid
        (d) TiBr4   **molecular nonpolar compound** - London forces only.
        (e) S_8 **molecular nonpolar compound** - London forces only.
        (f) BaH_2 **ionic compound** - attraction of full positive charges for negative charges.
        (g) AsH_3 **molecular, polar compound** - dipole-dipole attractions

11.   Lowest T_c ---> highest T_c

CO_2	C_3H_8	NH_3	CH_3COCH_3	CH_3CH_2OH
forces: London forces	London forces	H-bonding	polar, no H-bonds	H-bonding
g/mol: 44	44	17	52	42

The strongest factor is the strength of the forces and secondly, the molar mass.

12. A) **f**   B) **d**   C) **c**   D) **a**   E) **e**   F) **b**

13. A) (a) **B** (b) **G** c) **C**   B) (a) **A** (b) **E** (c) **F**   C) **D**   D) **H**
     E) **G**   F) (1) **A**   (2) **C**   (3) **E**   (4) **G**   (5) **F**

14. A) a) **sublimation and deposition**

b) $n_{I_2} = \dfrac{PV}{RT} = \dfrac{6.13 \times 10^{-4} \text{ atm } (0.050 \text{ L})}{0.08206 \frac{\text{L}-\text{atm}}{\text{K}-\text{mol}} (25 + 273) \text{ K}} = 1.25 \times 10^{-6} \text{mol } I_2 \times \dfrac{254.0 \text{ g } I_2}{1 \text{ mole } I_2} = \mathbf{3.18 \times 10^{-4} \text{ g } I_2}$

B) a) At **triple point so 3 phases, solid liquid and gas** present.

b) $n_{I_2} = \dfrac{PV}{RT} = \dfrac{0.1197 \text{ atm } (0.050 \text{ L})}{0.08206 \frac{\text{L}-\text{atm}}{\text{K}-\text{mol}} (25 + 273) \text{ K}} = 1.88 \times 10^{-4} \text{mol } I_2 \times \dfrac{254.0 \text{ g } I_2}{1 \text{ mole } I_2} = 0.0476 \text{ g } I_2$

   Mass of vapor = 0.0238 g, so **vapor darker since more I_2 in gas phase.**

c) **To heat solid:** $q = 0.050 \text{ mol } I_2 \times \dfrac{54.44 \text{ J}}{\text{mol } I_2-°C} \times (89°C) = 241 \text{ J}$

   To melt solid: $q_{\text{melt}} = 0.050 \text{ mol } I_2 \times \dfrac{15.52 \text{ kJ}}{1.0 \text{ mol } I_2} = 0.770 \text{ kJ} = 770 \text{ J} \approx \underline{\mathbf{3 \times (\text{heat for } \Delta T)}}$

C) **Pressure I_2 in flask = 129 atm> critical pressure, 116 atm, so supercritical fluid.**

   $P = \dfrac{nRT}{V} = \dfrac{(0.01 \text{ mol } I_2) \, 0.0821 \frac{\text{L}-\text{atm}}{\text{K}-\text{mol}} (785 \text{ K})}{0.050 \text{ L}} = 129 \text{ atm } I_2$

15. A) No. atoms in 1 cm³ Ag $= 10.5 \text{ g} \times \dfrac{1 \text{ mole Ag}}{107.8 \text{ g Ag}} \times \dfrac{6.02 \times 10^{23} \text{ atoms}}{1 \text{ mole Ag}} = 5.86 \times 10^{23} \text{ atoms}$

   Volume unit cell $= 4.0 \text{ } atoms \times \dfrac{1 \text{ cm}^3 \text{ Ag}}{5.86 \times 10^{22} \text{ atoms Ag}} = 6.82 \times 10^{-23} \text{ cm}^3$

   Edge unit cell $= \sqrt[3]{6.82 \times 10^{-23} \text{ cm}^3} = 4.09 \times 10^{-8} \text{ cm} = \underline{\mathbf{0.409 \text{ } nm}}$

B) $r = \sqrt{2} \times (0.409 \text{ nm}) = \underline{\mathbf{0.578 \text{ nm}}} = $ **radius of Ag atom**   C) **metallic forces**

16. A) mass, bcc unit cell $= 2.0 \text{ atoms} \times \dfrac{1 \text{ mole Fe}}{6.02 \times 10^{23} \text{ atoms Fe}} \times \dfrac{55.85 \text{ g Fe}}{1 \text{ mole Fe}} = \mathbf{1.85 \times 10^{-22} \text{ g Fe}}$

   mass, fcc unit cell $= 4.0 \text{ atoms} \times \dfrac{1 \text{ mole Fe}}{6.02 \times 10^{23} \text{ atoms Fe}} \times \dfrac{55.85 \text{ g Fe}}{1 \text{ mole Fe}} = \mathbf{3.71 \times 10^{-22} \text{ g Fe}}$

B) $d_{\text{bcc}} = \dfrac{1.85 \times 10^{-22} \text{ g Fe}}{\left(2.865 \times 10^{-8}\right)^3 \text{ cm}^3} = \mathbf{7.89 \text{ } g/cm^3}$, $d_{\text{fcc}} = \dfrac{3.71 \times 10^{-22} \text{ g Fe}}{\left(3.63 \times 10^{-8}\right)^3 \text{ cm}^3} = \mathbf{7.76 \text{ } g/cm^3}$

C) bcc: $4r = \sqrt{3}(0.2865 \text{ } nm)$, $r = 0.124 \text{ } nm$, fcc: $4r = \sqrt{2}(0.363 \text{ } nm)$, $r = 0.128 \text{ } nm$
   So radius of Fe atoms changes slightly when the crystal structure changes.

17. A)   B) a) Possible 1 atom unit cell: b) Possible 2 atom unit cells;

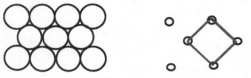
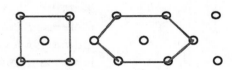

18. A) edge $= \dfrac{2 \text{ } r_- + 2 \text{ } r_+}{\sqrt{2}} = \dfrac{2(220) + 2(148) \text{ nm}}{\sqrt{2}} = \mathbf{368 \text{ nm}}$ D)

B) Each ion has **6 nearest neighbors**

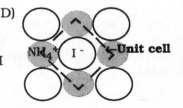

C) Unit cell: $(\frac{1}{8} NH_4^+ \text{ (corner)} \times 8) = 1 \text{ ion } NH_4^+ + (\text{ } I^- \text{ in center}) = NH_4I$

   so **unit cell contains empirical formula**

A - 40

Chapter 12

Answers to Key Terms Exercise:

Concerning Fuels and Carbon Dioxide Emissions:

1. petroleum fractions   2. catalyst   3. catalytic cracking
4. catalyst   5. catalytic reforming   6. catalyst   7. octane number
8. octane number   9. octane number   10. greenhouse effect
11. greenhouse effect   12. global warming   13. global warming

Functional Groups for Carbon Compounds:

14. aldehyde   15. carboxylic acid   16. ketone   17. aldehyde
18. carboxylic acid   19. ketone   20. carboxylic acid   21. monounsaturated acids
22. polyunsaturated acid   23. monounsaturated acids   25. partial hydrogenation
26. carboxylic acid   27. ester   28. ester   29. hydrolysis
30. carboxylic acid   31. ester   32. hydrolysis   33. saponification
34. amine   35. amine   36. carboxylic acid   37. carboxylic acid
38. amine   39. amide   40. amide linkage

Concerning Polymers and Reactions that Produce Them:

41. monomer   42. polymer   43. macromolecule   44. polymer
45. thermoplastics   46. thermosetting plastics   47. polymer   48. monomer
49. monomer   50. polymer   51. addition polymer   52. monomer
53. carboxylic acid   54. amine   55. polymer   56. condensation polymer
57. monomer   58. copolymer   59. addition polymer   60. condensation polymer
61. (condensation) polymer   62. carboxylic acid   63. (condensation) polymer
64. ester   65. polymer   66. polyester   67. carboxylic acid
68. amine   69. amide linkage   70. polyamides   71. condensation polymer

Concerning the Structure of Biological Macromolecules:

72. polyamides   73. monomer   74. amino acids   75. amine
76. carboxylic acid   77. amino acids   78. amine   79. carboxylic acid
80. alpha carbon   81. peptide linkage   82. alpha carbon   83. polyamide
84. polymer   85. polypeptide   86. amino acid   87. monomer
88. amino acid   89. amino acid   90. monomers   91. primary structure
92. amino acids   93. amino acids   94. secondary structure
95. tertiary structure   96. monomer   97. condensation polymer
98. glycosidic linkage   99. polymer   100. polymer   101. polyester
102. polyamides (polypeptides)   103. hydrolysis   104. monomer

Answers to True/False:

1.
   - A). **True**   B). **False**   Fractions contain a range of hydrocarbons.
   - C). **False**   Can be above 100, but not below zero.
   - D). **False**   Functions reversed, catalytic cracking breaks up the molecules and catalytic reforming makes the branched compounds.
   - E). **False**   Alcohols, not aldehydes, are used. Toluene is an aromatic compound.
   - F). **True**   G). **False**   The replacements (alkanes) are more volatile than aromatics.
   - H). **True**   I). **False**   Both primary alcohols, since 2 H's on C with OH group.
   - J). **False**   Secondary alcohols only produce ketones when oxidized and tertiary alcohols cannot be oxidized further.
   - K). **True**   L). **False**   Needs to be 160 proof, since proof two times the percentage.
   - M). **False**   It must be thermoplastic if reformed.
   - N). **False**   Need two acid or two alcohol groups per molecule.
   - O). **False**   Chains are branched and don't pack as well, but will contain many C's.
   - P). **False**   True of addition polymers.
   - Q). **False**   Made by addition polymerization   R). **True**
   - S). **True**   T). **True**   U). **True**   V). **False**   Fats have a lower percentage.

2.   A) Ethanol is not a fossil fuel, for two reasons: (1) fossil fuels are mixtures of many hydrocarbons and (2) fossil fuels are the product of long term conversion over millions of years, while ethanol forms quickly from fermentation of plant materials.

B) It is important since it is a precursor to many different substances and serves as a basic building block for countless number of compounds and polymers.

C)   (1) They lower CO emission and smog-producing potential

       (2) They are good octane enhancers as shown by the data.

D)   (a) an ester   (b) a polyamide, like nylon

E) If one was a primary alcohol and the other secondary, two very different products would result from the oxidation. Both type of alcohols have the same molecular formula.

F) Boiling point of acids are greater than boiling points of similar weight alcohols because of the increased number of H-bonds between molecules and increased polarity.

G)   (a) They can be remelted and reformed, unlike thermosetting plastics.

       (b) Since the temperatures will vary over a wide range, it would be much better to have a thermosetting plastic for the bench which will not become brittle or soften.

H)   (a) Branching produces softer, low density plastics, because the shapes do not allow the chains to align very well.

       (b) Cross-linking typically produces very rigid plastics. The density could become higher if the chains can be better aligned because of the cross-linking.

       (c) The side groups have a major role in determining the properties since they are sites for cross-linking and also produce noncovalent interactions between chains.

I) Cellulose has an trans (or a beta) linkage between glucose units.

J)   (a) H-bonding between the R (side) groups of the amino acids produce the secondary structure.

       (b) The -SH groups can make S-S bonds and disulfide linkages to cross-link chains.

3.   A) **a**   B) **e**   C) **c**   D) **f**   E) **d**   F) **b**

B) Low boiling fuels are burned at the top of the stack. They could act as pollutants or as greehhouse gases .

C) Jet fuel must be in the higher part of gasoline and lower part of kerosene fractions, so likely that: Jet fuel: C10- C14 range   boiling points $\approx$ 150 - 200°C .

4.   Lowest: **(b)** CH_3CH_3 < **(a)** CH_3OH < **(d)** CH_3CH_2OH < **(c)** CH_3CO_2H (highest)

5.   A) **(c)**   B) **(a)**   C) **(d)**   D) **(f)**   E) **(f)**   F) **(b)**   G) **(d)**   H) **(e)**   I) **(g)**

6.   A) primary: **(a), (b)**   secondary: **(c), (d)**   tertiary: **(e)**

B) Primary :

(a) $CH_3OH \rightarrow HCOH \rightarrow HCO_2H$ (b) $Cl(C_6H_4)CH_2OH \rightarrow Cl(C_6H_4)COH \rightarrow Cl(C_6H_4)CO_2H$

C) secondary:

(c) $C_6H_4C(CH_3)O$ =   (d) $CH_3COCH_2CONH_2$ =

7.   A) **vinyl acetate**, an **ester**   B) repeating units:

C) an **addition** polymer

D) The side groups are polar and would produce **dipole-dipole attractions**.

8. A)

B) condensation polymer

C) It is a thermoplastic, since it can be reformed into fibers and other forms when recycled.

9. A)

B) The polar C-Cl bonds will produce dipole- dipole attractions and the glass. The surface of glass is polar due to many O atoms at the surface.

C) Since it expands and then contracts with heating, it is acting like a thermoplastic. If it were a thermosetting plastic it would remainrigidly fixed in the bubble shape.

11. A) **a,d,e**   B) **c, f**   C) **b + c**   D) **f**   E) **e**   F) **d**   G) **f**

12. A) $10 \text{ mL CH}_3\text{OH} \times \dfrac{0.791 \text{ g}}{1 \text{ mL}} = 7.91 \text{ g CH}_3\text{OH}$; $90 \text{ mL C}_8\text{H}_{18} \times \dfrac{0.692 \text{ g}}{1 \text{ mL}} = 62.28 \text{ g C}_8\text{H}_{18}$

$\% \text{ O} = \dfrac{\text{wt O from CH}_3\text{OH}}{\text{total wt. mixture}} \times 100 = \dfrac{7.91 \text{ g CH}_3\text{OH} \times (16.0 \text{ gO}/32 \text{ g CH}_3\text{OH})}{70.19 \text{ g}} \times 100 = \textbf{5.63\% O}$

Yes, the mixture will exceed the minimum %O of 2.7 % required.

B) Reactions: $\text{CH}_3\text{OH} + 1.5 \text{ O}_2 \rightarrow \underline{1} \text{ CO}_2 + 2 \text{ H}_2\text{O}$; $\text{C}_8\text{H}_{18} + 12.5 \text{ O}_2 \rightarrow \underline{8} \text{ CO}_2 + 9 \text{ H}_2\text{O}$

1.0 gallon = 3.785 L, so 10% = 378.5 mL CH_3OH and 90% = 3.4065 L C_8H_{18}

No. mol CO_2 (CH_3OH) = $378.8 \text{ mL} \times \dfrac{0.791 \text{ g}}{1 \text{ mL}} \times \dfrac{1 \text{ mol}}{32.0 \text{ g}} \times \dfrac{1 \text{ mol CO}_2}{1 \text{ mol CH}_3\text{OH}} = 9.36 \text{ mol CO}_2$

No. mol CO_2 (C_8H_{18}) = $3406.5 \text{ mL} \times \dfrac{0.692 \text{ g}}{1 \text{ mL}} \times \dfrac{1 \text{ mol}}{114.0 \text{ g}} \times \dfrac{8 \text{ mol CO}_2}{1 \text{ mol C}_8\text{H}_{18}} = 165.4 \text{ mol CO}_2$

Vol. CO_2 = $174.78 \text{ mol} \times \dfrac{22.4 \text{ L}}{1 \text{ mol}} = \textbf{3915 L CO}_2$

b) less than, 3.785 L pure C_8H_{18} would produces 4147 L CO_2

C) From CH_3OH: 9.36 mol $\text{CO}_2 \times 1.5/1 = 14.0$ mol O_2
C_8H_{18}: 165.4 mol $\text{CO}_2 \times 12.5/8 = 258.4$ mol O, so total mol O_2 = 272.4 mol

Vol. O_2 = $272.4 \text{ mol} \times \dfrac{22.4 \text{ L}}{1 \text{ mol}} = \textbf{6103 L O}_2$

b) If pure C_8H_{18}, the volume of O_2 needed would **increase** since reaction requires more.

13. A) Taking the mass of 1unit to be the same as $\text{C}_6\text{H}_{12}\text{O}_6$, 180 g, have **about 380 units**

B) Figure (b) best since would be branched and more cis linkages than trans.

14. A)

B) 3 peptide linkages, see figure

C) alanylserylglycyltyrosine

D) No, need more than 50 units.

E) The OH group on the serine portion and the OH attached to the ring on the tyrosine portion could H-bond.

Chapter 13

Answers to Key Terms Exercise:

Describing Rates of Reaction:

1. rate   2. chemical kinetics   3. rate   4. homogeneous reaction
5. heterogeneous reaction   6. heterogeneous reaction   7. rate   8. heterogeneous catalyst   9. homogeneous reaction   10. reaction rate   11. reaction rate
12. average reaction rate   13. (reaction) rate   14. instantaneous reaction rate
15. initial rate   16. instantaneous rate of reaction   17. rate law   18. rate law
19. order of reaction   20. rate constant   21. rate constant   22. order of reaction
23. overall reaction order   24. half-life   25. half-life   26. rate constant

Concerning Reaction Mechanisms and Energy Profiles:

27. unimolecular reaction   28. bimolecular reaction   29. reaction mechanism
30. elementary reaction   31. reaction mechanism   32. unimolecular reaction
33. bimolecular reaction   34. order of reaction   35. reaction mechanism
36. rate-limiting step   37. order of reaction   38. intermediate
39. reaction intermediate   40. homogeneous catalysts   41. rate
42. transition state   43. activated complex   44. activation energy
45. activation energy   46. rate constant   47. reaction rate(law)
48. order of reaction   49. activation energy   50. rate constant
51. Arrhenius equation   52. Arrhenius equation   53. frequency factor
54. activation energy   55. frequency factor   56. steric factor
57. reaction rate   58. reaction mechanism   59. activation energy

Concerning Enzymes as Catalysts:

60. enzyme   61. enzyme   62. substrate   63. active site
64. substrate   65. enzyme   66. enzyme-substrate complex
67. activated complex   68. enzyme   69. substrate   70. induced fit
71. enzyme   72. cofactor   73. substrate   74. order of reaction
75. enzyme   76. substrate   77. enzyme   78. denaturation
79. active site   80. enzyme   81. enzyme-substrate complex
82. Inhibitor   83. enzyme   84. active site   85. enzyme
86. substrate

Answers to True/False:

1.   A). **False**   Molarity (mol/L) terms appears in rate law so volume very important.

     B). **False**   The energy profile is based on ΔH and Ea, not kinetic energies.

     C). **True**   D). **True**   E). **True**

     F). **False**   There is no reliable connection between them.

     G). **False**   It will double the rate.

     H). **False**   Only always true for the rate-limiting step only.

     I). **False**   The higher the barrier (Ea) the harder it is for reaction to occur.

     J). **True**   K). **True**   True no matter what order of reaction is.

     M). **False**   The intermediate do not affect the size of the Ea.

     N). **False**   Can't tell speed from ΔH alone.   O. **True**

2.   A) The purpose is to provide the energy to allow some particles to overcome the Ea barrier for the reaction so that the burning reaction can get started.

     B) The probability of the correct number and type of particles coming together and producing a successful collision goes down as the molecularity of the reaction increases.

     C) It depends on the order of reaction for that reactant. If first order or second order increasing a reactant concentration increases the rate, while in a zero order reaction, the changes will not affect the rate observed.

     D) Trapping intermediates is necessary to describe reaction mechanisms. The intermediates give the clues as to what particles are involved in the steps of the reaction. Using catalysts <u>may</u> cause the mechanism to change from when no catalysts are involved.

E) The inhibitor and substrate often share some structural or size characteristics so that the inhibitor also fits the active site for the enzyme, but they are unlikely to be completely alike.

3. **B**   4. **Fastest = D, slowest = A**   5. **A and C are true.**

6. A) (a) **decrease**   (b) **increase**   (c) **decrease**   (d) **no effect**   B) **(b)** since it doubles

7. A) Overall = **Second order**.

B) 4.89×10^{-5} M/s = **k** (0.100 M) (0.300 M) and **k $= 1.63 \times 10^{-3}$ M^{-1}s^{-1}**

C) Rate $= 7.85 \times 10^{-2} = 1.63 \times 10^{-3}$ [x][2x], so x = **[ICl] = 4.91 M**

D) Rate for ICl = 2 × (rate for H$_2$).

E) The orders indicate that one molecule of each reactant is involved in the rate-limiting step so it is very likely a bimolecular reaction.

8. A) (a) Average rate (0 → 1000) = **- 6.72×10^{-4} M/s**

(b) Average rate (0 → 3000) = **- 4.72×10^{-4} M/s**

(c) See lines marked (a) and (b) on the graph. The line for the average rate at the beginning of the reaction is the better fit to true behavior of rate.

(d) The rate decreases with time. (The negative sign in the rate indicates disappearance.)

B) The instantaneous rate is estimated using the triangles, to get the slope , marked on the graph.

At 1000 s $(\Delta y_1/\Delta x_1)$ = **3.13×10^{-4} M/s**

At 3000 s $(\Delta y_2/\Delta x_2)$ = **2.78×10^{-4} M/s**

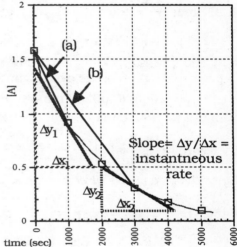

C) The data must replotted as ln[A] versus time and 1/[A] versus time, to find which is linear. In this case the ln[A] plot is linear, and the slope of that plot gives k = 6.30×10^{-4} s^{-1}.

D) The half life can most easily be gotten from the graph of concentration versus time above. The time at which the [A] is one-half of 1.59 M (≈ 0.8 M) is about 1400 s.

9. To determine order, need to find the % left = $[A]_t/[A]_0$, at each time. Applying the first order integrated equation to % left versus time, produces a constant value for k, so the reaction is must be first order.

Time:	1	5	10	20 days	
% left	79%	32%	10%	1%	B) Calculate average value of k,

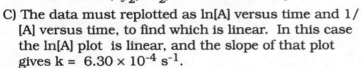

$k = \dfrac{\ln [A_t/A_0]}{t \text{ (da)}}$ 0.235 ·0.228 0.230   0.230   **k = 0.231 da^{-1}**

10. A) Need $\dfrac{1}{A_t} - \dfrac{1}{A_0} = kt$, $\dfrac{1}{A_t} = 1.67 + (0.080 \ M^{-1}s^{-1})$ 5 min $\times \dfrac{60 \text{ s}}{1 \text{ min}} = 25.67$; **A$_t$ = 0.390 M**

B) Number moles NOBr reacted = 0.60- 0.0390 = 0.561 mol;

so number mol Br$_2$ produced/L $= \frac{1}{2} \times$ (0.561) = 0.2805 mol Br$_2$ = **0.281 mol Br$_2$**

C) ΔH reaction = (2 (90.37 kJ) - 2 (82.2 kJ) = + 16.3 kJ

q = 0.281 mol Br$_2 \times \dfrac{16.3 \ kJ}{1 \text{ mol Br}_2} =$ **4.57 kJ**

11. A) Need: $\ln\left[\dfrac{A_t}{A_0}\right] = -\,kt$ and $t_{1/2} = 0.693/k$ (first order) and 25% decomposed = 75% left

$\ln\left[\dfrac{75}{100}\right] = -\,k(80\text{ s})$ with $k = 0.00360\text{ s}^{-1}$, $t_{1/2} = 0.693/(0.00360\text{ s}^{-1}) = 139\text{ sec}$ **= 3.21 min**

B) Since only one reactant <u>and</u> first order, the rate-limiting step must be **unimolecular**.

12. A) **First order** B) Need value of k from reaction data, $k = 1.93 \times 10^{-3}\text{ min}^{-1}$

so : $\ln\left[\dfrac{10\ \%}{40\ \%}\right] = -\,1.93 \times 10^{-3}\text{min}^{-1}\ (t)$ then **t = 574 min = 9.6 hours.**

B) (a) $\mathbf{H_2O_2(aq) \rightarrow 2\ H_2O + O_2(g)}$, same as the studied reaction
(b) catalyst = **Br-**   (c) intermediate = **BrO⁻**

13. A) **order for O_2 = 1, order for NO_2 = 2** B) **rate law = $k[NO_2]^2[O_2]$**
C) **k = 176 M⁻²s⁻¹**   D) rate $= 176\ M^{-2}s^{-1}\ (0.045)^2(0.22)\ M^3 =$ **0.00713 M/s**

14. A) Since $\dfrac{A_t}{A_0}$ = fraction left and $\ln\left[\dfrac{A_t}{A_0}\right] = -\,2.83 \times 10^{-3}\text{min}^{-1}(60\text{ min}) = -\,0.169$

% left = $e^{-0.169}$ = 0.844, 84.4% left, means **15.5% of sample decomposed**

B) The pressure will increase since when x moles of gas decomposes, 2x moles will be produced, so pressure overall must increases.

C) $t_{1/2} = 0.693/(0.00283\text{ min}^{-1}) = 245\text{ min}$ **= 4.08 hr** Fairly slow reaction since takes hours to reduce to one-half of original amount.

15. **Need**: $\ln\left[\dfrac{k_1}{k_2}\right] = \ln\left[\dfrac{\text{rate}_1}{\text{rate}_2}\right] = \dfrac{E_a}{R}\left[\dfrac{1}{T_2} - \dfrac{1}{T_2}\right]$

Since the rate varies directly with k, the rate constants

$\ln\left[\dfrac{0.498}{1.81}\right] = \dfrac{E_a}{8.314\ \text{J/mol}-\text{K}}\left[\dfrac{1}{627\ K} - \dfrac{1}{592\ K}\right]$

$-1.291(8.314) = E_a\ (-\ 9.43 \times 10^{-5})$,
$E_a = 113,\ 733\ \text{J/ mol}$ **= 114 kJ/mol**

B) $\Delta H = 2\ (92.37) - 2\ (33.18) =$ **114 kJ**   C)

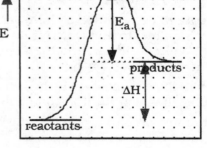

16. **Need**; $\ln\left[\dfrac{k_1}{k_2}\right] = \ln\left[\dfrac{t_{1/2}\ (at\ T_2)}{t_{1/2}\ (at\ T_1)}\right] = \dfrac{E_a}{R}\left[\dfrac{1}{T_2} - \dfrac{1}{T_1}\right]$

Since the half-life varies inversely with k, the rate constants

$\ln\left[\dfrac{t_{1/2}\ (at\ T_2)}{t_{1/2}\ (at\ T_1)}\right] = \dfrac{2.83 \times 10^{-3}J}{8.314\ \text{J/mol-K}}\ (-\ 8.63 \times 10^{-5}K) = -\ 2.346 \rightarrow t_{1/2}\ (at\ T_2) = 0.0958\ (78\ s) =$ **7.5 s**

B) Use same equation: $\ln\left[\dfrac{240\ s}{78\ s}\right] = \dfrac{2.26 \times 10^3 J}{8.314\ \text{J/mol-K}}\left[\dfrac{1}{x} - \dfrac{1}{723\ K}\right]\ x = T_2 = 701\ K =$ **428 °C**

17. A) Calculate ln (k) and 1/T(K) for each temperature and construct plot, ln k versus 1/T(K) or compare any two values to estimate Ea.

From plot, slope $= -\dfrac{E_a}{R} = -\ 4113\ K^{-1}$, so **Ea = 43.4kJ**   B) $A \approx e^{16.8}$ **= 2.0 × 10⁸**

C) $\dfrac{k_{45°C}}{k_{25°C}} = \dfrac{\text{rate}_{45°C}}{\text{rate}_{25°C}} = \dfrac{0.332}{0.101} =$ **3.3 times faster**   D) At some T higher than 37°C, the enzyme will undergo denaturation. At that point the rate will decrease, instead of increasing.

Chapter 14

Answers to Key Terms Exercise:

Concerning Chemical Equilibrium:

1. chemical equilibrium
2. chemical equilibrium
3. dynamic equilibrium
4. chemical equilibrium
5. equilibrium constant
6. product-favored
7. equilibrium constant
8. reactant-favored
9. equilibrium constant
10. equilibrium constant
11. equilibrium concentration
12. equilibrium constant expression
13. equilibrium concentration
13. equilibrium constant expression
15. equilibrium concentration
16. equilibrium constant expression
17. equilibrium constant expression

Shifting a Chemical Equilibrium:

18. reaction quotient concentration
19. equilibrium constant expression
20. equilibrium
21. chemical equilibrium
22. reaction quotient
23. chemical equilibrium
24. chemical equilibrium
25. shifting an equilibrium
26. reaction quotient
27. chemical equilibrium
28. reaction quotient
29. Le Chatelier's principle
30. shifting an equilibrium
31. chemical equilibrium
32. Le Chatelier's Principle
33. chemical equilibrium
34. equilibrium constant
35. equilibrium constant
36. equilibrium constant
37. equilibrium constant
38. equilibrium constant
39. entropy
40. equilibrium constant
41. product-favored
42. entropy
43. entropy
44. entropy
45. equilibrium constant
46. shifting an equilibrium
47. Haber-Bosch process
48. Haber-Bosch process
49. product-favored
50. equilibrium constant
51. Le Chatelier's Principle
52. equilibrium constant

Answers to True/False:

1. A). **True**   B). **False**   The ratio must be the rate constant times concentration(s) to their appropriate powers, not just rate constants.

   C). **False** Q is less than K at the start of the reaction.

   D). **True**   E). **False**   Ratios can be much larger than 1.0 and so can K.

   F). **False**   Not true for gases, since volume change changes molarity.

   G). **False**   Have no effect on value of K, just time needed to achieve the ratio.

   H). **True**   I). **False**   For exothermic, product favored at low temperatures.

   J). **True**   K). **True**   L). **False**   The value of K must be squared.

2.   A) Increasing T always makes the entropy effect greater, since the highly disordered states are favored at high temperatures and the high kinetic energy also increases the randomness of any collection of particles. The energy effect is related to the enthalpy and changes in bonding and changes in temperature can make either the forward or reverse reaction more probable.

   B) (a) The four possible types of changes are:
      (1) Decrease P (or increase V), K the same
      (2) Decrease T, K decreases (since reaction endothermic)
      (3) Remove some reactant (SO_2 or Cl_2), K same
      (4) Add some product (SO_2Cl_2), K same

   (b) Since ΔH is endothermic and the entropy change is negative, (two moles of gas to one mole), the reaction should be reactant-favored and the value of Kp should be less than 1.0 at 25°C.

3.   A) $N_2(g) + 2 O_2(g) <=> 2 NO_2(g)$ B) **(c)**
   C) **endothermic, + 67.8 kJ**   D) **(b)** The high pressure in the engine would **favor the second reaction** since 3 moles of gas converts to 2 moles of gas.

4. A) **stays the same** If $[O_2]$ the same, K the same and no shift to products will occur.
B) **increase** A shift to products will occur and Hg(l) is a product, even if not in the equilibrium expression as a concentration.
C) **decrease** (since $[O_2]$ decreases) D) **decreases**, (since less product formed at low T)
E) **increases** (since more O_2 must be formed)

5.   Reaction (a), CO reactant   Reaction (b), CO product   Reaction (c), CO product
• *exothermic reaction, Δn= -2*   • *endothermic reaction, Δn= 0*   • *exothermic reaction, Δn=+1*
   A):   **DEC**   **NC**   **DEC**
        P↑, *favor products*   P↑, *no effect*   P↑, *favor reactants*
   B):   **DEC**   **INC**   **DEC**
        T↑, K↓ *favor reactants*   T↑, K↑ *favor products*   T↑, K↓ *favor reactants*
   C):   **CT**   **DEC**   **INC**
T↓ K↑ *favor products*; P↓, *favor reactants* T↓, K↓   T↓, K↑ *favor products, same as* P↓
   D):   **DEC**   **DEC**   **CT**
   T↓ so K↑ *favor products, same as* P↑   T↓ so K↓   T↓, K↑ *favor products, opposite* P↑
   E):   **NC**   **NC**   **NC**
Catalysts cannot cause shifts since do not change equilibrium concentrations achieved.

6. A) **d** *Δn must be* (+)   B) **a,d** K < 1.0   C) **a,b,c** *Δn must be* 0
   D) **d**   E) **b,c,e**   F) **a, d**   G) **a,d** ΔH *must be* (+)   H) **e** *Δn must be* (-)

7. A)   (a) $Q> K$, shift to reactants   B)
        (b) $Q<K$, shift to products
        (c) $Q>K$, shift to reactants
        (d) $Q= K$, no shift

   C)   Point **(a)** $Q<K$, Point **(b)** $Q = K$
        Point **(c)** $Q> K$, Point **(d)** $Q = K$

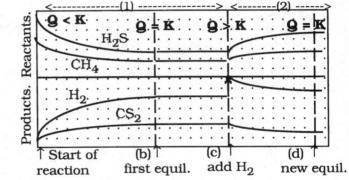

8. A) $Q = \dfrac{[C]^2}{[A]^2 [B]} = \dfrac{(1.30)^2}{(0.0075)^2 (0.0060)} = 5.0 \times 10^6$ Q **equals K for 430°C**
   B) Both entropy and energy effects are not favorable, since K increases with temperature, so reaction must be endothermic. The entropy effect should also become more favorable as temperature increases.

9. A) $K_p = \dfrac{(P\ N_2O)(P\ O_2)}{(P\ NO_2)(P\ NO)}$ B) $P_{NO_2} = P_{NO} = \dfrac{(2.0\ \text{mol})\ 0.08206\ \frac{L-atm}{K-mol}\ (298)\ K}{5.0\ L} = 9.78$ atm

	P NO_2	P NO	P N_2O	P O_2
Initial	9.78	9.78	0	0
Change	-x	-x	+x	+x
Equil.	9.78- x	9.78- x	x	x

C) $K_p = 0.914 = \dfrac{x^2}{(9.78 - x)^2}$ x = 4.78 atm
   $P_{NO} = (9.78- 4.78)$ atm = **5.00 atm**

10. A)

	P CCl_4	P Cl_2
Initial	1.00	0
Change	- x	+ 2x
Equil	1.00-x	2x

B) $K_p = \dfrac{((P\ Cl_2)^2}{(P\ CCl_4)} = \dfrac{(2x)^2}{(1.00 - x)}$

C) $P_{total} = P\ CCl_4 + P\ Cl_2 = 1.0- x + 2x = 1.0 + x = 1.35$
   so x = 0.35 atm, $K_p = \dfrac{(0.70)^2}{(1.00 - 0.34)} = \textbf{0.754}$

D) **Picture (b)** is the best, since just slightly more Cl_2 than CCl_4 particles at equilibrium.

11.

	[NOCl]	[NO]	[Cl$_2$]
Initial	1.67	0	0
Change	- 2x	+ 2x	+ x
Equil	1.67-2x	2x	x

A) $0.28(1.67)$ mol/L $= 2x$, $x = 0.234$ M

$$K_c = \frac{(2x)^2(x)}{(1.67 - 2x)^2} = \frac{(0.468)^2(0.234)}{(1.18)^2} = 0.0368$$

B) $\mathbf{K_p} = (0.0368)\left[0.08206 \frac{L-atm}{K-mol} (673)\ K\right]^1 = \underline{\mathbf{2.03}}$

C) *All pressures must be in atm:* $2.03 = \dfrac{(0.35)^2(0.10)}{x^2}$ and $x = \mathbf{P\ NOCl} = \underline{\mathbf{0.0777\ atm}}$

12. A) $K_p = \dfrac{((P\ H_2O)^3}{(P\ H_2)^3} = 8.11$ and $P\ H_2 = \dfrac{(0.50\ mol)\ 0.0821 \frac{L-atm}{K-mol} (1000)\ K}{2.0\ L} = 20.5\ atm$

$\dfrac{P\ H_2O}{P\ H_2} = \sqrt[3]{8.11}$ so $\dfrac{x}{20.5 - x} = 2.01$ gives $\mathbf{x} = P\ H_2O = \mathbf{13.7\ atm}$

B) $P\ H_2 = \mathbf{6.75\ atm}$ **so** no. mol H$_2$ left $= \dfrac{(6.75\ atm)\ 2.0L}{0.08206 \frac{L-atm}{K-mol} (1000)\ K} = 0.164$ moles left

No. moles H$_2$ reacted $= 0.50 - 0.176 = 0.336$ mol H$_2$ reacted

0.336 mol H$_2$ $\times \dfrac{1\ mol\ Fe_2O_3}{3\ mol\ H_2} \times \dfrac{160\ g\ Fe_2O_3}{1\ mol\ Fe_2O_3} = \mathbf{17.9\ g\ Fe_2O_3\ reacted}$

C) The pressure of H$_2$ at equilibrium is 6.75 atm, while that for H$_2$O is 13.7, indicating that the number of H$_2$O particles is nearly double the number of H$_2$ particles at equilibrium, the reaction can be considered product-favored, as expected. Although only about 20% of the Fe$_2$O$_3$ has been reacted, the amount of solid Fe$_2$O$_3$ is not relevant since it does not influence the value of K.

13. Starting with [N$_2$O$_4$] = 0.0294 M, need to use quadratic equation to solve for x

$K_c = \dfrac{(2x)^2}{(0.0294 - x)} = 4.63 \times 10^{-3}$ and $1.36 \times 10^{-4} + 4.63 \times 10^{-3}x - 4x^2 = 0$

Solving for x = 5.28 X 10^{-3} M $\quad$ **[NO$_2$] = 2x = $\underline{0.106\ M}$** and **[N$_2$O$_4$] = $\underline{0.0241\ M}$**

14. Let Y = initial concentration of HCN:

$K_c = \dfrac{[H_2][C_2N_2]}{[HCN]^2} = \dfrac{(x)^2}{(Y - 2x)^2} = 4.00 \times 10^{-4}$ where x = [H$_2$] = [C$_2$N$_2$]= 7.5 X 10^{-4}

$\dfrac{(7.5 \times 10^{-4})}{(Y - 1.5 \times 10^{-3})} = \sqrt{4.00 \times 10^{-4}}$ and **Y = initial concentration HCN = $\underline{0.039\ M}$**

15. A) $Ba(OH)_2(s) \Longleftrightarrow Ba^{+2}(aq) + 2\ OH^-(aq)$ $\quad$ B) **[Ba^{+2}]= 0.0488 M, [OH$^-$] = 0.0977 M**

C) $K_{sp} = [Ba^{+2}][OH^-]^2 = (0.0488)(0.0977)^2 = \mathbf{4.67\ X\ 10^{-4}}$

D) K_{sp} for KOH = (19.1 M)^2 = **365**, so $\approx$ **80,000 times larger** than K_{sp} for Ba(OH)$_2$

Chapter 15

Answers to Key Terms Exercise

Characteristics of the Solution Process and Solubility:

1. miscible	2. immiscible	3. enthalpy of solution	4. lattice energy
5. hydration	6. enthalpy of solution	7. saturated solution	8. solubility
9. solubility	10. unsaturated solution	11. solubility	12. supersaturated
solution	13. solubility	14. enthalpy of solution	15. enthalpy
of solution	16. solubility	17. enthalpy of solution	18. solubility
19. solubility	20. solubility	21. solubility	22. Henry's Law

Concentration Units Based on Mass Ratios:

23. mass fraction	24. mass fraction	25. mass percent
26. mass percent	27. part per million	28. parts per billion

Concerning the Colligative Properties:

29. colligative properties	30. Raoult's Law	31. boiling point elevation
32. freezing point lowering	33. molality	34. colligative properties
35. molality	36. semipermeable membrane	37. osmosis
38. osmotic pressure	39. colligative properties	40 osmotic pressure
41. osmotic pressure	42. reverse osmosis	43. reverse osmosis
44. semipermeable membrane	45. osmotic pressure	46. isotonic
47. hypertonic	48. hypotonic	49. hypotonic 50. molality

Concerning Colloidal Solutions:

51. colloid	52. dispersed phase	53. continuous phase
54. colloid	55. Tyndall effect	56. dispersed phase
57. continuous phase	58. colloid 59. emulsion	60. emulsion
61. surfactants	62. surfactants 63. detergents	64. emulsion

Answers to True/False:

1.   A). **True**

    B). **False**   The direction of change depends on the sign of $\Delta H_{solution}$

    C). **False**   The reverse is true, supersaturated solutions are unstable solutions.

    D). **False**   Magnesium chloride is ionic and does not dissolve in a nonpolar solvent.

    E). **True**

    F). **False**   The solubility would increase if the pressure of gas increases.

    G). **True**

    H). **False**   The molality is greater than or equal to molarity since the volume of solvent is less than (or equal to) volume of solution, depending on the density of the solution.

    I). **False**   The moles of each solute per kilogram of water would be different .

    J). **False**   100 ppm would equal 0.01% by mass solution

    K). **True**   L). **True**   M). **True**   N). **True**

    O). **False**   The water will flow to the side with the higher osmotic pressure.

    P). **False**   The solution temperature will increase.

    Q). **False**   Colloid particles are smaller than those that form a suspension.

    R). **True**

    K). **False**   They have similar structures, but detergents are not made from fatty acids, as soaps are, and have different groups for the polar end.

2.   A)   By changing the temperature of the solution in the direction that will decrease the value of the solubility or by allowing some solvent to evaporate.

    B)   The first process, $CaCl_2$ dissolving in water, has an exothermic enthalpy of solution, causing the temperature of the solution to increase. In the second process, two energy changes occur. The exothermic enthalpy of solution helps the ice to melt, since as the solution forms, the solid water absorbs heat (ΔH_{fusion}) while it melt, and the melting point is depressed from the added ions. The final solution is then liquid at a temperature below 0°C.

C)   The emulsifier acts at the surface boundary where a nonpolar liquid is in contact with the polar liquid, with different portions of the molecule in each liquid, breaking up large droplets into much smaller droplets making it possible to disperse one liquid phase in the other.

D)   Some of the essential vitamins (A, D and E) are fat soluble and in excess could (1) exceeds the body's ability to emulsify them or (2) accumulate in fatty tissues. Neither of these is likely in the short term, but could become a health problem in the long term.

E)   In reverse osmosis, an external pressure is applied to a aqueous solution that is much greater than the solution's osmotic pressure. This forces pure water to flow from solution, leaving the solutes behind, to the other side of the semipermeable membrane.   In a water softening process, the ions remain in the water, but Na^+ and Cl^- ions replace undesirable ions in the water, so that the water remains a solution and is not purified.

3.   A) **molarity**   B) **molality**   C) **molarity**

D) **mass percent, molality**   E) **mass percent**

4.   **D**   For hydrate: $0.91\,mol \times (248.2\ g/mol) = 226$ g; (1120 g solution)- 226 = 894 g water

5.   A) **False**   B) **False**   C) **True**   D) **False**

Although final molalities are the same, the moles of $Fe(NO_3)_2$ added is one-third the moles of fructose because of the van't Hoff factor. Different weights of solute have been added to for each 1000g of water, so the molarities and mole fractions are not the same.

6. Total mass solution = 932 g: *mass* LiBr = 2.30 mole $\times \dfrac{86.85\ g\ LiBr}{1.0\ mole\ LiBr} = 199.8$ g LiBr

A) $m = \dfrac{2.30\ mol\ LiBr}{0.732\ kg\ CH_3CN} = \underline{\textbf{3.14 } m}$ B) $\% = \dfrac{199.8\ g\ LiBr}{937\ g\ soln} \times 100 = \underline{\textbf{21.4\%}}$

C) LiBr is ionic and CH_3CN a very polar molecule, with a shape that is largely linear. Therefore a large enough number of the CH_3CN molecules can arrange themselves around the ions to produce strong enough noncovalent interactions to overcome the lattice energy of LiBr and the weaker dipoles of CH_3CN.

D) Nanoscale depiction of solution should have an arrangment something like shown in the figure on the right.

7. A) 5.0 ppm $= \dfrac{5.0\ g\ Cd \times (1 \times 10^6 \mu g\,/1.0\ g)}{1 \times 10^9\ g\ soln \times (1.0\ L\,/1000\ g)} = 5.0\ \mu g\,/L\ Cd \times 2.0\ L = \underline{\textbf{10 } \mu g\ \textbf{Cd}}$

B)   **3.3 packs of cigarettes**

8. A) $\dfrac{\Delta T_f}{K_f} = \dfrac{(-40°C - 0°C)}{-1.86\ °C/m} = m\ C_3H_8O_2$, m = 21.5 moles $C_3H_8O_2$/kg water

21.5 mole $\times \dfrac{76.1\ g\ C_3H_8O_2}{1.0\ mole\ C_3H_8O_2} = 1936$ g $\%\ C_3H_8O_2 = \dfrac{1636\ g\ C_3H_8O_2}{2636\ g\ soln} \times 100 = \underline{\textbf{62.1\%}}$

B) The OH groups allow for H-bonds to be formed between the water and the solute.

9. **Smallest ΔT_f, (B) < (A) < (C) < D, largest ΔT_f**   *Apply van't Hoff factors to ionic solutes:*
   effective molality: (B) 0.30 m CH_3OH ≈ 0.30m   (A) 0.20 m $NaNO_3$ ≈ 0.40m
                             (C) 0.15 m $HgCl_2$ ≈ 0.45m   (D) 0.10 m $Al_2(SO_4)_3$ ≈ 0.50m

10.   A) $M = \dfrac{0.0965 \text{ mol } C_{12}H_{22}O_{11}}{0.335 \text{ L soln}} = \textbf{0.271 M}$

      B) $\Pi = 0.271 \text{ M } (0.0821 \tfrac{L-atm}{K-mol}) \, (293 \text{ K}) = \textbf{6.54 atm}$

11.   A) $S_g = \dfrac{3.66 \times 10^{-3} \text{ mol } CO_2}{0.100 \text{ L soln}} = 0.0366 \text{ M} = k_H(760 \text{ mm Hg}) \text{ , so } \mathbf{k_H = 4.81 \times 10^{-5} \text{ M/mmHg}}$

      B) $S_g = k_H(P, \text{ mm Hg}) = 4.81 \times 10^{-5} \tfrac{M \, CO_2}{mm \, Hg} \, (5.50 \text{ atm} \times (760 \text{ mm Hg/atm}) = 0.2013 \text{ M } CO_2$

        mole CO_2 dissolved in 355 mL bottle = 0.2013 M (0.355 L) = 0.07146 mol CO_2

        Volume of CO_2 gas at STP = 0.07146 mol CO_2 (22.4 L/mol) = **1.60 L CO_2**

12. A) $m = \dfrac{0.200 \text{ mol methylbenzene}}{0.500 \text{ kg cyclohexane}} = 0.400 \text{ m}$ $\dfrac{\Delta T_f}{m} = \dfrac{(6.5°C + 2.04°C)}{(0.400 \text{ m})} = \mathbf{K_f = 20.1°C/m}$

     B) $\dfrac{\Delta T_f}{K_f} = \dfrac{8.04 \text{ °C}}{20.1°C/m} = (m \, C_6H_6) = \underline{0.40 \, m \, C_6H_6}$

     mass C_6H_6 = 0.40 mole (78.0 g/mole) = 31.2 g so   % $C_6H_6 = \dfrac{31.2 \text{ g } C_6H_6}{1031.2 \text{ g soln}} \times 100 = \textbf{3.02\%}$

13. A) $m = \dfrac{0.162 \text{ mole } CaCl_2}{0.082 \text{ kg } H_2O} = \textbf{1.98 m } CaCl_2$

     B) $\Delta T_f = K_f m(i) = -1.86°C/m \, (1.98 \text{ m}) \times (3) = \underline{\mathbf{-11°C}}$

     C) First convert the solubility to the mass percent allowed at 0°C to compare to determine whether 18% $CaCl_2$ is higher than the solubility.
          Solubility = 59.5 g/(159.5 g) = 37.3 % at 0°C.
     An 18% solution is unsaturated at 0°C and no solute would precipitate from solution. The solvent cannot solidify until a temperature of -11°C is reached, so **neither the solute nor the solvent will become a solid at 0°C.**

14. A) Molar mass $= \dfrac{K_b \, (\text{wt. solute, g})}{\Delta T_b \, (\text{wt. solvent, kg})} = \dfrac{0.52°C/m \, (1.90 \text{ g})}{(0.06 \text{ °C}) \, (0.04868 \text{ kg})} = \textbf{338 g/mol}$

     B) $\Pi = \dfrac{(\text{moles of solute added } (i)) \times RT}{V_{soln}} = \dfrac{(5.62 \times 10^{-3} \text{mol})(1) \, 0.0821 \tfrac{L-atm}{K-mol} \, (293) \text{ K}}{0.05058 \text{ L}} = \textbf{2.67 atm}$

     C) The measurement of the osmotic pressure is much more reliable. The pressure is much larger than the boiling point elevation for the same solution, so that the small errors encountered in normal laboratory procedures would not significantly affect the end result for the molar mass.

15.   A) Solution A, $M = \dfrac{0.0585 \text{ mol } C_{12}H_{22}O_{11}}{0.09234 \text{ L soln}} = 0.633 \text{ M}$

      $\Pi_A = C \, (i) \times RT = 0.633 \text{ M } (0.0821 \tfrac{L-atm}{K-mol} \, (293) \text{ K}) = \textbf{15.2 atm}$

      Solution B, $M = \dfrac{0.08554 \text{ mol } NaCl}{0.09652 \text{ L soln}} = 0.886 \text{ M}$

      $\Pi_B = C \, (i) \times RT = 0.886 \text{ M } (2) \, (0.0821 \tfrac{L-atm}{K-mol} \, (293) \text{ K}) = \textbf{43.4 atm}$

    B) The **water will flow from A→B**, to the region of higher solute concentration.

Chapter 16

General Characteristics of Acid and Base Equilibria:

1. acid-base reaction
2. Bronsted-Lowry acid
3. Bronsted-Lowry base
4. Bronsted-Lowry acid
5. Bronsted-Lowry base
6. Bronsted-Lowry base
7. amines
8. acid-base reaction
9. conjugate acid-base pair
10. Bronsted-Lowry acid
11. conjugate acid-base pair
12. Bronsted-Lowry base
13. Bronsted-Lowry acid
14. Bronsted-Lowry base
15. autoionization
16. ionization constant for water
17. ionization constant for water
18. neutral solution
19. Bronsted-Lowry acid
20. acidic solution
21. basic solution
22. Bronsted-Lowry base
23. pH
24. neutral solution
25. pH
26. acidic solution
27. pH
28. basic solution
29. pH
30. Bronsted-Lowry acid
31. acid ionization constant
32. acid ionization constant expression
33. acid ionization constant
34. acid ionization constant
35. Bronsted-Lowry base
36. base ionization constant
37. base ionization constant expression
38. base ionization constant

Concerning Molecular Structure and Acid-Base Behavior:

39. monoprotic acids
40. polyprotic acids
41. acid ionization constant
42. acid ionization constant
43. polyprotic acids
44. oxoacids
45. oxoacids
46. acid ionization constant
47. oxoacids
48. zwitterion
49. hydrolysis
50. hydrolysis
51. basic solution
52. hydrolysis
53. conjugate acid-base pair
54. acidic solution
55. amines
56. hydrolysis

Concerning Lewis Acid-Base Behavior:

57. acid-base reaction
58. Lewis acid
59. Lewis base
60. Lewis acid
61. Lewis base
62. coordinate covalent bond
63. Bronsted-Lowry acid
64. Bronsted-Lowry base
65. amphoteric
66. Lewis acid
67. coordinate covalent bond
68. Lewis acid
69. Lewis base

Answers to True/False:

1. A). **True**   B). **True**   C). **True**   D). **True**   E). **True**

F). **False**   The Ka will be very large, much greater than 1.0 for strong acid.

G). **False**   Amines are typically weak bases.

H). **False** CH_3CO_2H is a monoprotic acid.

I). **False**   The Ka gets smaller since it's harder to remove H^+ from a negative ion.

J). **False**   pH can below 0, as in 6.0M HCl and strong bases could go above 14.

K). **True**   L). **True**   M). **False**   Has a lower pH

N). **False**   Can't tell just from pH, need initial concentration to tell whether have dilute strong acid or whether weak acid in solution.

O). **False**   As with (N), can't tell from just the pH how much base has been dissolved in the solution, since only know how much dissociated.

P). **False**   The $K_b = 1.0 \times 10^{-14} / 1.4 \times 10^{-4} = 7.1 \times 10^{-11}$

Q). **True**   R). **False** $Fe(H_2O)_6^{+3}$ is Bronsted-Lowry and also Lewis acid.

S). **True**   T). **True**   U). **True**

V). **False**   Must have an amino acid, not just an amine group.

2. A)   (a) **OCl$^-$** + H_2O <=> HOCl + OH^- Should be **reactant-favored** since OCl$^-$ is a weak base, it will only partially dissociate in solution.

(b) **HNO$_3$** + NH_3 <=> NH_4^+ + NO_3^- Should be **product-favored** since the probability of the reverse reaction occurring is very small. HNO_3 is a strong acid so therefore NO_3^- is a very weak conjugate base and would not hydrolyze in water.

2. B) **Acid = H_2S, Base = CO_3^{-2}**   Reaction should be **product favored**. Since HCO_3^- is a much weaker acid than H_2S, the probability of the reverse reaction occurring should be much smaller than the forward reaction.

C) **$Zn(OH)_2$ is amphoteric and can act as either acid or base**. $Zn(OH)_2$ added to NaOH produces $Zn(OH)_4^{-2}$, which consumes OH^- making solution more acidic. When $Zn(OH)_2$ added to HCl, OH^- ions react with H^+ making solution less acidic.

3. A) (a) $C_6H_5CO_2H + H_2O \iff C_6H_5CO_2^- + H_3O^+$ $K_a = \dfrac{[C_6H_5CO_2^-]\,[H_2O^+]}{[C_6H_5CO_2H]}$

   (b) $H_2AsO_4^- + H_2O \iff HAsO_4^{-2} + H_3O^+$ $K_a = \dfrac{[HAsO_4^-]\,[H_2O^+]}{[H_2AsO_4^-]}$

   B) (a) $CH_3NH_2 + H_2O \iff OH^- + CH_3NH_3^+$ $K_b = \dfrac{[CH_3NH_3^+]\,[OH^-]}{[CH_3NH_2]}$

   (b) $IO_3^- + H_2O \iff OH^- + HIO_3$ $K_b = \dfrac{[HIO_3]\,[OH^-]}{[IO_3^-]}$

4.   (a) $HClO_4$ (b) NH_4^+ (c) C_6H_5N (d) HCN   (e) H_2 (f) S^{2-}

	(a)	(b)	(c)	(d)	(e)	(f)
A)	**strong acid**	**weak acid**	**weak base**	**weak acid**	**neither**	**strong base**
B)	ClO_4^-	NH_3	$C_6H_5NH^+$	CN^-	--	HS^-

5. A) **(d)** $BrCH_2CO_2H$ B) **(b)** HBrO
   C) (a) HCOO(H)   (b) (H)BrO   (c) $(CH_3)_2AsO_2$(H)   (d) $BrCH_2CO_2$(H)

6. **(c)** $pH = -\log [H_2O^+] = -\log (3.4 \times 10^{-5}) = 4.5$ **pOH** $= 14 - pH = 14 - 4.5 = \mathbf{9.5}$

7. **(d)**   0.10 M HF, 8.4% ionized means $[H_3O^+] = 0.084(0.10) = 0.0084$ M so **pH = 2.08**

8. **B, C, F** $pH = 3.0$, $[H_3O^+] = 1.0 \times 10^{-3} \to pH = 6.0$, 1.0×10^{-6}

9.   (a) $[H_3O^+] = 10^{-pH} = 10^{-2.30}$ (b) $[H_3O^+] = K_w / [OH^-]$ (c) $pH = 14 - pOH = 6.48$
   A) $[H_3O^+] =$ **5.01×10^{-3}**   **4.55×10^{-10}**   **3.31×10^{-7}**
   B)   **acidic**   **basic**   **acidic**

10. $M\ OH^- = \dfrac{0.463\ \text{g Ca(OH)}_2}{0.250\ \text{L}} \times \dfrac{1.0\ \text{mole Ca(OH)}_2}{74.0\ \text{g Ca(OH)}_2} \times \dfrac{2.0\ \text{mole OH}^-}{1.0\ \text{mole Ca(OH)}_2} = 0.0501\ M\ OH^-$

   $pOH = 10^{-0.0501} = 1.30$ so **pH** $= 14.0 - 1.30 = \underline{\mathbf{12.7}}$

11. A) $CH_3CH_2CO_2H + H_2O \iff CH_3CH_2CO_2^- + H_3O^+$ B) K_a, $K_a = \dfrac{[C_2H_5CO_2^-]\,[H_2O^+]}{[C_2H_5CO_2H]} = 1.4 \times 10^{-5}$

   C) $K_a = \dfrac{(x)(x)}{(0.15 - x)} = 1.4 \times 10^{-5} = \dfrac{x^2}{(0.15)}$ then $x = [H_3O^+] = 1.45 \times 10^{-3}$, **pH = 2.84**

   • Check: $0.15 - x = 0.15 - 0.00145 = 0.149$, so assumption to neglect reasonable. With quadratic result would be $x = 1.44 \times 10^{-3}$

12. A) **Conjugate= $(CH_3)_3NH^+$**   B) K_b: $(CH_3)_3N + H_2O \iff (CH_3)_3NH^+ + OH^-$
   C) $K_b = \dfrac{(x)(x)}{(0.25 - x)} = 6.25 \times 10^{-5} = \dfrac{x^2}{(0.25)}$ then $x = [OH^-] = 3.95 \times 10^{-3}$, **pH** $= 14.0 - 2.40 = \underline{\mathbf{11.6}}$

   • Check: $0.25 - x = 0.25 - 0.00395 = 0.246$, so assumption to neglect reasonable. With quadratic result would be $x = 3.92 \times 10^{-3}$ producing $pH = 11.6$.

13. $[H_3O^+]$ at equilibrium $= 10^{-2.50} = 1.51 \times 10^{-3}$ and $K_a = \dfrac{(3.16 \times 10^{-3})^2}{HA_{int} - 0.00316} = 1.4 \times 10^{-4}$

then $9.99 \times 10^{-5} = 1.4 \times 10^{-4}(HA_{int}) - 4.42 \times 10^{-7}$ and $HA_{int} = \dfrac{(1.044 \times 10^{-5})}{(1.4 \times 10^{-4})} = \underline{\textbf{0.0746 \textit{M}}}$

14. $[H_3O^+] = 3.16 \times 10^{-3}$ then $K_a = \dfrac{(0.00151)^2}{(0.15 - 0.00151)} = \underline{\textbf{1.54} \times 10^{-6}}$

15.

	Beaker A,	Beaker B:	Beaker C:	Beaker D:	Beaker E:
	0.202M HNO_3	0.202M HNO_2	2.02M KNO_3	0.202M KOH	2.02M KNO_2
	strong acid	*weak acid*	*salt*	*strong base*	*salt + anion, weak base*
(a)	**acidic,**	**acidic**	**neutral**	**basic**	**basic**

(b) **lowest pH** (c) **highest pH**

(d) **Need K$_a$, HNO$_2$** **Need K$_b$, NO$_2^-$**

(e) **Beaker E** K_b, NO_2^- less than K_a, HNO_2 so has largest number of undissociated molecules

(f) **Beaker C** KNO_3 = salt strong acid (HNO_3) + strong base (KOH)

(g) **Beaker E** KNO_2 = salt of weak acid (HNO_2) + strong base (KOH)

16. A) **KClO** Need a salt with anion that is a weak base. Only ClO^- would produce basic solution when dissolved by itself in water.

B) pOH = 5.25, $[OH^-] = 5.62 \times 10^{-6}$ $K_b = \dfrac{(5.23 \times 10^{-6})^2}{B_{int}} = 2.9 \times 10^{-7}$; B_{int} = **0.0109 M HClO**

17. $130 \text{ mL } C_2H_5OH \times \dfrac{0.785 \text{ g}}{1.00 \text{ mL}} \times \dfrac{1.0 \text{ mol } C_2H_5OH}{60.0 \text{ g } C_2H_5OH} \times \dfrac{2.0 \text{ mol } CH_3CO_2H}{3.0 \text{ mol } C_2H_5OH} = \underline{\textbf{1.13 mol CH}_3\textbf{CO}_2\textbf{H}}$

B) K_a, $CH_3CO_2H = \dfrac{(x)(x)}{(1.13 - x)} = 1.8 \times 10^{-5} = \dfrac{x^2}{(1.13)}$ and $x = [H_3O^+] = 4.52 \times 10^{-3}$, **pH = 2.35**

18.

Lewis acid = BF_3, accepts the electron pair to form the bond.

Lewis Base = F^-, donates the unshared electron pair to the bond.

Chapter 17

Answers to Key Terms Exercise:

Concerning Buffers, Titrations and Acid Rain:

1. buffer	2. buffer solution	3. buffer solution	4. buffer solution
5. Henderson-Hasselbalch equation		6. buffer	7. buffer solution
8. buffer capacity	9. buffer	10. buffer (solution)	11. buffer capacity
12. endpoint	13. titration curve	14. titrant	15. acid rain
16. Acid rain			

Concerning Solubility Equilibria:

17. solubility product constant   18. solubility product constant   19. ion product

20. solubility product constant   21. common ion effect   23. complex ion

23. formation constant   24. complex ion   25. common ion effect

Answers to True/False:

1.   A). **False**   Buffers are centered about pK_a, of acid form, not pK_b which gives pOH

   B). **False**   Adding strong base to weak acid makes a potential buffer.

   C). **True**

   D). **False**   Possible to make buffers with pH = 7.0, from weak acids or base, if this pH in the buffer region for weak acid or base.

   E). **True**   F). **False**   Are the same since moles of conjugate acid same.

   G). **False**   The pH will be lower, ratio falls to 0.5 and makes log term negative.

   H). **False**   The pH increases since larger value for ratio, conjugate base to acid.

   I). **True**   J). **True**   K). **True**   L). **True**   M). **True**

   N). **False**   The reverse would be true.

   O). **False**   The complex ion requires a Lewis base to react with metal, to donate the unshared pair of electrons to the metal.

2. A)   The conjugate acid concentration, from the ionization, will not be large enough to be nearly equal to the conjugate base concentration, so the condition for a buffer not met.

   B)   The equivalence point is the point of complete reaction between a substance and the titrant. The endpoint is when the color change of the indicator occurs, which requires a slightly larger volume of titrant than the equivalence point.

   C) The acid with the lower value of the pK_a is the stronger acid, since the negative of the log is taken. A pK_a of 4.74 is Ka of about 10^{-5} while a pK_a of 7.46 is a Ka of 10^{-8}.

   D)   Two major contributors are NO_2 and SO_2 gases. They react with water and oxidizers in rain drops to make the acids, HNO_3 and H_2SO_4. The limestone deposits in lakes can react to buffer the water and neutralize the added acid. Silicates are not as able to neutralize the acid, so the added acid from the rain can decrease the pH in these lakes.

3. A) **Increase**   B) **stays the same**   C) **decreases**   D) **stays the same**

4. **B and C**   *HCl is strong acid, so no equilibrium, but AgBr has K_{sp} and HClO, a K_a.*

5. **A, B and C** all buffers   ratio of $\dfrac{[\text{conj. base}]}{[\text{conj. acid}]}$ in final solution:

   A: *weak acid + salt of conjugate base* $\dfrac{\text{mol } HCO_2^-}{\text{mol } HCO_2H} = \dfrac{0.04 \text{ mol}}{0.03 \text{ mol}} = 1.33$

   B: *weak acid + strong base* $\dfrac{x}{\text{mol } HA_{int} - x} = \dfrac{0.01 \text{ mol added}}{0.03 - 0.01 \text{ mol}} = \dfrac{0.01 \text{ mol}}{0.02 \text{ mol}} = 0.50$

   C: *weak acid + salt of conjugate base* $\dfrac{\text{mol } HCO_2^-}{\text{mol } HCO_2H} = \dfrac{0.143 \text{ mol}}{0.50 \text{ mol}} = 0.286$

6.   A) **Acid**$= HCO_3^-$, **conjugate base** $= CO_3^{2-}$

   B). $HCO_3^- + H_2O <=> OH^- + CO_3^{2-}$

   C) $\mathbf{pK_a\ HCO_3^- = \underline{10.3}}$

7. $[SO_4^{2-}] = \dfrac{17.4\ g\ K_2SO_4}{0.100\ L} \times \dfrac{1\ mol\ K_2SO_4}{174.3\ g\ K_2SO_4} \times \dfrac{1\ mol\ SO_4^{2-}}{1\ mol\ K_2SO_4} = 1.00\ M\ SO_4^{2-}$, $[Ba^{2+}] = 0.50\ M$

   $[Ba^{2+}] \times [SO_4^{2-}] = 0.50 > K_{sp} = 1.1 \times 10^{-10}$ then: A) **False**   B) **False**

   C) **False** *KCl very soluble salt*   D) **True**   E) **True** $[K^+]= 2 \times [SO_4^{2-}]$ **F) False**

8. A) **b**   B) (a) **Point A**   (b) **Points B, C, D**   (c) **Point E**   (d) **Points F, G**

   C) Endpoint volume **about 22 mL**   D) **Methyl red**, since need the indicator to change
   color below pH of 4.

9. pH = $pK_a + \log \dfrac{[HCO_2^-]}{[HCO_2H]} = 3.75 + \log \dfrac{(0.63)}{(0.55)} = 3.75 + 0.059 = \mathbf{\underline{3.81 = pH}}$

10. A) Pair: (a) NH_4^+/NH_3 (b) HF/F^- (c) H_2S/HS^- (d) $B(OH)_3/B(OH)_4^-$
    Buffer range: **9.25 ± 1.0**   **3.14 ± 1.0**   **7.00 ± 1.0**   **9.14 ± 1.0**

   B) To test solution, add strong base or acid dropwise to the solution. Water would change it's
   pH significantly, with just a few drops added, but the buffer solution will not.

   C) $B(OH)_3 + H_2O <=> B(OH)_4^- + H^+$ To form the conjugate base, boric acid bonds with OH^-,
   leaving excess H^+ in solution and solution becomes acidic, just as if boric acid donated the
   proton itself. Boric acid is a Lewis acid and a the equilibrium is a K_a.

11. A) pH = $pK_a + \log \dfrac{[A^-]}{[HA]} = 4.87 + \log \dfrac{(0.10)}{(0.20)} = 4.87 - 0.301 = \mathbf{\underline{4.57 = pH}}$ B)

   (1) Add 0.01 mol H^+ so acid form increases, base form decreases by the added amount.

   (a) Ratio will decrease   **(b)** pH = $4.87 + \log \dfrac{(0.10 - 0.01)}{(0.20 + 0.01)} = 4.87 + \log \dfrac{(0.09)}{(0.21)} = \mathbf{\underline{4.51 = pH}}$

   (2) **(a) No effect**   *Water causes both concentrations to be diluted to same extent.*

   (3) Add 0.05 mole OH^- so acid form decreases, base form increases by the added amount.

   (a) Ratio will increase **(b)** pH = $4.87 + \log \dfrac{(0.10 + 0.05)}{(0.20 - 0.05)} = 4.87 + \log \dfrac{(0.15)}{(0.15)} = \mathbf{\underline{4.87 = pH}}$

12. A) pH = $pK_a + \log \dfrac{[C_6H_5CO_2^-]}{[C_6H_5CO_2H]}$, so $4.00 - 4.20 = \log \dfrac{[C_6H_5CO_2^-]}{(0.20)}$ and $[C_6H_5CO_2^-] = 0.126\ M$

   $0.750\ L \times \dfrac{0.126\ mol\ C_2H_5CO_2^-}{1.00\ L} \times \dfrac{1.0\ mol\ NaC_6H_5CO_2}{1.0\ mol\ C_6H_5CO_2^-} \times \dfrac{144.1\ g\ NaC_6H_5CO_2}{1.0\ mol\ NaC_6H_5CO_2} = \mathbf{\underline{13.6\ g}}$

   B) Both would furnish the needed benzoate ion, but the **lithium salt would be best** since it
   would not produce a competing equilibria and the NH_4^+ ion will set up its own acid-base
   equilibria (NH_4^+/ NH_3) in solution.

13. A)   **(d) Oxalic acid** with $pK_a = 1.23$ B) **(a) Acetic acid** with $pK_a = 4.74$

   C) $3.00 = 3.85 + \log \dfrac{(conj.base)}{(conj.\ acid)}$ then $\dfrac{(conj.base)}{(conj.\ acid)} = 10^{-0.85} = \mathbf{0.141}$ **Yes, solution a buffer**

14. A) K_a, HOCl $= \dfrac{(x)(x)}{(0.165 - x)} = 3.0 \times 10^{-8} = \dfrac{x^2}{(0.165)}$ then x = $[H_3O^+] = 7.0\ X\ 10^{-5}$ **pH = 4.15**

14.   B) Adding conjugate base, 0.10 mol $Ca(OCl)_2$ = 0.20 mol OCl^-, producing common ion effect. Then at equilibrium, $[OCl^-]_{eq} \approx 0.20$ M, $[HOCl]_{eq} \approx 0.165$ M, so that:

K_a, HOCl = 3.0×10^{-8} = $\dfrac{(x)(0.20 + x)}{(0.165 - x)}$ = $\dfrac{(x)(0.20)}{(0.165)}$ then x = $[H_3O^+]$ = 2.48×10^{-8} ,   **pH = 7.61**

C) Adding OH^- causes the base form to increase while the acid form decreases, so the **$[H_3O^+]$ decreases** to keep the equilibrium ratio at K_a.

15. (a) Combine **B + E for acidic buffer**, weak acid + strong base and pK_a HNO_2 = 3.35

    (b) Combine **D + E for basic buffer**, weak acid + strong base but pK_a HOBr = 8.60

16. A) M $Ce(IO_3)_4$ = $\dfrac{1.79 \times 10^{-4} \text{ mol } Ce(IO_3)_4}{0.100 \text{ L}}$ = **1.79×10^{-3} M**

   B) K_{sp} = $[Ce^{4+}][IO_3^-]^4$ = $(1.79 \times 10^{-3})(7.14 \times 10^{-3})^4$ = **4.67×10^{-12}**

17. A) $[Ba^{+2}]$ = s, $[OH^-]$ = 2s, K_{sp} = (s)$(2s)^2$ = $4s^3$ = 3.0×10^{-4} **s = molar solubility = 0.0422 M**

   B) $[OH^-]$ = 2s = 2(0.0422 M) = 0.0843 M   pOH = 1.07, so **pH = 12.9**

   C) Q = $[Ba^{+2}][OH^-]^2$ = $(0.05)(0.10)^2$ = 5.0×10^{-4} > K_{sp}, so **$Ba(OH)_2$ will precipitate**

18. A) Let s = molar solubility; PbC_2O_4, K_{sp} = $[Pb^{+2}][C_2O_4^{-2}]$ = (s)(s) = s^2 = 3.0×10^{-11}

            and for CaF_2 , K_{sp} = $[Ca^{+2}][F^-]^2$ = (s)$(2s)^2$ = $4s^3$ = 3.0×10^{-11}

   The molar solubility cannot be the same for both compounds. CaF_2 has the higher value with s = 1.95×10^{-4}, while s = 5.47×10^{-6} for PbC_2O_4

19. A) 500 mg = 0.500 g $C_6H_8O_6$, $\dfrac{0.500 \text{ g } C_6H_8O_6}{0.100 \text{ L}}$ $\times$ $\dfrac{1.0 \text{ mole } C_6H_8O_6}{176.0 \text{ g } C_6H_8O_6}$ = 0.0284 M $C_6H_8O_6$

K_a, $C_6H_8O_6$ = $\dfrac{(x)(x)}{(0.0284 - x)}$ = 7.94×10^{-5} = $\dfrac{x^2}{(0.0284)}$ then x = 1.50×10^{-3}, **pH = 2.82**

B) <u>At equivalence point:</u> $V_{NaOH} \times (0.100 \text{ M})$ = 0.0284 M (0.100 L) and V_{NaOH} = **28.4 mL**

C) <u>Halfway to endpoint</u>, one half of acid will be converted to its conjugate form, so:

   $[C_6H_8O_6]$ = 0.50 (0.0284) = 0.0142 M and $[C_6H_7O_6^-]$ = 0.0142 M, so **pH = pK_a = 4.10**

D) <u>pH at equivalence point</u>, all ascorbic acid converted to conjugate base, $C_6H_7O_6^-$, and the pH of solution determined by the K_b $C_6H_7O_6^-$ = $1.0 \times 10^{-14}/7.94 \times 10^{-5}$ = 1.26×10^{-10}

M, $C_6H_7O_6^-$ = $\dfrac{0.00284 \text{ moles}}{0.1285 \text{ L}}$ = 0.0221 M and K_b, $C_6H_7O_6^-$ = 3.0×10^{-8} = $\dfrac{x^2}{(0.0221)}$

so x = $[OH^-]$ = 1.69×10^{-6} pOH = 5.78, and **pH = 8.22 at equivalence point**

E) **Yes.** Phenolphthalein is red between 8.3 and 10, so it will change color at the proper pH.

20. A) K_f: $Au^+ + 2 CN^- \Longleftrightarrow Au(CN)_2^-$ and K_f = $\dfrac{[Au(CN)_2^-]}{[Au^+][CN^-]^2}$

   B) (a) $AuCl(s) + 2 CN^- \Longleftrightarrow Au(CN)_2^- + Cl^-$

       (b) $AuCl(s) \Longleftrightarrow Au^+ + Cl^-$ K_{sp}

         + $Au^+ + 2 CN^- \Longleftrightarrow Au(CN)_2^-$ K_f

       ———————————————————————

        $AuCl(s) + 2 CN^- \Longleftrightarrow Au(CN)_2^- + Cl^-$ (c)   **K for reaction = $K_{sp} \times K_f$**

Chapter 18

Answers to Key Terms Exercise:

Characteristics of Entropy and Gibbs Free Energy:

1. reversible process   2. reversible process   3. Third Law of Thermodynamics
4. Second Law of Thermodynamics   5. Gibbs free energy   6. Gibbs free energy
7. Gibbs free energy   8. Gibbs free energy   9. Gibbs free energy
10. Gibbs free energy   11. Gibbs free energy   12. extent of reaction
13. Gibbs free energy   14. Gibbs free energy   15. standard equilibrium constant
16. standard equilibrium constant   17. standard equilibrium constant

Free Energy and Biochemical Systems:

19. exergonic   20. Gibbs free energy   21. endergonic   22. Gibbs free energy
23. Metabolism   24. nutrients   25. Gibbs free energy   26. Metabolism
27. exergonic   28. endergonic   29. energy conservation   30. exergonic
31. endergonic   32. Photosynthesis   33. endergonic   34. exergonic

Answers to True/False:

1.   A). **True**   B). **False**   Not always, depends on the sign of $\Delta S°$ and temperature.

     C). **False**   Only perfect crystalline equal zero, everything else approximately zero.

     D). **True**   E). **True**   F). **False**   When signs alike, it can change.

     G). **False** $\Delta G°$ less than zero, means it is negative and K would greater than 1.0

     H). **False** $\Delta G°$ is fixed, but RT changes so K° changes.

     I). **False**   K° is always greater than 1.0, but will change value with temperature.

     J). **True**   K). **True**   L). **False**   Endergonic means + $\Delta G°$, not + $\Delta H°$

     M). **True**   N). **True**

2.   A)   An **endothermic (positive) ΔH and negative ΔS** will always produce a reactant favored reaction since $\Delta G°$ is positive no matter what the temperature is.

     B) (a) **entropy**   The ability to choose where you want to sit is having freedom of choice which leads to random, or disordered arrangements.

       (b) **enthalpy**   Having an attraction is like what atoms experience through bonding and intermolecular forces that induce a certain order.

       (c) **free energy**   Trying to balance the choices to get the best situation for yourself is what free energy represents for atoms.

     C) **Yes.**   Exergonic means that $\Delta G°$ is negative and negative values of $\Delta G°$ are definitely possible when $\Delta H°$ is exothermic (or has a negative sign).

     D)   Use the fact that ΔS for surroundings and system (your body) would be the same value, just opposite in sign. ΔS is defined by (heat lost)/T for reversible process, so

$$\Delta S_{surroundings} = -\Delta S_{body} = \frac{q}{T(K)} = \frac{-(-100\ J/s\ (50\ min\ )\ (60\ s/min))}{293\ K} = 1024\ J/K$$

3.   **A, B**   *C, D either go to less moles of gas as products or a more ordered state*

4.   A) **c**   *$\Delta H° = [\ 2(-285.83)+2(-296.83)] - [2(-20.63)] = -1124\ kJ$*

     B) $\Delta S°$ **should be negative**   *5 moles of gas → 4 moles of gas, means less disorder*

5.   **D**   *Product-favored at low T, but not at high is a $\Delta H(+)$, $\Delta S(-)$ combination*

6.   A) **Negative ΔS.**   Producing less moles of gas and there are fewer choices for particle arrangement since atoms becoming grouped.

     B) **Negative ΔS**   Producing a liquid or solid-like state from more random gas state.

     C) **Positive ΔS**   Producing a much more random gas state from solid-like state.

7. $\Delta G^0 = -141.8$ kJ *then* $\ln K^0 = \dfrac{-(-1.418 \times 10^4 J)}{(8.314\ J/mol - K)(298\ K)} = 57.23$ $K^0 = 7.18 \times 10^{24} = B$

8. **C, F, G** $\Delta S° = [\ 2(192)\ ] - [\ (192 + 3(130)\ ]\ J/K = 198\ J/K$, so $\Delta S° (-)$, $\Delta H°(-)$ so $\Delta G° (+\ or\ -)$

9. (a) **Arrow E**   (b) **Point C**   (c) **Points A, B**   (d) **Point D**

10. (a) Water, **H-bonding** , Mercury, **metallic bonding**, and Br_2 , **London forces**.

    (b) Mercury has strong interactions which induces a high degree of order and a low entropy, but the H-bonding in water creates a very structured (ordered) arrangement in liquid water, producing slightly less freedom of movement (or arrangement) than in mercury. In contrast, the weak forces in liquid bromine allow for much more movement of the particles producing the highest entropy.

11. A) $\Delta H° = + 285$ **kJ**, $\Delta S° = $ **-137.4 J/K**, $\Delta G° = $ **326 kJ**   B) **reactant-favored**

    C) $K° = 7.2 \times 10^{-58}$ D) Colder temperatures will make K° even smaller.

12. A) $\Delta S° = $ **- 533 J/K**   B) $\Delta G° = $ **- 1010 kJ**   C) **(a)**   D) **(c)**

13. A) (a) $2\ NO(g) \Longleftrightarrow 1\ N_2O(g) + \frac{1}{2}\ O_2\ (g)$ (b) $2\ NO(g) + O_2(g) \Longleftrightarrow 2\ NO_2(g)$

    B)

	$\Delta H°(kJ)$	$\Delta S°(J/K)$	$\Delta G°(kJ)$	C) **both are product favored**
Reaction (a)	**- 98.5**	**- 99.2**	**- 68.9**	
Reaction (b)	**-114**	**- 147**	**- 70.5**	

    D) **c**   E) Reaction: $N_2O(g) + \frac{3}{2}\ O_2\ (g) \Longleftrightarrow 2\ NO_2(g)$ $\Delta G° = $ **-1.58 kJ**

14. A) $\Delta G^0 = \Big[\ 2(-111.25) + 86.55\Big] - \Big[(-237.1) + 3(51.31)\Big]\ kJ = +$ **52.78 kJ**

    B) $\ln K^0 = \dfrac{-52.78 \times 10^6 J}{(8.314\ J/mol - K)(298\ K)} = -21.30$, $e^{-21.30} = K° = $ **5.6 $\times 10^{-10}$**

15. A) $\Delta G^0 = -RT \ln K^0 = -(8.316\ J/mol - K)(973\ K)\ln(5.10) = -13,182\ J = $ **- 13.2 kJ**

    B) $\Delta G = \Delta G^0 + RT \ln Q$ If extent of reaction 0.33 then, $Q = 0.33 \times (5.10) = 1.68$

    $\Delta G = (-13.18\ kJ) + (8.314\ J)(973) \ln (1.68) = $ **- 9.38 kJ**

16. A) $\Delta H° = $ **-471.2 kJ**, $\Delta S° = $ **-10.79 J/K**, so $\Delta G° = -471.56 + 0.533\ kJ = $ **- 468 kJ**

    B) **The $\Delta H°$ factor dominates,** producing the nearly all of the free energy.

    C) **Yes.** Since signs of $\Delta H°$ and $\Delta S°$ are both negative, decreasing temperature will make the reaction less product-favored and lower the value of K°.

17. A) $\Delta G^0 = -RT \ln K_c^0 = -(8.316\ J/mol - K)(298\ K)\ln(1.8 \times 10^{-4}) = + 21,369\ J = $ **+ 21.4 kJ**

    B) The sign of $\Delta H°$ **is positive** ($\Delta H° = + 92.7$ kJ)

    C) **Yes**, since both $\Delta S°$ and $\Delta H°$ are positive, changing the temperature can make the reaction product-favored. The **temperature needs to be made higher than 25°C**, so that the $T\Delta S°$ will get larger than $\Delta H°$ and the $\Delta G°$ can become negative.

    D) $K_c^0 = 1.8 \times 10^{-4} = x^2$ $x = 0.0134$ M = no. moles NH_4HS converted in 1.0 L

    so free energy needed = 0.0134 mol (21.4 kJ) = 0.2868 kJ = **287 J**

18. Can use K to calculate $\Delta G°$ from: $\Delta G^0 = -RT \ln K_c^0$ and then $\Delta S^0 = (\Delta H^0 - \Delta G^0)(1000)/T$

Reaction:	(a)	(b)	(c)	(d)	(e)
$\Delta G°$ (kJ)	- 673.7	- 530.5	+ 28.6	- 32.8	+ 158.0
$\Delta S°(J/K)$	+1899	+ 1537	- 222.8	- 194.0	+ 74.5

    A) **c, e**   B) **a, b, d**   C) **a, b, c**   D) **c, d** (*need negative $\Delta H°$ and $\Delta S°$*)

    E) **e** (*need positive $\Delta H°$ and $\Delta S°$*)   F) **a, b** (*need negative $\Delta H°$ and positive $\Delta S°$*)

Chapter 19

Answers to Key Terms Exercise:

Components and Types of Electrochemical Cells:

1. Electrochemistry   2. half reaction   3. half reaction
4. electrochemical cell   5. half reaction   6. half cell   7. half cell
8. anode   9. cathode   10. half cell   11. electrode
12. salt bridge   13. half cell   14. voltaic cell   15. anode
16. cathode   17. battery   18. voltaic cell   19. electromotive force
20. emf   21. volt(V)   22. half cell   23. ampere
24. coulomb   25. electrolysis   26. electrolysis   27.electrochemical cell
28. voltaic cell (battery)   29. ampere   30. Faraday constant   31. electrolysis

Measuring Voltages in Electrochemical Cells:

32. cell voltage   33. half cell (reaction)   34. electrochemical cell
35. standard voltages(E°)   36. standard conditions   37. half reaction
38. half reaction   39. standard hydrogen electrode   40. half cell (reaction)
41. standard reduction potential   42. half reaction   43. anode
44. standard reduction potential   45. half reaction   46. cathode
47. standard reduction potential   48.voltaic cell   49. electrochemical cell
50. cell voltage   51. standard conditions   52. Faraday constant
53. Nernst equation   54. cell voltage   55. standard conditions
56. Nernst equation   57. cell voltage   58. half cell(reaction)
59. concentration cell   60. cell voltage   61. pH meter
62. concentration cell   63. neurons

Concerning Applications of Electrochemical Cells:

64. voltaic cell   65. primary battery   66. secondary battery
67. electrochemical cell   68. battery   69. fuel cell   70. secondary battery
71. fuel cell   72. corrosion   73. anode   74. anode   75. corrosion
76. anodic inhibition   77. cathodic protection   78. standard reduction potential
79. corrosion   80. cathode   81. standard reduction potential

Answers to True/False:

1.   A). **False**   The reactions always occur together, so equal numbers.

B). **True**   C). **True**   D). **True**   E). **True**

F). **False**   Reduction is the gain of electrons.

G). **True**   H). **False**   Total the same, but electrons per mole can be different.

I). **True**   J). **True**

K). **False**   The measured cell voltage is always accurate, only half reaction values need to set by a reference.

L). **False**   Need a positive value for $E°_{cell}$.

M). **True**   N). **False**   Faraday constant is the charge for 1.0 mole e's.

O). **False** $E°_{cell}$ will be positive and greater than zero, but can be less than 1.0 V.

P). **False**   Example of nonreversible reaction, so primary battery.

Q). **False**   It's reversible since the products of both half-reactions stick to the electrodes and remain in contact.

R). **False**   Oxygen is reduced, not H^+. The function of water is as a salt bridge.

S). **True**   A). **True**

2.   A) Oxidation is the loss of electrons by an element causing reduction of another element. The element that causes oxidation, the oxidizing agent, has to be the one reduced.

B) (1) H oxidized, O reduced (2) H is the reducing agent, O the oxidizing agent

(3) OH^- ions participate in each half reaction, so they are needed when reactions are separated, but appear to cancel out in the overall reaction.

2. C) A spontaneous product-favored reaction is needed, so the $E°_{cell}$ must be negative and that only applies to some combinations of half cells. Also practicality of design and safety considerations of cell reactions must also be considered.

D) A primary battery uses an irreversible reaction, where even reversing the flow of electrons will not reproduce reactants from products. In a secondary battery, the cell reaction can be reversed, and the battery can be recharged.

E) Metals such as platinum or solid graphite rods are often used. The basic criteria is that the electrode material be chemically inert, have such a high value for E° that it will not be oxidized, and also electrically conducting.

F) There are not enough ions in pure water that could alleviate the charge buildup in a running cell.

G) No. Copper is above Fe and has a higher reduction potential, so Cu would be reduced, not Fe.

3. **A, C, D, E**   • *No current flows when at equilibrium, since E° = 0 and E°$_{cell}$ never 0.*

4. Use Nernst equation: $E_{cell} = E°_{cell} - \frac{0.0257 \text{ V}}{n} \ln \frac{[F^-]^2}{[Br^-]^2}$ A) **less than** $(\ln \frac{[F^-]^2}{[Br^-]^2} = \text{positive})$

   B) **greater than** $(\ln \frac{[F^-]^2}{[Br^-]^2} = \text{negative})$ C) **same as E°** since $(\ln \frac{[F^-]^2}{[Br^-]^2} = \ln(1) = 0)$

5. **C, D, F**   *Since Zn oxidized at anode and E°$_{cell}$ = + 0.78 V, Cl⁻ spectator ion*

6.   A) **Silver metal would deposit** on the Fe strip, as Ag^+ reduced, the Fe oxidized.

   B) **Droplets of liquid elemental mercury would appear** on the strip, as mercury ions are reduced and Al is oxidized.

   C) **No reaction** would occur, since Mg^{+2} ions cannot oxidize the Ag metal.

7. **A) False**, *Zn(s) is in anode* **B) True**, *E° positive*   **C) False**, *need double line for salt bridge*
D) False, *ln Q will be negative, so $E_{cell} > E°$*   **E) False**, *$E_{cell} = E°$ when $[Zn^{+2}] = [Pb^{+2}]$*

8. **A is True.** *Combinations B-D will not result in a positive value for E°$_{cell}$.*

9. A) **Li(s)** , **Br₂** would be produced   b) **Br₂** , **H₂** produced and OH⁻ ions appear in solution

10. A) (a) = **Mn** (b) initial = **+ 4**, final = **+ 2**   B) (a) = **N** (b) initial = **+ 2**, final = **+ 5**
   (c) $MnO_2(s) + 4H^+ + 2e^- \rightarrow Mn^{2+} + 2 H_2O$ (c) $NO(g) + 2 H_2O \rightarrow NO_3^- + 4H^+ + 3e^-$
   C) $3 MnO_2(s) + 4H^+ + 2NO(g) \rightarrow Mn^{2+} + 2 H_2O + 2NO_3^-$ **E°$_{cell}$** = (1.23 - 0.96) = **0.27 V**

11.

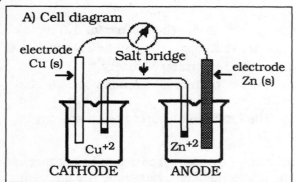

B) $E_{cell} = E°_{cell} - \frac{0.0257 \text{ V}}{n} \ln \frac{[Zn^{2+}]}{[Cu^{2+}]}$

$= 1.12 \text{ V} - \frac{0.0257 \text{ V}}{2} \ln \frac{[0.001]}{[3.0]}$

$= 1.12 \text{ V} - \frac{0.0257 \text{ V}}{2} (-8.01) = \underline{\textbf{1.23 V}}$

C) $E°_{cell} \times \frac{n}{0.0257 \text{ V}} = \ln K$

then **K = $\underline{7.13 \times 10^{37}}$**

The cell must be close to the start of the reaction, since the ratio of $[Zn^{2+}] / [Cu^{2+}]$ is small in the situation described and will be very large at equilibrium, based on value of K.

12. A) Overall Reaction for Voltaic cell:

$2\ Ag(s) + 2\ I^- + F_2(g) \rightarrow 2F^- + 2\ AgI(s)$

B) $E°_{cell} = 2.87 + 0.15\ V = \textbf{3.02 V}$

$$\Delta G° = -\ nFE°_{cell}$$

$\Delta G° = -\ 2\ mol\ (\ 9.65 \times 10^4 \frac{coul}{mol})(3.02\ V)\ \frac{1\ joule}{1\ coul - V}$

$\Delta G° = -\ 582,860\ J = -\ 583\ kJ$

C) Cell diagram

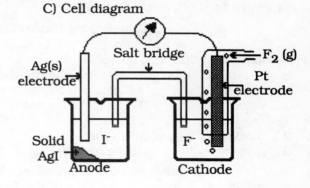

13. A) The best oxidizing agent, should have highest reduction potential = **$Cr_2O_7^{2-}$**

B) Best reducing agent, should have lowest reduction potential = **Cu^+**

C) Overall reaction with highest $E°_{cell}$:

   $Cr_2O_7^{2-} + 14\ H^+ + 6\ Cu(s) \rightarrow 6\ Cu^+ + 2\ Cr^{+3} + 7\ H_2O$ **$E°_{cell} = 1.17V$**

14. A) Cathode reaction: **$2\ e^- + H_2O \rightarrow H_2(g) + 2\ OH^-$** $E° = -\ 0.8277\ V$

   Anode reaction: **$2\ Cl^- \rightarrow Cl_2(g) + 2\ e^-$** $E° = +\ 1.358\ V$

B) **$E°_{cell} = -\ 0.8277 - 1.358\ V = -\ \textbf{2.19 V}$**

C) *No. mole electrons* $= \frac{I \times t}{F} = \frac{(2.0\ coul/s) \times (8.0\ hr) \times (3600\ s/hr)}{9.65 \times 10^4\ coul/mol\ electrons} = 0.597\ mole$

   $0.597\ mole\ electrons\ \frac{(\ 2\ mole\ OH^-\ )}{(2\ mole\ electrons)} = 0.597\ mole\ OH^-\ produced$

   since volume of solution 1.0 L, the [OH-] = 0.597 M , **pH = 13.78**

15. A) Electrolysis of Cu reaction: $Cu^{+2} + 2\ e^- \rightarrow Cu(s)$, so

   $2.542\ g\ Cu \times \frac{1\ mole\ Cu}{63.55\ g} \times \frac{2\ mole\ electrons}{1\ mole\ Cu} = \textbf{0.080 mole electrons used}$

B) Electrolysis of Cr reaction: $Cr^{+3} + 3\ e^- \rightarrow Cr(s)$

   $0.080\ mole\ electrons \times \frac{1\ mole\ Cr}{3\ mole\ electrons} \times \frac{52.00\ g}{1\ mole\ Cr} = \textbf{1.387 g Cr}$

C) *No. mole electrons* $= \frac{I \times t}{F}$ *rearranges to:* $t = \frac{F \times (no.\ mole\ electrons)}{I}$

   $t = \frac{9.65 \times 10^4\ coul/mol\ e's \times (0.080\ mol\ e's)}{10.0\ coul/s} = 772\ sec = \textbf{12.9 minutes}$

Chapter 20

Answers to Key Terms Exercise:

Concerning Natural Decay of Radioactive Atoms:

1. nuclear reactions   2. alpha radiation   3. beta radiation
4. alpha radiation   5. alpha (α) particles   6. Beta radiation   7. beta (β) particles
8. gamma radiation   9. Gamma radiation   10. alpha (α) particles   11. beta (β) particles
12. positron   13. electron capture   14. nucleons   15. nucleons
16. nucleons   17. nuclear reactions   18. radioactive series   19. nuclear reactions
20. nucleons   21. nucleons   22. binding energy   23. binding energy
24. binding energy per nucleon   25. binding energy   26. nucleons
27. activity   28. becquerel (Bq)   29. curie (Ci)   30. nuclear reactions

Concerning Other Nuclear Reactions and Units for Exposure:

31. nuclear fission   32. nuclear fusion   33. binding energy
34. nuclear reactor   35. binding energy   36. nuclear fission   37. critical mass
38. nuclear reactor   39. nuclear reactor   40. critical mass   41. Nuclear fusion
42. nuclear fission   43. Nuclear fusion   44. nucleons   45. plasma
46. binding energy   47. nuclear fusion (reactions)   48. background radiation
49. roentgen(R)   50. rad   51. gray (Gy)   52. rad   53. rad
54. rem   55. beta radiation (particles)   56. gamma radiation
57. alpha radiation (particles)   58. rem   59. rem   60. background radiation
61. rem   62. sievert (Sv)   63. rem   64. nuclear medicine
65. tracers   66. tracers

Answers to True/False:

1.
     A). **False**   There is no change in the radioactive properties.

     B). **True**   C). **False**   All three can cause ionization of atoms.

     D). **False**   Does not change the mass of the atom.

     E). **False**   The final product must a stable, nonradioactive atom and is lead-206.

     F). **True**   G). **True**   H). **True**

     I). **False**   Reverse true, generally, neutrons must be greater than protons in atom.

     J). **True**   K). **True**

     L). **False**   Unstable (radioactive) atoms are produced by both fission and fusion.

     M). **False**   Transuranium elements are artificial, but are in the actinides.

     N). **False**   Uranium-238 not fissionable part of the fuel, Uranium-235 is.

     O). **False**   Can't be pure or reaction less controlled.

     P). **True**   Q). **True**   R). **True**

     S). **False**   Activity is a rate, rate = kN, so need the number of atoms at the start as well as the half life, which gives the decay constant,k.

     T). **False**   Need higher temperatures for plasma of nucleons than reactor produces.

     U). **True**   V). **True**   W). **True**   X). **True**

     Y). **False**   An alpha particle is 2 protons and 2 neutrons.

2. A)   (a) Alpha least penetrating, stopped by paper or skin; beta penetrates further, 1- 2 cm in tissue; but gamma most penetrating, since need thick concrete or lead to stop.

     (b) Alpha is the most dangerous if emitted inside the body, because of the much larger size of the particle than beta, or gamma radiation (which has no size). This is evidenced by the larger quality factor applied to the dose in rads for alpha radiation.

B)   (a) Nuclear reactions release millions of kilojoules per mole while chemical reactions produce thousands at best.

     (b) A 100°C would have no effect on rate of nuclear reaction, but has a dramatic effect on chemical reaction rates.

2. B)   (c) The nuclear binding energy between nucleons is the source for nuclear reactions, whereas chemical bonding energy is the source in chemical reactions.

C)   The main factors are the neutron/proton ratio and the binding energy per nucleon.

D)   Transmutation is the result of bombarding a nucleus with a small particle so that a new, larger new nucleus is formed. In fission, the nucleus splits apart after absorbing the particle. In decay, no bombarding particle is needed and the nucleus gets smaller.

3. **A, D**   4. **Other product = proton** (hydrogen nucleus)

5.A) $^{226}_{88}$Ra $\rightarrow$ $^{4}_{2}$He + $^{222}_{86}$ **Rn, alpha** decay   B) $^{34}_{17}$Cl $\rightarrow$ $^{0}_{-1}$e + $^{34}_{18}$ **Ar, beta** decay

6. **B, D**   7. A) $^{157}_{63}$Eu $\rightarrow$ $^{0}_{-1}$e + $^{157}_{64}$Gd   B) $^{75}_{35}$Br $\rightarrow$ $^{0}_{+1}$e + $^{75}_{34}$Se

C) $^{114}_{48}$Cd + $^{0}_{-1}$e $\rightarrow$ $^{114}_{47}$Ag   D) $^{239}_{94}$Pu $\rightarrow$ $^{4}_{2}$He + $^{235}_{92}$U

8.   Atomic number = **15**, atomic mass = **30**, element = **phosphorus**

9.   A) **decreases**   B) **increases**

C) **increases**   *Although both go down by 2 particles, the ratio will change: 100/50 is not same as 98/48*

10. Use $\ln \dfrac{A_t}{A_o} = -\dfrac{0.693\,(t)}{t_{1/2}}$ so $\ln \dfrac{5.28 \times 10^4}{5.12 \times 10^5} = -2.272 = \dfrac{-0.693\,(6.0\text{ hr})}{t_{1/2}}$ $t_{1/2}$ **= 1.83 hr**

11. $\ln \dfrac{N_t}{N_o} = \ln \left[\dfrac{\text{wt. left}}{\text{wt. start}} \right] = -\dfrac{0.693\,(t)}{t_{1/2}}$ *so that* $\dfrac{\text{wt. left}}{80.0\text{ g}} = 0.0359$ *and* **wt. left = 2.87 g**

12. $\ln \dfrac{A_t}{A_o} = -\dfrac{0.693\,(4\text{ da})(24\text{ hr/da})}{15\text{ hr}} = -4.435$ so $A_t = 0.01185(A_o) = $ **2.96 X 10^7 Bq**

13. Volume of room = volume of air = $640\text{ ft}^3 \times \dfrac{28.3\text{ L}}{1.0\text{ ft}^3} = 1.81 \times 10^4\text{ L}$

$1.81 \times 10^4\text{ L} \times \dfrac{4.0 \times 10^{-12}\text{ Ci}}{1.0\text{ L}} \times \dfrac{3.7 \times 10^{10}\text{ dps}}{1.0\text{ Ci}} = $ **2680 atoms/sec disintegrating**

14. A) $1.5 \times 10^{-3}\text{ L} \times \dfrac{2.5 \times 10^{-9}\text{ mole}}{1.0\text{ L}} \times \dfrac{6.02 \times 10^{23}\text{ atoms}}{1.0\text{ mole}} = $ **2.26 ×10^{13} atoms**

B) 6.0 hrs = $t_{1/2}$, One γ per decayed atom: $\frac{1}{2}$ (2.26 X 10^{13}) = **1.13 X 10^{13} gamma rays**

C) 1.13×10^{13} rays $\times \dfrac{1.0\text{ mole rays}}{6.02 \times 10^{23}\text{ rays}} \times \dfrac{1.35 \times 10^{10}\text{ J}}{1.0\text{ mole rays}} = 0.253\text{ J}$

$\dfrac{0.253\text{ J given}}{65.0\text{ kg}} \times \dfrac{0.01\text{ J absorbed}}{1.0\text{ J given}} \times \dfrac{1\text{ rad}}{0.01\text{ J/kg}} = $ **0.0039 rads**

15. A) $\ln \dfrac{N_t}{N_o} = -\dfrac{0.693\,(2\text{ yr})(365.25\text{ da/yr})}{138.4\text{ da}} = -3.657$, $\dfrac{N_t}{N_o} = 0.02578$ *and* **2.6% of weight left**

B) **The amount of solid will appear about the same**, since the product of the decay is lead-206, a very stable element and nearly all the mass of Po is converted into Pb.

16. A) **A= activity = kN**, where k = decay constant and N = number of atoms in sample

B) **k = 7.82 X 10^{-10} s^{-1}**   C) A = 7.82×10^{-10}s^{-1} (6.689 $\times 10^{21}$ atoms) = **5.23 ×10^{12} Bq**

Initial activity in curies = 141 Ci

D) $\ln \dfrac{A_t}{A_o} = \ln \dfrac{N_t}{N_o} = -\dfrac{0.693\,(100\text{ yr})}{28.3\text{ yr}} = -2.45$ so $\dfrac{A_t}{A_o} = 0.0864$ **8.64% of activity remains**

Chapter 21

Answers to Key Terms Exercise:

Concerning Formation of Elements and Processes to Obtain or Convert Some Elements:

    1. nuclear burning   2. hydrogen burning   3. helium burning
4. minerals   5. Ores   6. mineral   7. cryogen
8. nitrogen fixation   9. Nitrogen fixation   10. Frasch process
11. Frasch process   12. chlor-alkali process   13. chlor-alkali process

Answers to True/False:

1.   A). **False**   Two carbons nuclei would only bring 6 protons, need 14 protons.
        B). **False**   Fusion not chemical bond formation, a new atom with 8 protons formed.
        C). **True**   D). **False**   Most abundant is aluminum
        E). **False**   Clays have 4 O's bonded to each Si, 2 atoms of O for each Si.
        F). **True**   G). **False**   Clays are aluminosilicates
        H). **True**   I). **True**   J). **True**
        K). **True**   L). **False**   The nitrogen comes mainly form sodium azide
        M). **False**   S in +4 oxidation state in SO_2 and needs to go to +6 for H_2SO_4, so needs to be oxidized to SO_3, to form H_2SO_4 by combination reaction with water.
        N). **True**   O). **False**   Not an aloof metal since found combined in nature.
        P). **False**   Both pure forms of elemental phosphorus, not compounds, and difference due to bonding arrangement in the solids.

2.   A) The similarities are that both processes involve the helium nucleus and both are nuclear reactions. The differences would be that helium burning occurs only at extremely high temperatures and is a nuclear fusion process, while alpha emission occurs at room temperature and is a nuclear disintegration.

        B) The considerable savings of electrical energy that the recycling of Al(s) generates is the main reason.

        C) The substances are: Na, Cl_2, Al, Mg, Br_2 and I_2.

        D) NaN_3 • decomposes to produce the N_2 gas and heat to inflate the bag.

            KNO_3 • reacts with the Na(s) produced from the NaN_3 decomposition to form K_2O, Na_2O and more N_2. The heat released from the reactions forms the glass.

            SiO_2 • fuses with K_2O and Na_2O into a glass, so that potential hazard from exposure to these by-products is removed.

        E) NH_3 has N in the -3 oxidation state, which is its lowest oxidation number. NO_3^- ion has the N in a +5 state, the highest possible oxidation number.

        F) Elemental N_2 is a gas and forms many gaseous oxides, so it is frequently found in the atmosphere. Elemental phosphorus and its oxides are mainly solids, so it cannot enter the atmosphere easily.

3. There are only two acidic H's, bonded to O in H_3PO_3, while there are 3 acidic H's in the H_3PO_4 molecule. *(See the structures on the right)*

4. A) **2**   B) **N is reduced, O is oxidized**

5. A) **7**   B) The reaction is very **product-favored**, with a $\Delta G°$ **of about - 2900 kJ**

6. A) $C_6H_{12}O_6 + 6\ H_2SO_4 \rightarrow 6\ C(s) + 6\ H_2SO_4 \cdot H_2O$

B) **Yes, the reaction is the same** because paper and cotton are largely composed of cellulose which is made up of glucose units.

7. A) , B) *Lewis structure + Resonance forms:*   *Lewis structure + Resonance forms:*

$$|\overset{..}{\underset{..}{O}}-\overset{..}{N}=\overset{..}{\underset{.}{O}} \leftrightarrow \overset{..}{\underset{.}{O}}=\overset{..}{N}-\overset{..}{\underset{..}{O}}| \leftrightarrow |\overset{..}{\underset{..}{O}}-\overset{..}{N}=\overset{..}{\underset{.}{O}}$$

$$\overset{..}{\underset{.}{O}}=\overset{|\overset{..}{\underset{..}{O}}|}{N}-\overset{|\overset{..}{\underset{..}{O}}|}{N}=\overset{..}{\underset{.}{O}} \leftrightarrow |\overset{..}{\underset{..}{O}}-\overset{|\overset{..}{\underset{..}{O}}|}{N}-\overset{|\overset{..}{\underset{..}{O}}|}{N}-\overset{..}{\underset{..}{O}}| \leftrightarrow \overset{..}{\underset{.}{O}}=\overset{|\overset{..}{\underset{..}{O}}|}{N}-\overset{|\overset{..}{\underset{..}{O}}|}{N}-\overset{..}{\underset{..}{O}}|$$

C) $\Delta H° = -85.9$ kJ, $\Delta S° = -270.9$ J/K   The **entropy factor is unfavorable** and making the reaction reactant-favored at room temperature.

8.   % P in $Ca_3(PO_4)_2 = \dfrac{62.0}{310.2} \times 100 = \underline{\mathbf{20.0\%}}$, %P in superphosphate $= \dfrac{62.0}{432.4} \times 100 = \underline{\mathbf{14.3\%}}$

Let x = solubility: K_{sp} $Ca_3(PO_4)_2 = (3x)^3(2x)^2 = 108\,x^5 = 1 \times 10^{-25}$ and $x = 3 \times 10^{-5}$ M

$$K_{sp}\ Ca(H_2PO_4)_2 = x\,(2x)^2 = 4\,x^3 = 1 \times 10^{-3} \text{ and } x = 0.063 \text{ M}$$

So definitely the solubility, not the % P, is why the compound is called superphosphate.

9.   A) $\Delta G° = [-760 + 3(-131.23)] - [-272.3 + 3(-237.13)]$ kJ = -160.4 kJ

B) Since $\Delta G°$ is negative, $K > 1.0$, since can assume complete conversion of PCl_3 to products. Can assume $[H_3O^+]$ is from HCl only since, HCl is a strong acid and will fully ionize and impede the dissociation of weak acid, H_3PO_3, by Le Chatelier's Principle. Then:

$$\frac{15.0 \text{ g PCl}_3}{0.500 \text{ L}} \times \frac{1.0 \text{ mole PCl}_3}{137.5 \text{ g PCl}_3} \times \frac{3.0 \text{ mole HCl}}{1.0 \text{ mole PCl}_3} \times \frac{1.0 \text{ mole H}_3O^+}{1.0 \text{ mole HCl}} = 0.654 \text{ M}, \underline{\mathbf{pH = 0.18}}$$

10. A) The first reaction is product-favored since the sign of $\Delta G°$ will be negative at all temperatures. The positive H° with the positive entropy, results in the reaction being reactant-favored at low temperatures.

B) $T = \dfrac{\Delta H°}{\Delta S°} = \dfrac{124 \text{ kJ}}{0.626 \text{ kJ/K}} = \underline{\mathbf{198 \text{ K}}}$ *So at -75°C, the reaction starts to become product-favored from the thermodynamics standpoint.*

C)   The compound would have considerable kinetic stability since this is an ionic solid and the energy of activation should be high. However, the compound is very thermodynamically unstable so once if this reaction starts, it will go to completion very quickly.

11. A) Total moles of Cl^- = 0.5374 moles in 1.0 kg of solution, so

$$\frac{0.5374 \text{ mole Cl}^-}{1000 \text{ g soln}} \times \frac{1040 \text{ g soln}}{1.0 \text{ L soln}} = \underline{\mathbf{0.559 \text{ M Cl}^-}}$$

B) Calculate moles of electrons needed to oxidize the Cl^-:

$$10.0 \text{ L soln} \times \frac{0.559 \text{ mole Cl}^-}{1.0 \text{ L soln}} \times \frac{2.0 \text{ mole e's}}{1.0 \text{ mole Cl}^-} = \underline{\mathbf{11.18 \text{ mole } e's}}$$

$$t = \frac{F \times (no.\ mole\ electrons)}{I} = \frac{9.65 \times 10^4 \text{ coul/mol e's} \times (11.18 \text{ mole e's})}{4 \times 10^4 \text{ coul/s}} = \underline{\mathbf{27.0 \text{ sec}}}$$

Chapter 22

Answers to Key Terms Exercise
Concerning Properties of Transitions Metals:

1. lanthanide contraction   2. pyrometallurgy   3. pyrometallurgy
4. steel   5. ligands   6. coordinate covalent bond   7. ligands
8. coordination compound   9. coordinate covalent bond
10. coordination number   11. ligands   12. monodentate   13. bidentate
14. coordinate covalent bond   15. coordinate covalent bond   16. polydentate
17. chelating ligands   18. chelating ligands   19. hexadentate

Answers to True/False:

1.
- A). **False**   The transition elements are more dense.
- B). **False**   The transition metals have partially filled d sublevels.
- C). **True**
- D). **False**   Oxidation of the carbon occurs to lower the %C.
- E). **False**   Blister copper is the anode to dissolve the Cu^{2+}.
- F). **True**   G). **True**   H). **True**
- I). **False**   The number of coordinate covalent bonds gives the coordination number.
- J). **False**   Complex ions with six ligands have octahedral geometry.
- K). **True**   L). **True**
- M). **False**   Linkage isomerism describes a monodentate ligand in which uses either of two ends of the molecule and does not produce cis-trans isomerism.

2.
   A)   The similarity is that both involve the sharing of an electron pair to produce a bonding orbital and then a definite geometry is produced for the molecule. The difference is in the formation of the bond. Typical covalent bonds between main group elements result from each atom contributing equal numbers of electrons to the bond, but in coordinate covalent bonds, a single atom contributes both electrons to the bond.

   B)   No, it cannot. Although it has two atoms with unshared pairs, the atoms are next to each other. The donating atoms must be separated by at least two other atoms to produce the needed geometry for a bidentate ligand.

   C)   The CO-Fe^{2+} coordinate covalent bond is about 200 times stronger than the O_2-Fe^{2+} coordinate covalent bond, so the energy needed to reverse the process is very high.

   D)   The Cl^- ion from HCl, instead of H_3O^+, is critical to changing the solubility since the H_3O^+ from other acids has no effect. The AgCl forms a complex ion when combined with excess Cl^- which then causes the solid to dissolve.

3. The precipitate is not $NiCl_2$, since it is not yellow, but is an insoluble compound of Ni^{2+} and the OH^-, $Ni(OH)_2$ which is light green. In excess NH_3, the $Ni(OH)_2$ is converted to the $Ni(NH_3)_6^{+2}$ complex ion which has a deep blue color.

4.
A) 4s [↑↓] 3d [↑][↑][↑][↑][ ] 4p [ ][ ][ ]   B) 4s [ ] 3d [↑][↑][↑][↑][ ] 4p [ ][ ][ ]

C) 4s [ ] 3d [↑][↑][↑][ ][ ] 4p [ ][ ][ ]   D) 4s [↑↓] 3d [↑↓][↑↓][↑↓][↑↓][↑↓] 4p [ ][ ][ ]

E) 4s [ ] 3d [↑↓][↑↓][↑↓][↑↓][↑↓] 4p [ ][ ][ ]   F) 4s [ ] 3d [↑↓][↑↓][↑↓][↑↓][↑↓] 4p [ ][ ][ ]

5. Answers for:

	(A)	(B)	(C)	(D)	(E)
(a)	**cobalt**	**+3**	**6**	**Cl^-**	**monodentate**
(b)	**iron**	**+3**	**6**	**$C_2O_4^{2-}$**	**bidentate**
(c)	**iron**	**+2**	**4**	**CO**	**monodentate**
(d)	**ruthenium**	**+3**	**6**	**CN^-, Cl^-**	**monodentates**

6.   A) $Na[CoCl_4(NH_3)_2]$ B) $[CoBr_2(NH_3)_4]Cl$ C) $[PtCl(NH_3)_5]Cl_3$

      D) $[Fe(NCS)]Cl$ E) $K[Fe(CN)_4(H_2O)_2]$

7.   (a)   (b)   (c)

(A) complex ion: $[Co(en)_2F_2]^+$ $[Mn(H_2O)_4(OH)_2]$ $[CuCl_3(NH_3)]^-$

(B) need: one Cl⁻   none needed   one K⁺

(C) Coordination compound:

                          $[Co(en)_2F_2]Cl$ $[Mn(H_2O)_4(OH)_2]$ $K[CuCl_3(NH_3)]$

(D) Name compound:

      (a)   **diethylenediaminedifluorocobalt(III) chloride**

      (b)   **tetraaquadihydroxymanganese(II)**

      (c)   **potassium triamminetrichlorocuprate(II)**

8. A) **triamminecyanomaganese(II) chloride**   B) **potassium tetrachloronickelate(II)**

   C) **diamminedichloropalladium(II)**   D) **potassium hexafluorocobalt (IV)**

9.   A) $\left[Cl-Au-Cl\right]^-$ **no geometric**   B) $\left[\begin{smallmatrix}Cl & & Cl \\ & Cu & \\ Cl & & Cl\end{smallmatrix}\right]^{2+}$ **no geometric**
                                     isomers   **isomers**

      C) **cis isomer:**   **trans isomer:**

      D) **cis isomer:**   **trans isomer:**

      E)

no geometric isomers

10. A) **dichloroaurate (II) ion**

    B) $2.1 \text{ mg Ag} \times \dfrac{1.0 \text{ g Ag}}{1000 \text{ mg}} \times \dfrac{1 \text{ mole Ag}}{107.9 \text{ g}} \times \dfrac{1 \text{ mole Au}}{1 \text{ mole Ag}} \times \dfrac{1097.0 \text{ g Au}}{1 \text{ mole Au}} \times \dfrac{1000 \text{ mg}}{1.0 \text{ g Au}} = \underline{\textbf{3.8 mg Au}}$

11. • Calculate the mass Cr needed to plate object:

$$1750 \text{ in}^2 \times (3.0 \times 10^{-3} \text{in}) \times \frac{16.38 \text{ cm}^3}{1.0 \text{ in}^3} \times \frac{7.2 \text{ g Cr}}{1.0 \text{ cm}^3} = 61.94 \text{ g Cr plated out}$$

   • If only 70% efficient, to get 61.54 g Cr as actual yield need enough current for **0.70 (x) = 61.54 and x = 88.5 g Cr** (the theoretical yield).

   • Convert 88.5 g Cr to moles - from half reaction $Cr^{+4} + 4e^- \rightarrow Cr(s)$ need 4 moles e's per Cr^{+4}

$$88.5 \text{ g Cr} \times \frac{1.0 \text{ mole Cr}}{52.0 \text{ g}} \times \frac{4.0 \text{ mole e's}}{1.0 \text{ mole Cr}} = 6.807 \textit{ mole e's}$$

   • Calculate the minimum current needed from: $I = \dfrac{(\textit{No. mole electrons}) \times F}{(\textit{time, sec})}$

$$I = \frac{(6.807 \text{ mole e's}) \times (9.65 \times 10^4 \text{ coul/mol e's})}{3600 \text{ sec}} = \underline{\textbf{\textit{182 amperes}}}$$

http://chemistry.brookscole.com

The Wadsworth Group is the publisher of the following imprints:
Brooks/Cole, Duxbury, Heinle & Heinle, Schirmer, Wadsworth, and West.

BROOKS/COLE

THOMSON LEARNING ™

Visit us online at **http://chemistry.brookscole.com**

Visit Brooks/Cole online at **www.brookscole.com**

ISBN 0-03-032401-7